IET POWER AND ENERGY 93

Cogeneration and District Energy Systems

Other volumes in this series:

Volume 1	**Power Circuit Breaker Theory and Design** C.H. Flurscheim (Editor)
Volume 4	**Industrial Microwave Heating** A.C. Metaxas and R.J. Meredith
Volume 7	**Insulators for High Voltages** J.S.T. Looms
Volume 8	**Variable Frequency AC Motor Drive Systems** D. Finney
Volume 10	**SF_6 Switchgear** H.M. Ryan and G.R. Jones
Volume 11	**Conduction and Induction Heating** E.J. Davies
Volume 13	**Statistical Techniques for High Voltage Engineering** W. Hauschild and W. Mosch
Volume 14	**Uninterruptible Power Supplies** J. Platts and J.D. St Aubyn (Editors)
Volume 15	**Digital Protection for Power Systems** A.T. Johns and S.K. Salman
Volume 16	**Electricity Economics and Planning** T.W. Berrie
Volume 18	**Vacuum Switchgear** A. Greenwood
Volume 19	**Electrical Safety: a guide to causes and prevention of hazards** J. Maxwell Adams
Volume 21	**Electricity Distribution Network Design, 2nd Edition** E. Lakervi and E.J. Holmes
Volume 22	**Artificial Intelligence Techniques in Power Systems** K. Warwick, A.O. Ekwue and R. Aggarwal (Editors)
Volume 24	**Power System Commissioning and Maintenance Practice** K. Harker
Volume 25	**Engineers' Handbook of Industrial Microwave Heating** R.J. Meredith
Volume 26	**Small Electric Motors** H. Moczala *et al.*
Volume 27	**AC-DC Power System Analysis** J. Arrillaga and B.C. Smith
Volume 29	**High Voltage Direct Current Transmission, 2nd Edition** J. Arrillaga
Volume 30	**Flexible AC Transmission Systems (FACTS)** Y-H. Song (Editor)
Volume 31	**Embedded generation** N. Jenkins *et al.*
Volume 32	**High Voltage Engineering and Testing, 2nd Edition** H.M. Ryan (Editor)
Volume 33	**Overvoltage Protection of Low-Voltage Systems, Revised Edition** P. Hasse
Volume 36	**Voltage Quality in Electrical Power Systems** J. Schlabbach *et al.*
Volume 37	**Electrical Steels for Rotating Machines** P. Beckley
Volume 38	**The Electric Car: Development and future of battery, hybrid and fuel-cell cars** M. Westbrook
Volume 39	**Power Systems Electromagnetic Transients Simulation** J. Arrillaga and N. Watson
Volume 40	**Advances in High Voltage Engineering** M. Haddad and D. Warne
Volume 41	**Electrical Operation of Electrostatic Precipitators** K. Parker
Volume 43	**Thermal Power Plant Simulation and Control** D. Flynn
Volume 44	**Economic Evaluation of Projects in the Electricity Supply Industry** H. Khatib
Volume 45	**Propulsion Systems for Hybrid Vehicles** J. Miller
Volume 46	**Distribution Switchgear** S. Stewart
Volume 47	**Protection of Electricity Distribution Networks, 2nd Edition** J. Gers and E. Holmes
Volume 48	**Wood Pole Overhead Lines** B. Wareing
Volume 49	**Electric Fuses, 3rd Edition** A. Wright and G. Newbery
Volume 50	**Wind Power Integration: Connection and system operational aspects** B. Fox *et al.*
Volume 51	**Short Circuit Currents** J. Schlabbach
Volume 52	**Nuclear Power** J. Wood
Volume 53	**Condition Assessment of High Voltage Insulation in Power System Equipment** R.E. James and Q. Su
Volume 55	**Local Energy: Distributed generation of heat and power** J. Wood
Volume 56	**Condition Monitoring of Rotating Electrical Machines** P. Tavner, L. Ran, J. Penman and H. Sedding
Volume 57	**The Control Techniques Drives and Controls Handbook, 2nd Edition** B. Drury
Volume 58	**Lightning Protection** V. Cooray (Editor)
Volume 59	**Ultracapacitor Applications** J.M. Miller
Volume 62	**Lightning Electromagnetics** V. Cooray
Volume 63	**Energy Storage for Power Systems, 2nd Edition** A. Ter-Gazarian
Volume 65	**Protection of Electricity Distribution Networks, 3rd Edition** J. Gers
Volume 66	**High Voltage Engineering Testing, 3rd Edition** H. Ryan (Editor)
Volume 67	**Multicore Simulation of Power System Transients** F.M. Uriate
Volume 68	**Distribution System Analysis and Automation** J. Gers
Volume 69	**The Lightening Flash, 2nd Edition** V. Cooray (Editor)
Volume 70	**Economic Evaluation of Projects in the Electricity Supply Industry, 3rd Edition** H. Khatib
Volume 72	**Control Circuits in Power Electronics: Practical issues in design and implementation** M. Castilla (Editor)
Volume 73	**Wide Area Monitoring, Protection and Control Systems: The enabler for Smarter Grids** A. Vaccaro and A.Zobaa (Editors)
Volume 74	**Power Electronic Converters and Systems: Frontiers and applications** A. M. Trzynadlowski (Editor)
Volume 75	**Power Distribution Automation** B. Das (Editor)
Volume 76	**Power System Stability: Modelling, analysis and control** B. Om P. Malik
Volume 78	**Numerical Analysis of Power System Transients and Dynamics** A. Ametani (Editor)
Volume 79	**Vehicle-to-Grid: Linking electric vehicles to the smart grid** J. Lu and J. Hossain (Editors)
Volume 86	**Advances in Power System Modelling, Control and Stability Analysis** F. Milano (Editor)
Volume 101	**Methane and Hydrogen for Energy Storage** R. Carriveau & David S-K. Ting
Volume 905	**Power system protection, 4 volumes**

Cogeneration and District Energy Systems

Modeling, Analysis and Optimization

Marc A. Rosen and
Seama Koohi-Fayegh

The Institution of Engineering and Technology

Published by The Institution of Engineering and Technology, London, United Kingdom

The Institution of Engineering and Technology is registered as a Charity in England & Wales (no. 211014) and Scotland (no. SC038698).

First published 2016

The Institution of Engineering and Technology
Michael Faraday House
Six Hills Way, Stevenage
Herts, SG1 2AY, United Kingdom

www.theiet.org

British Library Cataloguing in Publication Data
A catalogue record for this product is available from the British Library

ISBN 978-1-78561-126-1 (hardback)
ISBN 978-1-78561-127-8 (PDF)

Typeset in India by MPS Limited
Printed in the UK by CPI Group (UK) Ltd, Croydon

To my wife, Margot, and my children, Allison and Cassandra, for their inspiration and love.
And to my parents for their love and support.
Marc A. Rosen
&
To my husband, Ali, and my mother, for their inspiration and love.
And to the memory of my father.
Seama Koohi-Fayegh

Contents

Preface

District energy (DE) systems use central heating and/or cooling facilities to provide heating and/or cooling services for communities. The advantages of district energy over conventional heating and cooling include improved efficiency, reliability and safety, reduced environmental impact, and for many situations better economics. DE systems can be particularly beneficial when integrated with cogeneration plants for electricity and heat, i.e., with combined heat and power (CHP) plants. One of the main impediments to increased use of cogeneration-based district energy is a lack of understanding of the behavior of integrated forms of such systems. This book is aimed at providing information on district energy and cogeneration technologies, as well as systems that combine them.

In this book, therefore, cogeneration and district energy systems are described, with a focus on their modeling, analysis and optimization. The modeling includes characterizing the general configurations and components of cogeneration and DE systems, and the development, verification and validation of models for such systems. The analysis is focused on the application of thermodynamic analysis and economic assessment techniques to these systems. The optimization involves the identification of objective functions and constraints, and the development and application of optimization tools and computer codes.

The book contains several main parts, which indicate its scope. At the start, background information is provided, the motivation, objectives and rationale for the book are discussed, and the approach taken in the book is described (Chapter 1). Then, relevant background material on thermodynamic analysis is presented, with a comprehensive treatment of exergy analysis, including a procedure for applying energy and exergy analyses (Chapter 2).

In the next part of the book (Chapters 3–5), models and descriptions are presented for the core technologies relevant to the book:

- cogeneration, or combined heat and power,
- heating and district heating, and
- chilling and district cooling.

Configurations for integrating cogeneration and district energy technologies are then described and assessed (Chapter 6). To illustrate the material discussed in the book, and to help demonstrate the merits of different integrated systems for cogeneration and district energy, a comparison of several such systems is presented (Chapter 7).

To address the importance of economics in design and decision-making, the economics of cogeneration and district energy technologies are discussed, as are appropriate optimization procedures (Chapter 8).

In the next part of the book (Chapters 9 and 10), the focus shifts to the growing concerns with environmental issues in general and climate change in particular. The environmental impact of cogeneration systems is described, with a focus on the allocation wastes for cogeneration systems (Chapter 9), while the contributions of cogeneration to climate change are discussed, with an emphasis on carbon dioxide emissions (Chapter 10).

Next, a more detailed routine and computer code for comparing and optimizing the benefits of such systems while accounting for constraints such as environmental emission limits is presented (Chapter 11).

Developments and advances in cogeneration and district energy, and related technologies, such as trigeneration, multigeneration and distributed energy systems, are then described, drawing heavily on recent research and development activities throughout the world (Chapter 12).

Finally, closing remarks and future considerations are presented, including perspectives on the state of the art of the technologies of cogeneration and district energy, and their integration, as well as thoughts on future directions (Chapter 13).

Examples and case studies are included throughout the book to illustrate the material covered, and to demonstrate the importance, benefits and value of existing cogeneration and district energy technologies in achieving sustainable and efficient energy systems. The book also contains many references, including some related to books and articles on various aspects of cogeneration and district energy systems that are not fully covered. Some links to websites with basic freeware for ground-source heat transfer modeling and building heating loads are referenced throughout the book.

The book is research oriented, thereby ensuring that new developments and advances in cogeneration and DE systems are covered. This is particularly evident in Chapter 12, where many recent developments and advances in cogeneration, district energy and related technologies are described.

The book strives to provide clear information on cogeneration, district energy and related technologies and the many advances occurring in the field in a way that makes it understandable for students, practitioners, researchers and policy-makers. The book is intended for use by engineers and researchers in cogeneration, district energy and related technologies. Readers are likely to be from several engineering disciplines, including energy engineering, mechanical engineering, chemical engineering, environmental engineering, process engineering and industrial engineering. The book is also likely to be of use in research centers, institutes and labs focusing on the areas mentioned above, by related learned societies and professional associations, and in industrial organizations and companies interested in cogeneration, district energy and their applications. The book is also intended for use by advanced undergraduate or graduate students in various engineering disciplines, including those mentioned above. The information included in this book is sufficient for energy, environment and sustainable development courses. Furthermore, the book offers a valuable and readable reference source for anyone interested in learning about cogeneration, district energy systems and related technologies.

The main features of the book follow:

- comprehensive coverage of cogeneration, district energy and related technologies, and their integration to form cogeneration-based district energy systems
- detailed descriptions of methods to determine potential environmental and ecological impacts of cogeneration and district energy systems and their benefits for combating climate change
- extensive coverage of the economics of cogeneration and district energy, as well as optimization of such systems
- an emphasis on utilizing exergy methods, so as to improve the thermodynamic treatment of technologies and systems for cogeneration and district energy
- presentations of the most up-to-date information in the area
- suitability as a good reference for cogeneration and district energy
- a research orientation to provide coverage of the state of the art and emerging trends and recent developments
- numerous illustrative examples and case studies
- clarity and simplicity of presentation of cogeneration and district energy systems

We hope this book allows cogeneration and district energy to be used more widely for the provision of electrical, heating and cooling services in a sustainable manner, using both existing and conventional equipment and systems as well as new and advanced technologies. By exploiting the benefits of applying these energy technologies and systems, we believe that they can be made more efficient, clean and sustainable, and help humanity address many of the challenges it faces.

Marc A. Rosen and Seama Koohi-Fayegh
September 2016

About the Authors

Marc A. Rosen is a Professor at the University of Ontario Institute of Technology in Oshawa, Ontario, Canada, where he served as founding Dean of the Faculty of Engineering and Applied Science. A former President of the Engineering Institute of Canada and the Canadian Society for Mechanical Engineering, he is a registered Professional Engineer in Ontario. He has served in many professional capacities, including Editor-in-Chief of several journals and a member of the Board of Directors of Oshawa Power and Utilities Corporation. He is an active teacher and researcher in energy, sustainability, geothermal energy and environmental impact. Much of his research has been carried out for industry, and he has written numerous books. He was a member of an International Energy Agency Experts Group on cogeneration/district cooling, and the Canadian Experts Group on the same topic. He has worked for organizations such as Imatra Power Company in Finland, Argonne National Laboratory near Chicago, and the Institute for Hydrogen Systems near Toronto. He has received numerous awards and honors, including an Award of Excellence in Research and Technology Development from the Ontario Ministry of Environment and Energy, the Engineering Institute of Canada's Smith Medal for achievement in the development of Canada and the Canadian Society for Mechanical Engineering's Angus Medal for outstanding contributions to the management and practice of mechanical engineering. He is a Fellow of the Engineering Institute of Canada, the Canadian Academy of Engineering, the Canadian Society for Mechanical Engineering, the American Society of Mechanical Engineers, the International Energy Foundation and the Canadian Society for Senior Engineers.

Seama Koohi-Fayegh is a Post-doctoral Fellow at the Department of Mechanical Engineeering at the University of Ontario Institute of Technology (UOIT) in Oshawa, Canada. She received her PhD in Mechanical Engineering at UOIT under supervision of Prof. Marc A. Rosen. Her PhD thesis topic was proposed by Ontario ministry of environment and was focused on thermal sustainability of geothermal energy systems: system interactions and environmental impacts. She did her master study in Mechanical Engineering (Energy Conversion) at Ferdowsi University of Mashhad, Iran, and worked on entropy generation analysis of condensation with shear stress on the condensate layer. Her thesis research won multiple awards at the school level and at the Iranian Society of Mechanical Engineering in 2009. Her research interests include heat transfer, sustainable energy systems and energy technology assessment.

Acknowledgments

The inspiration and thought-provoking insights provided by many of our colleagues helped greatly in the development of this book, and is gratefully acknowledged. Some of the material in this book is derived from research that we have carried out with numerous collaborators over the years. Particular thanks in this regard are offered to several former faculty and students of Ryerson University, Toronto, Ontario, Canada, especially Judith Dimitri, June Wu, Minh Le, Melanie Sherwood and Jack Simpson.

Elements of the material covered in this book were assisted by ideas that germinated through collaborations with CanmetENERGY, a research and technology organization in the field of clean energy within Natural Resources Canada. Specific thanks for the valuable insights and advice are offered to Michael Wiggin, Paul Sears, Cliff Adams and Ingvar Larsson.

Last but not least, the authors warmly thank their families, for their endless encouragement and support throughout completion of this book. Their patience and understanding is most appreciated.

Marc A. Rosen and Seama Koohi-Fayegh
September 2016

Nomenclature

AFUDC	allowance for funds used during construction
BAU	business as usual
C	waste; CO_2 emissions; vector of unit costs associated with supply variables
c	unit economic value of the product of the cogeneration system
COP	coefficient of performance
D	matrix for demand requirement constraints
DC	direct cost
e	specific energy; electricity
E	energy; annual rate of CO_2 emissions
er	CO_2 emission quantity
Ex	exergy
F	fuel consumption
f	figure of merit; fraction of wastes
FC	fuel cost
FCI	fixed-capital investment
h	specific enthalpy; specific base enthalpy
HHV	higher heating value
I	exergy consumption
IC	indirect cost
ke	specific kinetic energy
LHV	lower heating value
LRD	licensing, research and development costs
M	accumulation of CO_2 in atmosphere
m	mass
N	moles
OFSC	off-site cost
ONSC	on-site cost
pe	specific potential energy

P	absolute pressure; vector of prices; price
P_{cost}	sum of discounted costs incurred by producers
PEC	purchased-equipment cost
PFI	plant facility investment
Q	heat interaction; vector of demands; quantity
R	discount factor per period
r	discount rate
s	specific entropy; scenario
SUC	start-up cost
T	temperature
t	time
th	thermal
TCI	total capital investment
u	specific internal energy
V	vector of dual variable for demand requirement constraint
v	specific volume
W	present worth of partial social welfare
w	work
WC	working capital
X	quantity of energy flow
Z	vector of all supply activities

Greek symbols

Δ	change
δ	parameter indicating natural reduction of CO_2 accumulated in atmosphere
ε	specific exergy
$\in$	exergy
$\in^Q$	thermal exergy transfer
η	energy efficiency
μ	chemical potential
Π	entropy production
σ	rate of CO_2 absorption by oceans, plants and other absorbing media
τ	exergetic temperature factor
φ	waste coefficient for fuel
ψ	exergy efficiency

Subscripts

a	air
b	boiler
c	curtailed
ch	chiller
CHP	combined heat and power (i.e., cogeneration)
COP	coefficient of performance
E	electrical
e	exit
elec	all-electric mode of operation
ex	export
f	fuel
H	heating
i	inlet
im	import
j	species *j*
kin	kinetic component
load	load
n	nuclear
net	net
o	environmental state; chemical exergy
oo	dead state
p	product
pp	power plant
ph	physical component
pot	potential component
Q	thermal
r	region of heat interaction
sys	system
w	waste

Superscripts

·	rate with respect to time

Acronyms

BFW	boiler feed water
CFC	chlorofluorocarbon
DE	district energy
GT	gas turbine
HVAC	heating, ventilating and air conditioning
ORC	organic Rankine cycle
O&M	operating and maintenance
PFBC	pressurized fluid-bed combustion
TES	thermal energy storage

Chapter 1
Introduction

Overview

An introduction is provided to the material covered in the book, starting with the groundwork related to cogeneration and district energy systems, and their modeling, analysis and optimization. The motivation for the book is also provided, in order to explain its rationale and the needs it seeks to satisfy – in terms of the technologies involved as well as the emphasis on modeling, analysis and optimization. The aims of the book are also given, complementing the material given in the motivation. The approach taken in preparing the book is described, to ensure that it provides appropriate and comprehensive coverage of cogeneration and district energy systems as well as their modeling, analysis and optimization. As part of that coverage, the approach involving an emphasis on modeling, analysis and optimization is detailed and explained. The scope of the book is explained to make it clear to the reader what is and is not covered, and a detailed outline of the book is provided.

1.1 Groundwork

District energy systems, which can include both district heating and district cooling systems, use central heating and/or cooling facilities to provide heating and/or cooling services for groups of buildings or communities. The advantages of district energy systems over conventional heating and cooling systems include improved efficiency, reliability and safety, reduced environmental impact, and for many situations better economics.

Cogeneration is the simultaneous generation of two products in a single device. Cogeneration usually refers to the production of electricity and useful heat. As a consequence, it is also commonly referred to as combined heat and power (CHP). Cogeneration constitutes an advantageous and useful technology, particularly because it is usually much more efficient that separate processes for the co-products.

District energy systems and CHP plants, i.e., cogeneration plants for electricity and heat, are in general complementary technologies. If the thermal energy outputs of a cogeneration plant are needed at a distance from the cogeneration plant (tens of meters to tens of kilometers away), rather than onsite, a district energy system is a useful addition. In fact, district energy systems can be particularly beneficial when

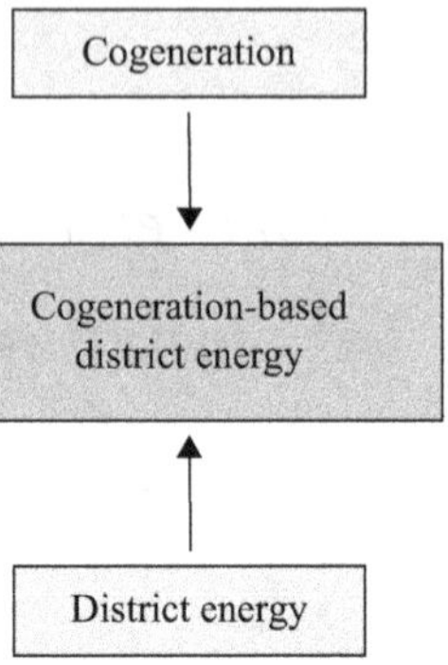

Figure 1.1 Conceptual illustration of integration of cogeneration and district energy, which are complementary technologies when the thermal energy outputs of cogeneration are needed at a distance from the cogeneration site, to form cogeneration-based district energy systems

integrated with cogeneration plants for electricity and heat. A simple illustration of this pairing of technologies is shown in Figure 1.1.

Cogeneration and district energy systems have been examined and applied broadly, and are increasingly identified as potentially beneficial in many applications. Although the number of cogeneration and/or district energy systems is relatively small at present compared to the potential number of applications, the role of such systems is growing.

1.2 Motivation

There exist challenges and barriers that limit the expansion of the use of cogeneration and district energy. One of the main impediments to increased use of cogeneration-based district energy systems is a lack of experience with and understanding of the behavior of integrated forms of such systems. A larger base of knowledge exists for each of the relevant technologies for circumstances when they are applied independently. However, the behavior and characteristics of these separate technologies when they are combined to form integrated energy systems is not always predictable in a straightforward manner and often can be complex and confusing.

Many important and interesting thermodynamic implications arise from the integration of cogeneration systems with district energy technologies (including technologies for heating, chilling, and district heating and cooling). When integrating cogeneration and chilling technologies, for example, it is not necessarily true that the efficiency of such an integrated system is maximized if the efficiencies of the components from which it comprised are each maximized, even though one may intuitively feel on first consideration that that would be the case.

To illustrate this point, consider the following two cases. In case A, the product heat from a cogeneration plant drives a single-effect absorption chiller to satisfy a fixed cooling requirement, and excess electricity is exported. Case B is similar, except that the chiller is replaced with a double-effect device, which has a higher efficiency but also requires heat at a higher temperature. Since the electrical output of a cogeneration plant generally decreases as the temperature of the product heat increases, the cogeneration plant in case B could generate less electricity than that in case A (depending on the quantity of heat required to drive each chiller), while satisfying the same cooling requirement. Hence an integrated system with a more efficient chiller (case B) can be less efficient than one with a less efficient chiller (case A), and this can be demonstrated to occur in many situations.

1.3 Aims

This example highlights that the appropriate integration of cogeneration and district energy (specifically district cooling/chiller technologies in the above example) requires a good understanding of the systems involved and how they can be integrated. This book provides such information by investigating the implications of alternatives for integrating cogeneration and district energy technologies, and aims at improving understanding of cogeneration and district energy systems and their integration, as well as fostering increased utilization of these technologies in beneficial ways.

The utilization of district energy and cogeneration could be greatly increased if more information on these technologies could be made available. Thus, we hope that this book makes significant contributions to forwarding these technologies, by helping to satisfy this need for information. Beyond discussing existing technology for cogeneration and district energy, ongoing developments and advances are also covered, in part to reflect the need for increased research efforts to advance the technology for cogeneration-based district energy systems.

1.4 Approach

Descriptions are provided of various cogeneration and district energy technologies, as well as integrated energy systems that combine these technologies advantageously. The descriptions cover technical, environmental, economic and other factors.

A good understanding of these technologies and their applications relies heavily on modeling, analysis and optimization. Hence, a strong focus is placed in this book, on the modeling, analysis and optimization of cogeneration-based district energy systems.

The modeling includes characterizing the general configurations and components of cogeneration and district energy (DE) systems, and the development, verification and validation of models for such systems.

The analysis focuses on the application of thermodynamic analysis, especially second-law analysis, and economic assessment techniques applied to these systems. The optimization work involves the identification of objective functions and constraints, and the development and application of optimization tools and computer codes.

The reasons for considering second-law analysis may not be obvious to all readers, and thus merit elaboration. In the development of energy technologies, their thermodynamic performances are often investigated using energy analysis. A better understanding is attained, however, when a more complete thermodynamic view is taken. Exergy analysis (or second-law analysis) provides an additional thermodynamic perspective and, in conjunction with energy analysis, permits the performance of more complete thermodynamic analyses. In simple terms, exergy is that portion of energy which is available for performing useful tasks. For conventional energy technologies, energy and exergy analyses have been performed and have yielded useful results. For advanced energy technologies, the use of energy and exergy analyses can be expected to provide meaningful insights into performance that assist in achieving optimal designs. Consequently, both energy- and exergy-based approaches to thermodynamic analysis are considered here.

In addition to the focused material on modeling, analysis and optimization, recent research reported in the literature and elsewhere is examined, to ensure the reader has an appreciation of new developments and possible future advances.

1.5 Scope

The focus of this book is on cogeneration and district energy systems. Considering the possible ways these technologies can be broken down into smaller parts, or integrated into more complex forms, provides a good perspective on the scope of the book. Figure 1.2 presents such a summary in diagrammatic form.

Cogeneration and district energy technologies are shown in Figure 1.2. The following main points can be seen:

- Cogeneration can involve simple production of electricity and thermal energy in the form of heat, or it can involve production of electricity and thermal

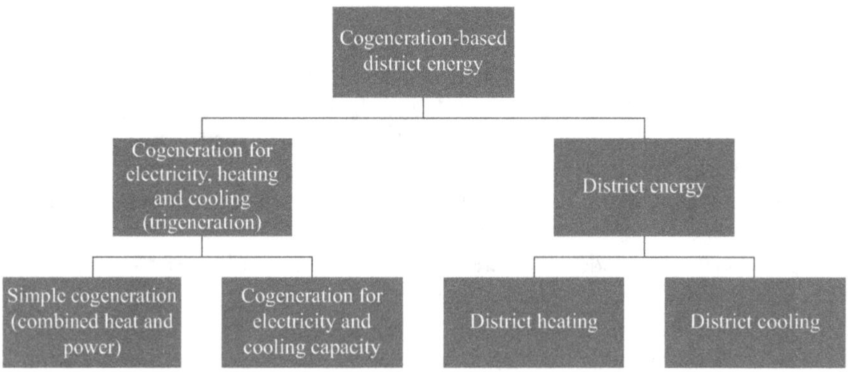

Figure 1.2 Cogeneration and district energy technologies, showing a breakdown into their main parts and their integration to form cogeneration-based district energy systems

energy in the form of "cool." In its extended or broader sense, cogeneration can involve the production of electricity, heat and "cool." In the latter case, it is often referred to as trigeneration or as combined cooling, heating and electricity (CCHP).

- Also presented is district energy and its breakdowns into its two main parts, i.e., district heating and district cooling. District heating comprises heating technology and piping technology for hot fluids (hot water, steam, etc.), while district cooling is made up of cooling technology and piping technology for cool fluids (cold water, cold brine, cold water slurry, etc.).
- The integration of these technologies leads to the formation of cogeneration-based district energy, which is presented at the top of the diagram.

All of the technologies illustrated in Figure 1.2 together encompass the scope of this book.

1.6 Outline of the book

This book contains 13 chapters (in addition to this introductory chapter), which indicate its scope and intent. An outline of the structure of the book and its primary content is provided here.

The first two chapters are focused on introductory and background material and are intended to ensure the reader has the requisite knowledge for the material covered throughout the book. Specifically:

- Chapter 1 provides an introduction to the material covered in the book. The motivation is provided to explain its rationale and the needs it seeks to satisfy. The approach taken to ensure the book provides appropriate and comprehensive coverage of cogeneration and district energy systems as well as their modeling, analysis and optimization, is described.
- Chapter 2 provides relevant background material on thermodynamic analysis, with a description of the concepts involved in exergy analysis. The fundamentals of exergy are covered, and a procedure for applying energy and exergy analyses is presented and illustrated. Various energy and exergy efficiencies are defined and the implications are explained of energy and exergy analyses, particularly aspects relevant to cogeneration and/or district energy.

Chapters 3–5, respectively, describe and discuss the core technologies covered in the book: cogeneration, heating and district heating, and chilling and district cooling. That is:

- Chapter 3 describes cogeneration, starting with fundamentals and definitions and extending to benefits, uses, operation and applications. Cogeneration is compared with thermal electrical generation, highlighting how cogeneration is an extension, and the trade-off between thermal and electrical outputs of cogeneration is explained. The interface between cogeneration and energy storage is introduced and the categories of heat demands that can normally be satisfied with cogeneration are described: residential, commercial, institutional and industrial processes. The selection of the size and type of a cogeneration system to match the thermal and electrical demands is discussed, as are matching schemes.

- Chapter 4 discusses systems for heating and district heating. A general model for heating and district heating systems is provided, with energy and exergy analyses of it. Characteristics of district heating systems are provided, including advantages, operation, applications and energy sources. The types of district heating systems covered include fossil fuel, geothermal, solar, biomass and waste-to-energy-based district heating and typical energy conversion efficiencies are given for heating and district heating.
- Chapter 5 covers systems for chilling and district cooling. A general model for cooling and district cooling systems is provided to facilitate engineering descriptions and thermodynamic assessments. The coverage for cooling systems includes applications, types, classifications, operation, performance and efficiency. Characteristics of district cooling systems are provided, including advantages, operation, applications, distribution and capacity.

Chapters 6 and 7 focus on integrated systems for cogeneration and district energy, including descriptions, comparisons and illustrations:

- Chapter 6 describes and discusses configurations for integrating cogeneration and district energy technologies. A general model for cogeneration-based district energy for engineering and thermodynamic evaluations is presented. Measures of system efficiency are also provided and examined in detail, emphasizing the need for proper measures of merit for systems for cogeneration-based district energy. It is also explained how district cooling using absorption chillers often complements district heating when both use cogenerated heat by balancing seasonal loads.
- Chapter 7 presents a comparison of several configurations for integrating cogeneration and district energy technologies and systems. The comparison is intended to illustrate the material discussed previously in the book, and to help demonstrate the merits of different integrated cogeneration and district energy systems.

Chapters 8–10 cover economics as well as environmental factors, particularly greenhouse gas (GHG) emissions relevant to cogeneration systems and the allocations of wastes. Specifically:

- Chapter 8 discusses the economics of cogeneration and district energy technologies, beginning with fundamentals and general economic considerations such as methods for estimating total capital investment (TCI), performing economic evaluations, and calculating revenue requirements and extending to optimization procedures. Economic considerations involved in integrated cogeneration-based district energy systems are examined, for generators, distributors and consumers. An analogy is presented between allocating wastes for cogeneration among products, and allocating economic costs among products.

- Chapter 9 covers environmental factors relevant to cogeneration systems, with a focus on wastes from cogeneration and their allocations to products. Various allocation methods for cogeneration wastes are described, including methods based on techno-economic and other factors. This material can benefit society through the better design and utilization of cogeneration technologies based on environmental considerations, and through better decision and policy-making.
- Chapter 10 parallels the previous chapter, but focusing on GHG emissions relevant to cogeneration systems and climate change. The allocation of carbon dioxide emissions for cogeneration is described, and the use of that information for developing carbon dioxide emissions credits for trading purposes from switching to cogeneration from equivalent independent plants is described.

The book is brought to a close with three final chapters, focusing on optimization of cogeneration-based district energy systems, as well as recent developments and advances. That is:

- Chapter 11 presents a more detailed routine and computer code, developed by the authors for comparing and optimizing the benefits of such systems while accounting for constraints such as environmental emission limits. The modeling and optimization of cogeneration-based district energy systems help illustrate and demonstrate the comparative merits of different integrated systems for cogeneration and district energy.
- Chapter 12 describes developments and advances in cogeneration and district energy, and related technologies such as trigeneration, multigeneration and distributed energy systems, drawing heavily on recent research and development activities throughout the world.
- Chapter 13 presents closing remarks, including perspectives on the state of the art of the technologies of cogeneration and district energy, as well as their integration, and thoughts on future directions.

Chapter 2

Thermodynamic analysis: fundamentals, energy and exergy

Overview

The fundamentals of thermodynamics and thermodynamic analysis are provided, as they are central to the contents of the book. The centrality of thermodynamics can be seen by noting that it constitutes the study of the concepts and laws describing energy and its conversion in processes and systems. The means by which thermodynamics permits the behavior, performance and efficiency of energy systems to be described, particularly via energy and exergy analyses, is described at length. Details on aspects of thermodynamics most relevant to energy and exergy analyses are presented and illustrated, along with descriptions of energy and exergy analyses themselves. Thermodynamic balances and basic quantities in them are described, including the exergy of matter, heat, work and electricity. The reference environment used in exergy analysis is described and models for it (e.g., natural-environment-subsystem, reference-substance, equilibrium and constrained-equilibrium, and process-dependent models) are detailed. Various energy and exergy efficiencies are defined and properties for energy and exergy analyses are provided. The implications are explained of energy and exergy analyses, particularly on research and development. A step-by-step procedure is given for energy and exergy analyses, and a detailed example is given to illustrate how the analyses are applied. Finally, exergy values for typical commodities encountered in cogeneration and district energy (e.g., electricity, work, heated and cooled substances) are provided, and extensions are explained of exergy methods beyond thermodynamics to areas such as economics and environmental studies. The coverage in this chapter is kept general where possible, although points relevant to cogeneration and/or district energy are raised as appropriate.

2.1 Introduction

Thermodynamics is the study of energy, including the various concepts and laws describing the conversion of energy from one form to another, and the various systems used in the conversions. Thermodynamics permits the behavior, performance and efficiency of engineering systems, and particularly energy systems, to be described.

Numerous investigations and applications have been reported for energy analysis (e.g., [1–4]), since it is the conventional method of thermodynamic analysis.

In the past decades, exergy analysis – an alternative but complementary approach to energy analysis – has received increasing recognition by researchers in industry, academia and government, and the number of reported applications has grown considerably [5–20].

It is noted that, although efficiencies based on exergy provide a measure of how nearly system performance approaches the ideal (a desirable objective from a thermodynamic perspective), such ideality does not in general lead to optimal economic performance. Clearly, a trade-off must be made between thermodynamics (based on energy or exergy) and economics to obtain optimal economic performance.

In sections 2.4–2.11, as well as in many references [1–90], details on aspects of thermodynamics most relevant to energy and exergy analyses are presented and illustrated. The coverage in this chapter is kept intentionally general in most parts, although points of direct or special relevance to cogeneration and/or district energy are raised where appropriate. For instance, the final section in this chapter, exergy values for typical commodities encountered in cogeneration and district energy are described and assessed.

2.2 Energy analysis

Conventional thermodynamic analysis is based primarily upon the first law of thermodynamics, which states the principle of conservation of energy. This law has been presented in a variety of ways (e.g., energy is always conserved when it is transformed from one form to another, energy can neither be created nor be destroyed, the sum total of all energy remains constant), all in essence have the same fundamental meaning.

An energy analysis of an energy conversion system is essentially an accounting of the energies entering, exiting and stored within the system. The exiting energy can be broken down into products and wastes. Efficiencies, normally evaluated as ratios of energy quantities, are often used to assess and compare systems.

However, energy analysis has several shortcomings:

- It is the usefulness or quality of an energy quantity, rather than simply the energy quantity itself, that is of value. For example, the heat rejected from the condensers of an electrical generating station, although great in quantity, is of little usefulness since its temperature is only a few degrees above that of the surrounding water or air (i.e., the thermal energy is of low quality). Focusing an analysis on energy quantities, therefore, can be misleading in that it puts the emphasis of the analysis on the wrong quantity.
- Energy efficiencies are often misleading, in that they do not always provide a measure of how nearly the performance of a system approaches ideality (or reversibility, in thermodynamic terms). For example, the thermal efficiency of a Carnot (ideal) power plant operating between heat reservoirs at 300 and 600 K, respectively, is 50 per cent, even though the meaning of efficiency would lead one to believe that the efficiency of any ideal device should be 100 per cent. Consequently, energy analysis often indicates a state of technological efficiency different than actually exists.

- The thermodynamic losses which occur within a system (i.e., those factors which cause performance to deviate from ideality) often are not accurately identified and assessed with energy analysis. Although energy is conserved, and therefore energy losses must be associated with waste energy emissions from a system, energy quality is not conserved and can be degraded during a process (even if there are no energy losses). For example, the energy efficiency for electrical resistance heating is almost 100 per cent as there are almost no energy losses; however, the quality of the energy is greatly degraded in converting it from electricity to heat at relatively low temperature, and this "loss" is not accounted for with energy analysis. Energy quality losses can be associated with waste emissions of energy and with energy quality degradation within a system. Consequently, energy analysis often indicates the main inefficiencies to be within the wrong sections and associated with the wrong processes of a system.

2.3 Exergy analysis

Exergy analysis permits many of the shortcomings of energy analysis to be overcome. Exergy is the maximum work obtainable as a system comes to equilibrium with a reference environment, or, more simply, exergy is that portion of energy which is available for performing useful tasks. Exergy analysis identifies the causes, locations and magnitudes of process inefficiencies, and is founded upon the second law of thermodynamics. The second law states that, although energy cannot be created or destroyed, it can be degraded in quality, eventually reaching a state in which it is in complete equilibrium with the surroundings and hence of no further use for performing tasks. The terms "energy crisis" and "energy conservation" are incomplete in that the concern is not with the conservation of energy, as first law assures, but with the conservation of the energy's ability to perform tasks (i.e., its exergy).

For electrical generating stations, for example, exergy analysis allows one to determine the maximum potential for electricity production associated with the incoming fuel. This maximum is only achieved if the fuel is utilized in processes that ultimately bring it to complete thermodynamic equilibrium with the environment, while producing electricity in a reversible manner. No further electricity can be extracted by allowing a system and its environment to interact if they are in equilibrium. Losses in the potential for electricity generation occur in the real world because actual processes are always irreversible. Exergy analysis states the theoretical limitations imposed upon any electrical generating station, clearly pointing out the portion of the input energy that can be converted to electricity. Also, exergy analysis determines losses as a direct quantitative measure of lost potential for electricity.

Several of the key ways in which exergy analysis permits the shortcomings of energy analysis to be overcome are illustrated in terms of thermodynamic parameters in Table 2.1. The main factors or measures considered are quality of energy, efficiencies, thermodynamic losses and the basis for the method. The advantages of exergy analysis, and the ways these advantages overcoming challenges with energy analysis, are explicitly highlighted through this comparative table.

Table 2.1 Thermodynamic rationale for exergy analysis over energy analysis

Factor or measure	Energy analysis shortcoming	Exergy analysis advantage
Quality	Ignores quality (or usefulness or value) of energy quantities	Explicitly accounts for quality (or usefulness or value) of energy quantities
Efficiency	Can provide misleading efficiencies that do not reflect how efficiently a system or process performs	Always provides meaningful efficiencies that measure how efficiently a system or process performs relative to ideality (i.e., reversibility)
Loss	Does not in general identify clearly the causes, locations and magnitudes of thermodynamic losses for a process or system	Identifies clearly the types, causes, locations and magnitudes of thermodynamic losses for a process or system in terms of detractions from efficiency, and separates losses into external waste exergy emissions and internal exergy destructions
Basis	Limited to the first law of thermodynamics (which embodies the principle of conservation of energy)	Comprehensively exploits thermodynamics, being premised primarily on both the first law of thermodynamics (which embodies the principle of conservation of energy) and the second law of thermodynamics (which embodies the principle of nonconservation of entropy)

2.3.1 The exergy method of analysis

As the thermodynamic analysis tool exergy analysis is central to this book, the exergy methodology is described and illustrated in depth.

The exergy of an energy form or a substance is a measure of its usefulness or quality. Exergy is based on the first and second laws of thermodynamics, and combines the principles of conservation of energy and nonconservation of entropy.

Exergy is defined as the maximum amount of work which can be produced by a system or a flow of matter or energy as it comes to equilibrium with a reference environment. Exergy is a measure of the potential of the system or flow to cause change, as a consequence of not being completely in stable equilibrium relative to the reference environment. Unlike energy, exergy is not subject to a conservation law (except for ideal, or reversible, processes). Rather exergy is consumed or destroyed, due to irreversibilities in any real process. The exergy consumption during a process is proportional to the entropy created due to irreversibilities associated with the process. Exergy and exergy analyses are discussed further elsewhere [1,2,4–6,22–29].

As a simple example, consider an adiabatic system containing fuel and air at ambient conditions. The fuel and air react to form a mixture of hot combustion gases. During the combustion process, the energy in the system remains fixed because it is adiabatic. But the exergy content declines as combustion proceeds due to the irreversibilities associated with the conversion of the high-quality energy of

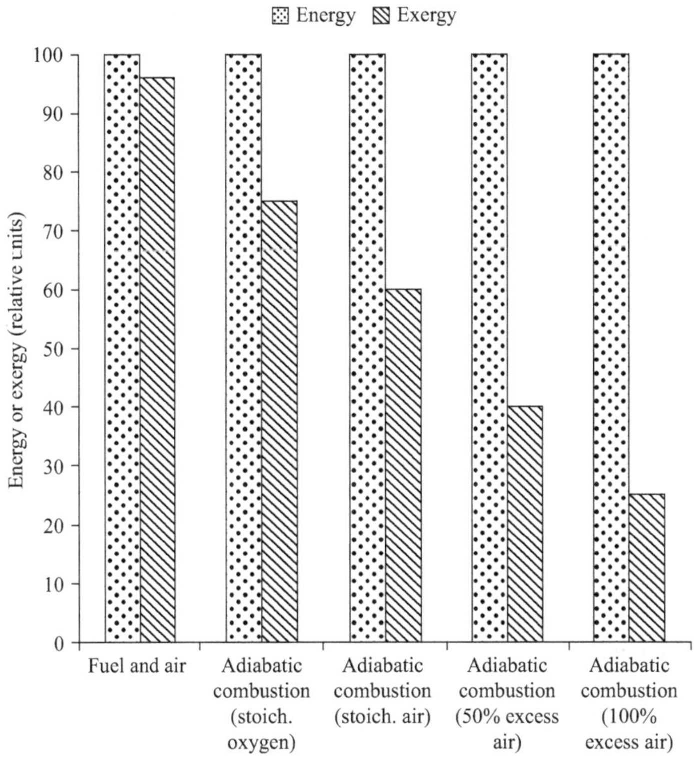

Figure 2.1 Qualitative comparison of the changes in energy and exergy quantities during the adiabatic combustion of a fossil fuel. The energy content of the fuel and air mixture prior to combustion (at the left) is seen to remain constant as adiabatic combustion occurs for a range of scenarios (represented by the rightmost four pairs of bars). But the exergy content of the fuel and air mixture prior to combustion is seen to decline as combustion occurs, and the decrease in exergy is seen to become more pronounced as the combustion occurs with an increasing amount of matter beyond the stoichiometric oxygen requirement

fuel to the lower quality energy of combustion gases. The different behaviors of energy and exergy during this process are illustrated qualitatively in Figure 2.1.

For exergy analysis, the characteristics of the reference environment must be specified completely. This is commonly done by specifying the temperature, pressure and chemical composition of the reference environment. The results of exergy analyses, consequently, are relative to the specified reference environment, which in most applications are modeled after the actual local environment. The exergy of a system is zero when it is in equilibrium with the reference environment. The tie between exergy and the environment has implications regarding environmental impact has been investigated previously [1,5,30–40].

Exergy analysis is a methodology that uses the conservation of mass and conservation of energy principles together with the second law of thermodynamics for the analysis, design and improvement of energy and other systems. The exergy method is useful for improving the efficiency of energy-resource use, for it quantifies the locations, types and magnitudes of wastes and losses. In general, more meaningful efficiencies are evaluated with exergy analysis rather than energy analysis, since exergy efficiencies are always a measure of the approach to the ideal. Therefore, exergy analysis identifies accurately the margin available to design more efficient energy systems by reducing inefficiencies. Many engineers and scientists suggest that thermodynamic performance is best evaluated using exergy analysis because it provides more insights and is more useful in efficiency-improvement efforts than energy analysis.

Many applications of exergy analysis to processes and systems have been reported [1,41–47]. Some of those have included investigations by one of the present authors of cogeneration using exergy methods [1,48,49].

2.3.2 Improving efficiency with exergy analysis: illustration for electricity generation

The use of exergy methods to analyze a device so as to permit its performance to be better understood and its efficiency improved is demonstrated for the process of electricity generation using a coal-fired steam power plant.

The plant considered is the coal-fired Nanticoke generating station, which has been operating since 1981 in Ontario, Canada. Each of the eight units in the station has a net electrical output of 505 MW. A single unit of the electrical generating station is illustrated in Figure 2.2, and consists of four main sections [44,45,48]:

(a) **Steam generation**. This section contains the steam generators. Pulverized-coal-fired natural-circulation steam generators combust coal to produce primary and reheat steam. Air is supplied to the furnace by motor-driven forced-draft fans, and regenerative air preheaters are used. The flue gas passes through an electrostatic precipitator and exits the plant via multi-flued chimneys.

(b) **Power production**. This section contains the turbine generators and transformers. The steam produced passes through a turbine generator, which is connected to a transformer. Each turbine generator has one single-flow high-pressure cylinder, one double-flow intermediate-pressure cylinder and two double-flow low-pressure cylinders. Steam exhausted from the high-pressure cylinder is reheated in the steam generator. Several steam extractions from the turbines preheat feed water (FW) in low- and high-pressure heat exchangers and one spray-type open de-aerating heat exchanger. The low-pressure turbines exhaust to the condenser.

(c) **Condensation**. This section includes the condensers. Cooling water condenses the steam exhausted from the turbines.

(d) **Preheating**. This section contains various types of reheating heat exchangers and pumps. The temperature and pressure of the condensed steam are increased in a series of pumps and heat exchangers.

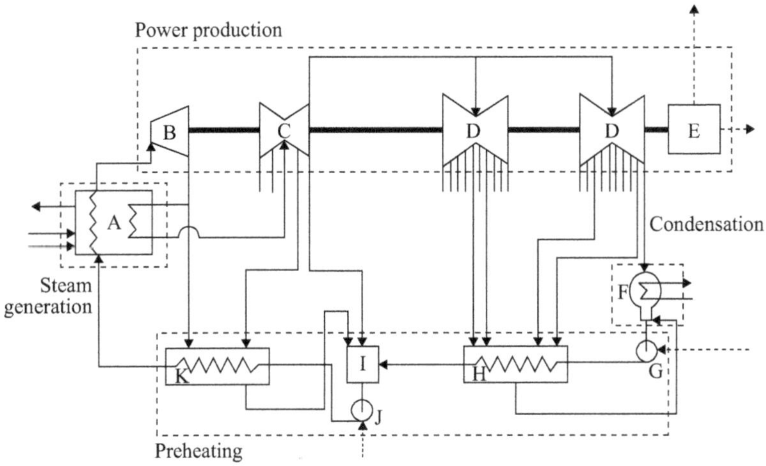

Figure 2.2 *Breakdown of the electrical generating station unit considered into four main sections. Material flows are represented by solid lines with arrows, electricity flows by dotted lines with arrows and heat flows by dashed lines with arrows*
Notes:

- *For unit A, the external inputs are coal and air, and the outputs are stack gas and solid waste.*
- *For unit E, the external outputs are electricity and waste heat.*
- *For units G and J, the input is electricity.*
- *For unit F, cooling water enters and exits.*

Legend:

A *Steam generator (boiler plus superheater and reheater)*
B *Steam turbine (high pressure)*
C *Steam turbine (medium pressure, dual flow)*
D *Steam turbine (low pressure, dual flow)*
E *Electrical generator and electrical transformer*
F *Condenser*
G *Pump (hot well)*
H *Heat exchanger (several closed, low-pressure units cascaded in series)*
I *Heat exchanger (open, deaerating)*
J *Pump (boiler feed)*
K *Heat exchangers (several closed, high-pressure units cascaded in series)*

Exergy and energy analyses of the station have been performed [44,45,48], and the results are now described. Overall, balances of exergy and energy for the station are illustrated in Figure 2.3. The main findings, which improve understanding of the thermodynamic behavior of the plant and help identify areas

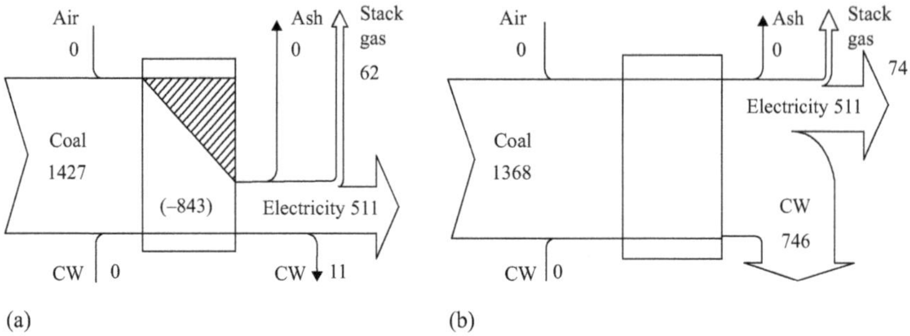

Figure 2.3 Overall energy and exergy balances for the station, represented by the rectangle in the center of each diagram. (a) Exergy balance showing flow rates (positive values) and consumption rate (negative value, denoted by hatched region) of exergy (in MW). (b) Energy balance showing flow rates of energy (in MW). Widths of flow lines are approximately proportional to the relative magnitudes of the represented quantities. Legend: CW denotes cooling water

having significant efficiency-improvement potential, follow, broken down by plant section:

- **Overall plant**. The energy efficiency (ratio of net electrical energy output to coal energy input) for the overall plant was found to be 37 per cent, and the corresponding exergy efficiency 36 per cent.
- **Steam generators**. The energy and exergy efficiencies were evaluated for the steam generators, considering the increase in energy or exergy of the water as the product. The steam generators appear significantly more efficient on an energy basis (95 per cent) than on an exergy basis (50 per cent). Physically, this discrepancy implies that, although most of the input energy is transferred to the preheated water, the energy is degraded as it is transferred. Most of the exergy losses in the steam generators are associated with internal consumptions (mainly due to combustion and heat transfer).
- **Condensers**. A large quantity of energy enters (about 775 MW for each unit) the condensers of which close to 100 per cent is rejected; and a small quantity of exergy enters (about 54 MW for each unit), of which about 25 per cent is rejected and 75 per cent internally consumed.
- **Other devices**. In other plant devices, energy losses were found to be very small (about 10 MW total), and exergy losses were found to be moderately small (about 150 MW total). The exergy losses are almost completely associated with internal consumptions.

2.3.3 Illustration of exergy analysis for electrical resistance space heating

An electrical resistance space heater converts electricity to heat at a temperature suitable for keeping a room at a comfortable temperature. Figure 2.4 shows a schematic of an electrical resistance heater. A wire or coil having electric resistance heats

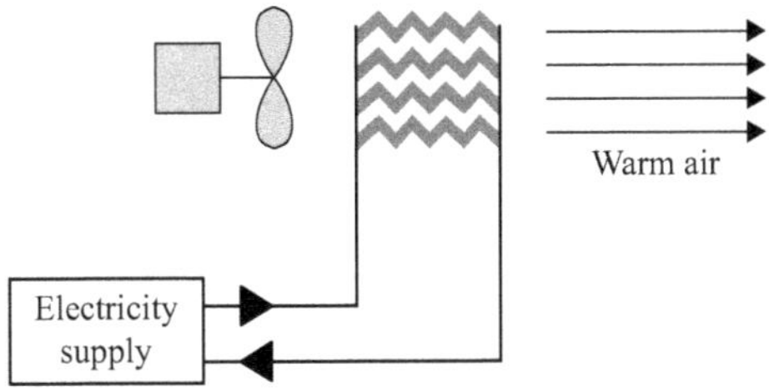

Figure 2.4 Schematic of an electric heater

up when electricity is passed through it, and air is warmed as it passes over the wire or coil. A fan may be employed to move the air. The energy efficiency of electric resistance space heating is often quoted to exceed 99 per cent. The implication clearly is that the maximum possible energy efficiency for electric resistance heating is 100 per cent, corresponding to the most efficient device possible.

This understanding is erroneous, however, as energy analysis ignores the fact that in this process high-quality energy (electricity) is used to produce a relatively low-quality product (warm air). Exergy analysis recognizes this difference in energy qualities, and indicates the exergy of the heat delivered to the room to be about 5 per cent of the exergy entering the heater. Thus, the efficiency, based on exergy, of electric resistance space heating is found to be about 5 per cent.

We therefore obtain useful information from the exergy results. Since thermodynamically ideal space heating has an exergy efficiency of 100 per cent, the same space heating can in theory be achieved using as little as 5 per cent of the electricity used in conventional electric resistance space heating. In practical terms, one can achieve space heating with a greatly reduced electricity input using a high-efficiency electric heat pump, using 15 per cent of the electricity that electric resistance heating would require, for a heat pump with a "coefficient of performance" (COP) of 7.

2.3.4 *Illustration of exergy analysis for thermal energy storage*

A thermal energy storage (TES) receives thermal energy and holds the energy until it is required. TESs can store energy at temperatures above or below the environment temperature, and come in many types (e.g., tanks, aquifers, ponds, caverns). A simple schematic of a tank energy storage is shown in Figure 2.5. Heat is stored in the tank, which contains a storage medium, often water, and is used later.

The evaluation of a TES system requires a measure of performance that is rational, meaningful and practical. The conventional energy storage efficiency is an inadequate measure. A more perceptive basis for comparison is needed if the true usefulness of a thermal storage is to be assessed, and so permit maximization of its economic benefit. Efficiencies based on ratios of exergy do provide rational measures of performance, since they can measure the approach of the performance of a system to the ideal.

That the energy efficiency is an inappropriate measure of thermal storage performance can best be appreciated through a simple example. Consider a perfectly insulated thermal storage containing 1,000 kg of water, initially at 40°C. The

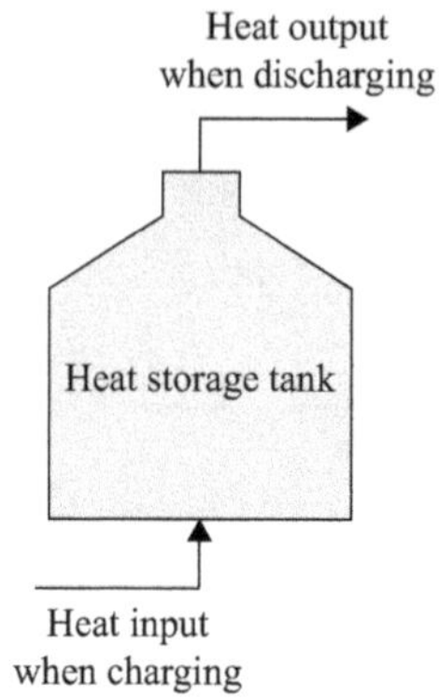

Figure 2.5 Thermal energy storage in a tank

ambient temperature is 20°C, and the specific heat of water is taken to be constant at 4.2 kJ/kg K.

A quantity of 4,200 kJ of heat is transferred to the storage through a heat exchanger from an external body of 100 kg of water cooling from 100°C to 90°C. This heat addition raises the storage temperature 1.0°C, to 41°C. After a period of storage, 4,200 kJ of heat are recovered from the storage through a heat exchanger which delivers it to an external body of 100 kg of water, raising the temperature of that water from 20°C to 30°C. The storage is returned to its initial state at 40°C.

For this storage cycle, the energy efficiency, the ratio of the heat recovered from the storage to the heat injected, is 4,200 kJ/4,200 kJ = 1, or 100 per cent. But the recovered heat is at only 30°C, and of little use, having been degraded even though the storage energy efficiency was 100 per cent. The exergy recovered in this example is evaluated as 70 kJ, and the exergy supplied as 856 kJ. Thus, the exergy efficiency, the ratio of the thermal exergy recovered from storage to that injected, is 70/856 = 0.082, or 8.2 per cent, a much more meaningful expression of the achieved performance of the TES.

Consequently, a device that appears to be ideal on an energy basis is correctly shown to be far from ideal on an exergy basis, clearly demonstrating the benefits of using exergy analysis for evaluating TESs.

2.4 Thermodynamic nomenclature and terminology

A relatively standard nomenclature and terminology has evolved for conventional classical thermodynamics.

But a standard nomenclature and terminology has not as of yet developed for exergy analysis. A diversity of symbols and names exist for most exergy-related quantities (e.g., exergy is often called available energy, availability, work capability, essergy) [21]. The nomenclature and terminology used in the present book were selected to be as consistent as possible with those in the literature, particularly those proposed [51] as a standard exergy-analysis nomenclature.

2.5 Thermodynamic balance equations and basic quantities

Thermodynamic balance equations for various quantities, and the basic quantities in the balances are provided and explained in this section.

2.5.1 Balance equations

General balance equations for commodities are provided, along with balance equations for specific thermodynamic quantities.

2.5.1.1 General balance equations

A general balance for a quantity in a system may be written as

$$\text{Input} + \text{Generation} - \text{Output} - \text{Consumption} = \text{Accumulation} \tag{2.1}$$

Each of the terms in this equation has an important meaning:

- Input and output refer, respectively, to quantities entering and exiting through system boundaries.
- Generation and consumption refer, respectively, to quantities produced and consumed (or destroyed) within the system.
- Accumulation refers to buildup (either positive or negative) of the quantity within the system.

2.5.1.2 Thermodynamic balance equations

Versions of the general balance equation may be written for mass, energy, entropy and exergy. Mass and energy, being subject to conservation laws (neglecting nuclear reactions), can be neither generated nor consumed. Consequently, the general balance (2.1) written for each of these quantities becomes

$$\text{Mass Input} - \text{Mass Output} = \text{Mass Accumulation} \tag{2.2}$$

$$\text{Energy Input} - \text{Energy Output} = \text{Energy Accumulation} \tag{2.3}$$

Before giving the balance equation for exergy, it is useful to examine that for entropy:

$$\text{Entropy Input} + \text{Entropy Generation} - \text{Entropy Output} = \text{Entropy Accumulation} \tag{2.4}$$

Entropy is created during a process due to irreversibilities. It is a rough measure of the order of a flow or system, and can be utilized in second-law-based assessments of efficiency for processes and systems. However, the results of entropy analysis are difficult to interpret and apply since the meaning of entropy is often not clear. Exergy analysis is a more useful and simple technique for second-law-based assessments of efficiency, and, since exergy is a measure of the work equivalent of any flow, the results of such analyses are more readily understood. By combining the conservation law for energy and nonconservation law for entropy, the exergy balance can be obtained

$$\text{Exergy Input} - \text{Exergy Output} - \text{Exergy Consumption} = \text{Exergy Accumulation} \quad (2.5)$$

Exergy is consumed due to irreversibilities. Exergy consumption is proportional to entropy creation. Equations (2.3) and (2.5) demonstrate an important main difference between energy and exergy: energy is conserved while exergy, a measure of work potential, can be consumed.

Many forms of the above balance equations may be written. For continuous processes, it is common to write rate balances. Then, the terms in (2.1)–(2.5) are written as rates, and (2.1), for example, becomes

$$\begin{aligned} &\text{Input Rate} + \text{Generation Rate} - \text{Output Rate} - \text{Consumption Rate} \\ &\quad = \text{Accumulation Rate} \end{aligned} \quad (2.6)$$

The rate balance describes what is happening in a system at a given instant of time. For steady-state continuous processes, the accumulation rate terms in all the balances are zero.

Rate forms of the balances for mass, energy and entropy in (2.2)–(2.4), respectively, can be written mathematically for a steady-state process in an open system as follows:

$$\sum_i \dot{m}_i - \sum_e \dot{m}_e = 0 \quad (2.7)$$

$$\sum_i (e + Pv)_i \dot{m}_i - \sum_e (e + Pv)_e \dot{m}_e + \sum_r \dot{Q}_r - \dot{W} = 0 \quad (2.8)$$

$$\sum_i s_i \dot{m}_i - \sum_e s_e \dot{m}_e + \sum_r \frac{\dot{Q}_r}{T_r} + \dot{\Pi} = 0 \quad (2.9)$$

Here, $\dot{m}_i$ and $\dot{m}_e$ denote, respectively, the rates mass is input across port i and output across port e; $\dot{Q}_r$ denotes the rate heat is transferred into the system across region r on the system boundary; $\dot{W}$ denotes the rate shaft work is transferred out of the system; $\dot{\Pi}$ denotes the rate entropy is created in the system; and e, s, P, T and v denote specific energy, specific entropy, absolute pressure, absolute temperature and specific volume, respectively. The shaft work rate, $\dot{W}$, done by a system excludes flow work, but includes all forms of work that could have been used to raise a weight (i.e., mechanical work, electrical work, etc.). The specific energy, e, is given by

$$e = u + ke + pe \quad (2.10)$$

where u, ke and pe denote, respectively, specific internal, kinetic and potential (due to conservative force fields) energies. For irreversible processes $\dot{\Pi} > 0$, and for reversible processes $\dot{\Pi} = 0$; $\dot{\Pi}$ can never be negative.

The left sides of (2.7)–(2.9) represent the net rates mass, energy and entropy are transferred into (and in the case of entropy created within) the system, while the right sides represent the rates these quantities are accumulated within the system.

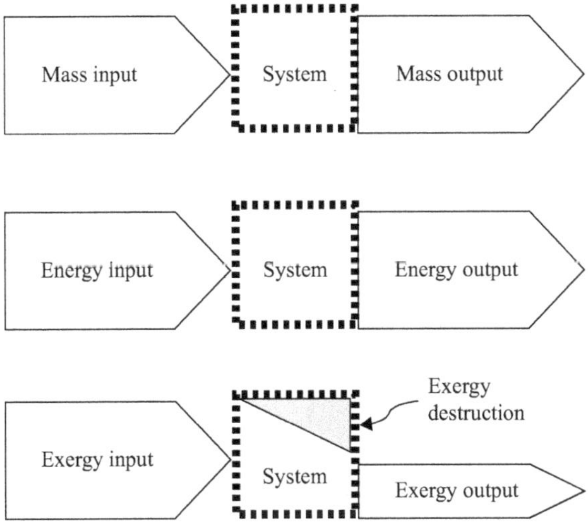

Figure 2.6 Illustration of general balances of mass (top), energy (middle) and exergy (bottom) for a system (denoted by dashed box in center of each diagram) at steady state. Line thicknesses for flows are proportional to their magnitudes. Energy and mass are seen to be conserved quantities, with inputs equaling outputs, while exergy is seen to be nonconserved due to the exergy destruction (shaded region)

An analogous balance to those given in (2.7)–(2.9) can be written for exergy as follows:

$$\sum_i \varepsilon_i \dot{m}_i - \sum_e \varepsilon_e \dot{m}_e + \sum_r \dot{\epsilon}_r^Q - \dot{W} - \dot{I} = 0 \tag{2.11}$$

where ε denotes the specific flow exergy; $\dot{\epsilon}_r^Q$ denotes the exergy transfer rate associated with $\dot{Q}_r$ and $\dot{I}$ denotes the rate exergy is consumed within the system. The first two terms on the left side of (2.11) represent the net input rate of exergy associated with matter, the third term the net input rate of exergy associated with heat, the fourth term the net input rate of exergy associated with work and the fifth term the exergy consumption rate. The balance demonstrates that exergy is subject to a nonconservation law (recall (2.5)).

Steady-state balances of mass, energy and exergy for a system are illustrated in Figure 2.6. The figure highlights the conservation of energy and mass and the nonconservation of exergy and cost.

2.5.2 Quantities in balance equations

The quantities involved in the exergy balance are described and discussed further in this section. They relate to flowing quantities as well as quantities in systems.

Exergy quantities can in general be divided into material and non-material types. A breakdown of exergy quantities into these types, along with the main types of exergy in each category, is shown in Figure 2.7.

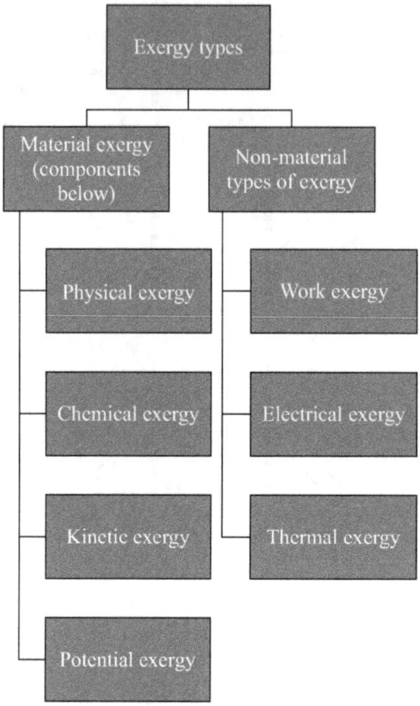

Figure 2.7 Breakdown of exergy into material and non-material types. Different types of non-material are listed, as are components of material exergy

2.5.2.1 Exergy of a flowing stream of matter

The exergy flow rate of a flowing stream of matter is given by the following expression:

$$\dot{\epsilon} = \dot{m}\varepsilon = \dot{\epsilon}_{ph} + \dot{\epsilon}_{o} + \dot{\epsilon}_{kin} + \dot{\epsilon}_{pot} \tag{2.12}$$

where

$$\dot{\epsilon}_{pot} = \dot{m}pe \tag{2.13}$$

$$\dot{\epsilon}_{kin} = \dot{m}ke \tag{2.14}$$

$$\dot{\epsilon}_{o} = \sum_{j} (\mu_{jo} - \mu_{joo})\dot{N}_j \tag{2.15}$$

$$\dot{\epsilon}_{ph} = \dot{m}[h - h_o - T_o(s - s_o)] \tag{2.16}$$

Here, h_o and s_o are the values of h and s, respectively, for the flow evaluated at T_o and P_o; μ_{jo} is the chemical potential of species j in the flow evaluated at T_o and P_o; μ_{joo} is the chemical potential of species j in the flow evaluated in the reference environment (or as if it exists in the reference environment) and $\dot{N}_j$ is the molar flow rate of species j. The terms on the far right side of (2.12) represent, respectively, the physical, chemical, kinetic and potential components of the exergy flow rate of a flowing stream of matter.

2.5.2.2 Exergy of heat

The flow rate of exergy associated with the heat transfer rate $\dot{Q}$ at a constant temperature T is denoted by $\dot{\epsilon}^Q$, and is given by

$$\dot{\epsilon}^Q = \tau\dot{Q} \tag{2.17}$$

where τ is the exergetic temperature factor. This factor is a dimensionless measure of the fraction of any quantity of heat transferred at temperature T which is entirely convertible to work, and can be expressed as

$$\tau = 1 - T_o/T \tag{2.17a}$$

2.5.2.3 Exergy of work and electricity

The exergy associated with shaft work rate $\dot{W}$ is by definition $\dot{W}$. Similarly, the exergy associated with electricity is equal to the energy.

2.5.2.4 Exergy consumption

For a steady-state process in an open system, the difference between the total exergy flow rates into and out of the system is the exergy-consumption rate, $\dot{I}$, and is given by the following expression:

$$\dot{I} = T_o\dot{\Pi} \tag{2.18}$$

2.6 The reference environment

Exergy is evaluated with respect to a reference environment (or a reference-environment model). The intensive properties of the reference environment determine the exergy of a stream or system. The exergy of the reference environment is zero. The exergy of a stream or system is zero when it is in equilibrium with the reference environment. Many researchers have examined the characteristics of and models for reference environments as well as the impact of variations in dead-state properties on exergy analysis results [18,20,52–63].

A brief discussion of the theoretical characteristics of the reference environment is now provided. The reference environment is in stable equilibrium. For this to occur, the reference environment must simultaneously be in thermal equilibrium (characterized by temperature), mechanical equilibrium (characterized by pressure) and chemical equilibrium (characterized by chemical potentials). The equilibrium characteristic of the reference environment is conceptually illustrated in Figure 2.8, where overall stable equilibrium is represented by the intersection of thermal, mechanical and chemical equilibria at the center of the figure.

Further details on the theoretical characteristics of the reference environment follow. All parts are at rest relative to one another. No chemical reactions can occur between the environmental components. The reference environment acts as an infinite system, and is a sink and source for heat and materials. It experiences only

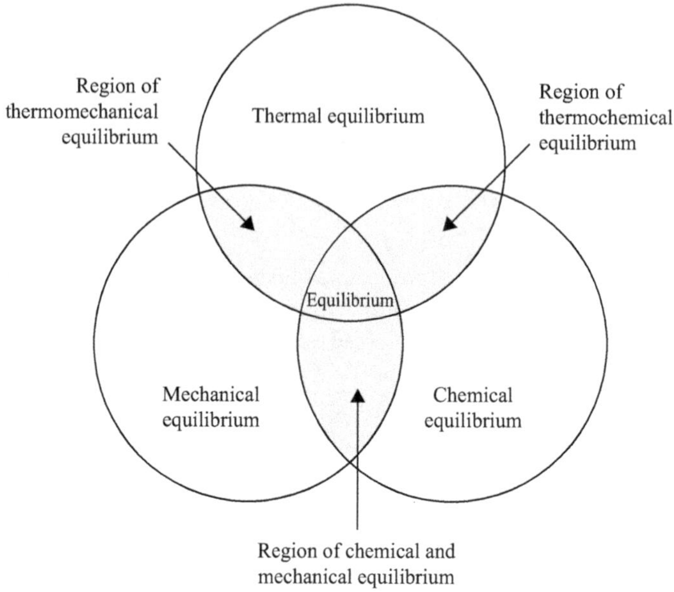

Figure 2.8 Conceptual illustration of the reference environment used in exergy analysis, showing how the reference environment must be in overall equilibrium, represented by the center of the diagram. For this to occur, the reference environment must simultaneously be thermal equilibrium (characterized by temperature), mechanical equilibrium (characterized by pressure) and chemical equilibrium (characterized by chemical potentials)

internally reversible processes in which its intensive state remains unaltered (i.e., its temperature T_o, pressure P_o and the chemical potentials, μ_{joo}, for each of the j components present remain constant).

The natural environment does not have the theoretical characteristics of a reference environment. The natural environment is not in equilibrium, and its intensive properties exhibit spatial and temporal variations. Many chemical reactions in the natural environment are blocked because the transport mechanisms necessary to reach equilibrium are too slow at ambient conditions. Thus, the exergy of the natural environment is not zero; work could be obtained if it were to come to equilibrium.

In developing reference-environment models, a compromise is made between theoretical requirements of the reference environment and the actual behavior of the natural environment. Several reference-environment models have been proposed. The most significant of these are grouped into classes and described below.

2.6.1 Natural-environment-subsystem models

Natural-environment-subsystem models attempt to simulate realistically subsystems of the natural environment. A model consisting of saturated moist air and liquid

Table 2.2 A reference-environment model (adapted from Gaggioli and Petit [12])

Temperature (T_o)	**298.15 K**		
Pressure (P_o)	**1 atm**		
Composition	(i) Atmospheric air saturated with H_2O at T_o and P_o	Air constituents	Mole fraction
		N_2	0.7567
		O_2	0.2035
		H_2O	0.0303
		Ar	0.0091
		CO_2	0.0003
		H_2	0.0001
	(ii) Condensed phases at T_o and P_o Water (H_2O) Limestone ($CaCO_3$) Gypsum ($CaSO_4{\cdot}2H_2O$)		

water in phase equilibrium was proposed by Baehr and Schmidt [55]. An extension of the above model which allowed sulfur-containing materials to be analyzed was proposed by Gaggioli and Petit [20] and Rodriguez [56]. The temperature and pressure of this reference environment (see Table 2.2) are normally taken to be 25°C and 1 atm, respectively, and the chemical composition is taken to consist of air saturated with water vapor, and the following condensed phases at 25°C and 1 atm: water (H_2O), gypsum ($CaSO_4{\cdot}2H_2O$) and limestone ($CaCO_3$). The stable configurations of C, O and N, respectively, are taken to be those of CO_2, O_2 and N_2 as they exist in air saturated with liquid water at T_o and P_o; of hydrogen is taken to be in the liquid phase of water saturated with air at T_o and P_o; and of S and Ca, respectively, are taken to be those of $CaSO_4{\cdot}2H_2O$ and $CaCO_3$ at T_o and P_o.

In essence, the natural-environment-subsystem model attempts to mimic realistically the three main subsystems of the natural environment: lithosphere (the land), hydrosphere (the water) and atmosphere. This idea is illustrated in Figure 2.9, where the overall reference environment is represented by the intersection of the three main subsystems of the natural environment, in the center of the diagram.

2.6.2 Reference-substance models

With reference-substance models, a "reference substance" is selected and assigned zero exergy for every chemical element.

A model in which the reference substances were selected as the most valueless substances found in abundance in the natural environment was proposed by Szargut [57]. The criteria for selecting such reference substances is consistent with the notion of simulating the natural environment, but is primarily economic in nature, and is vague and arbitrary with respect to the selection of reference substances. Part of this environment is the composition of moist air, including N_2, O_2, CO_2, H_2O and the noble gases, gypsum (for sulfur) and limestone (for calcium).

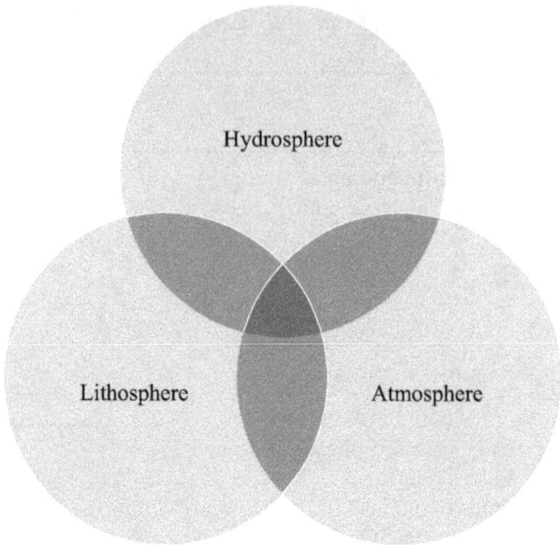

Figure 2.9 The natural-environment-subsystem model. The overall reference environment for this model represented by the intersection of the three main subsystems of the natural environment: lithosphere (the land), hydrosphere (the water) and atmosphere

Another model, in which reference substances are selected arbitrarily, was proposed by Sussman [53]. This model is not similar to the natural environment. Consequently, absolute exergies evaluated with this model do not relate to the natural environment, and cannot be used rationally to evaluate efficiencies. Since exergy-consumption values are independent of the choice of reference substances, they can be rationally used in analyses.

2.6.3 Equilibrium and constrained-equilibrium models

Equilibrium and constrained-equilibrium models have been proposed, as a way of obtaining realistic and physically meaningful models of the reference environment.

A model in which all the materials present in the atmosphere, oceans and a layer of the crust of the earth are pooled together and an equilibrium composition is calculated for a given temperature was proposed by Ahrendts [54]. The selection of the thickness of crust considered is subjective and is intended to include all materials accessible to technical processes. Ahrendts considered thicknesses varying from 1 to 1,000 m, and a temperature of 25°C. For all thicknesses, Ahrendts found that the model differed significantly from the natural environment Exergy values obtained using these environments are significantly dependent on the thickness of crust considered, and represent the absolute maximum amount of work obtainable from a material. Since there is no technical process available which can obtain this work from materials, Ahrendts' equilibrium model does not give meaningful exergy values when applied to the analysis of real processes.

Ahrendts [54] also proposed a modified version of his equilibrium environment in which the calculation of an equilibrium composition excludes the possibility of the formation of nitric acid (HNO_3) and its compounds. That is, all chemical reactions in which these substances are formed are in constrained equilibrium, and all other reactions are in unconstrained equilibrium. When a thickness of crust of 1 m and temperature of 25°C were used, the model was similar to the natural environment.

2.6.4 Process-dependent models

Process-dependent models take a pragmatic approach to defining a reference environment, by explicitly linking it to the process under consideration in a given analysis. A model which contains only components that participate in the process being examined in a stable equilibrium composition at the temperature and total pressure of the natural environment was proposed by Bosnjakovic [58]. This model is dependent on the process examined, and is not general. Exergies evaluated for a specific process-dependent model are relevant only to the process; they cannot rationally be compared with exergies evaluated for other process-dependent models.

2.7 Efficiencies

Various efficiency definitions can be defined which generally answer different questions. Many researchers [4,5,53,64] have examined efficiencies and other measures of performance.

2.7.1 Conventional energy and exergy efficiencies

An energy efficiency (η) is often written for steady-state processes occurring in systems as

$$\eta = \frac{\text{Energy in product outputs}}{\text{Energy in inputs}} = 1 - \frac{\text{Energy loss}}{\text{Energy in inputs}} \tag{2.19}$$

A corresponding exergy efficiency (ψ) is often written for steady-state processes occurring in systems as

$$\psi = \frac{\text{Exergy in product outputs}}{\text{Exergy in inputs}} = 1 - \frac{\text{Exergy loss plus consumption}}{\text{Exergy in inputs}} \tag{2.20}$$

Exergy efficiencies often give more illuminating insights into process performance than energy efficiencies because (i) they weigh energy flows according to their exergy contents and (ii) they separate inefficiencies into those associated with effluent losses and those due to irreversibilities. In general, exergy efficiencies provide a measure of potential for improvement.

2.7.2 *Alternative efficiencies*

Two other common exergy-based efficiencies for steady-state devices are as follows:

$$\text{Rational efficiency} = \frac{\text{Total exergy output}}{\text{Total exergy input}} = 1 - \frac{\text{Exergy consumption}}{\text{Total exergy input}} \tag{2.21}$$

$$\text{Task efficiency} = \frac{\text{Theoretical minimum exergy input required}}{\text{Actual exergy input}} \tag{2.22}$$

These efficiencies essentially both are different types of exergy efficiencies. That is because they incorporate exergy terms in the definitions. Like conventional exergy efficiencies, these alternative exergy-based efficiencies provide a measure of potential for improvement.

2.8 Properties for energy and exergy analyses

Many material properties are needed for energy and exergy analyses of processes. Sources of conventional property data are abundant for many substances (e.g., combustion gases [65], steam [66] and chemical substances [67]).

Energy values of heat and work flows are absolute, while the energy values of material flows are relative. Enthalpies are evaluated relative to a reference level. Since energy analyses are typically concerned only with energy differences, the reference level used for enthalpy calculations can be arbitrary. For the determination of some energy efficiencies, however, the enthalpies must be evaluated relative to specific reference levels [e.g., for energy conversion processes, the reference level is often selected so that the enthalpy of a material equals its higher heating value (HHV)].

If, however, the results from energy and exergy analyses are to be compared, it is necessary to specify reference levels for enthalpy calculations such that the enthalpy of a compound is evaluated relative to the stable components of the reference environment. Thus, a compound which exists as a stable component of the reference environment is defined to have an enthalpy of zero at T_o and P_o. Enthalpies calculated with respect to such conditions are referred to as "base enthalpies" [56]. The base enthalpy is similar to the enthalpy of formation. While the latter is the enthalpy of a compound (at T_o and P_o) relative to the elements (at T_o and P_o) from which it would be formed, the former is the enthalpy of a component (at T_o and P_o) relative to the stable components of the environment (at T_o and P_o). For many environment models, the base enthalpies of material fuels are equal to their HHVs. Base enthalpies for many substances, corresponding to the reference environment in Table 2.2, are listed by Rodriguez [56] and are reproduced in Table 2.3.

It is necessary for chemical exergy values to be determined for exergy analysis. Many researchers have developed methods for evaluating chemical exergies and tabulated values (e.g., [56,57]). Included are methods for evaluating the chemical exergies of solids, liquids and gases. For complex materials (e.g., coal, tar, ash), approximation methods have been developed. By considering environmental air

Table 2.3 Base enthalpy and chemical exergy values of selected species[1]

Species	Base enthalpy (kJ/g mol)	Chemical energy[2] (kJ/g mol)
Ammonia (NH_3)	382.585	2.478907 ln x + 337.861
Argon (Ar)	0.000	2.478907 ln x + 11.650
Benzene (C_6H_6)	3,301.511	2.478907 ln x + 3,253.338
Carbon (graphite) (C)	393.505	410.535
Carbon dioxide (CO_2)	0.000	2.478907 ln x + 20.108
Carbon monoxide (CO)	282.964	2.478907 ln x + 275.224
Carbon oxysulfide (COS)	891.150	2.478907 ln x + 848.013
Ethane (C_2H_6)	1,564.080	2.478907 ln x + 1,484.952
Hydrogen (H_2)	285.851	2.478907 ln x + 235.153
Hydrogen sulfide (H_2S)	901.757	2.478907 ln x + 803.374
Methane (CH_4)	890.359	2.478907 ln x + 830.212
Methanol (CH_3OH)	764.018	2.478907 ln x + 721.500
Nitrogen (N_2)	0.000	2.478907 ln x + 0.693
Oxygen (O_2)	0.000	2.478907 ln x + 3.948
Phenol (C_6H_5OH)	3,122.226	2.478907 ln x + 3,090.784
Sulfur (rhombic) (S)	636.052	608.967
Sulfur dioxide (SO_2)	339.155	2.478907 ln x + 295.736
Water (H_2O)	44.001	2.478907 ln x + 8.595

[1]Compiled from data in Rodriguez [56] and Gaggioli and Petit [20].
[2]x represents the mole fraction for each of the respective species.

and gaseous process streams as ideal gas mixtures, chemical exergy can be calculated for gaseous streams using component chemical exergy values [i.e., values of $(\mu_{jo} - \mu_{joo})$ for species j] listed in Table 2.3 after Rodriguez [56]. The base enthalpy and the component chemical exergy were calculated for methanol and argon by the present authors, following the procedures outlined elsewhere [20,56].

2.9 Implications of energy and exergy analyses on related research and development

2.9.1 Correlating energy and exergy analyses with allocations of research efforts

The results of energy and exergy analyses have direct implications on where research and development (R&D) efforts should be focused for energy systems. Further, it is mainly the results of exergy analyses that provide insights into the "best" directions for R&D effort. Here, "best" is loosely taken to mean "most promising for significant efficiency gains." There are two main reasons for this statement:

- Exergy losses represent true losses of the potential that exists to generate the desired product from the given driving input. This is not true in general for energy losses. Thus, focusing on exergy losses permits R&D to focus on reducing losses that will affect the objective.

- Exergy efficiencies always provide a measure of how nearly the operation of a system approaches the ideal, or theoretical upper limit. This is not in general true for energy efficiencies. By focusing R&D effort on those plant sections or processes with the lowest exergy efficiencies, the effort is being directed to those areas which inherently have the largest margins for efficiency improvement. By focusing on energy efficiencies, on the other hand, one can expend R&D effort on topics for which little margins for improvement, even theoretically, exist.

It is noted that R&D allocation decisions should not be based exclusively on the results of energy and exergy analyses, even though these results provide useful information to assist in such decision-making. Other factors must also be considered, such as economics, environmental impact, safety, and social and political implications.

The results of assessments of energy systems often suggest that R&D efforts should concentrate more on internal rather than external exergy losses. In order of priority, the "best" areas to devote R&D efforts, based on thermodynamic considerations, are those processes having the larger exergy losses. But the above statement does not indicate that R&D should not be devoted to those processes having low-exergy losses. Rather, the large-exergy loss processes are those for which margins for improvement are greatest. If, in fact, a simple and cost-effective way is found to increase efficiency by reducing a small exergy loss, it should certainly be considered.

2.9.2 Measures to reduce exergy losses

Methods to address those items identified as having high R&D priority are only limited by the creativity and knowledge of the engineer/designer. In terms of practical measures, there are many causes of exergy losses, and each can be addressed in a different way.

The following can be considered for two of the more significant losses in many energy systems (combustion and heat transfer across large temperature differences):

- Combustion losses can be reduced by improving combustion efficiency, usually through minor adjustments to the combustion chamber and related devices. Larger reductions in combustion losses are attainable if metallurgical temperature limits for materials used in the processes are improved so that high temperatures are permitted in the combustion chamber and related devices. This is particularly true in electricity generation if a topping cycle which utilizes the higher temperature heat to generate additional electricity is applied. Also, if electricity generation is the final objective, significant combustion losses are eliminated if the entire process is reconfigured so that electricity is generated electrochemically (e.g., via a fuel cell) rather than thermomechanically. Such a measure, however, involves a total redesigning of the process.
- Heat-transfer losses can be reduced by using smaller heat-exchanger temperature differences and more efficient heat exchangers. This measure also involves ensuring that heat flows of appropriate temperatures are used to heat cooler flows.

2.10 Steps for energy and exergy analyses

A simple step-by-step procedure for performing energy and exergy recommended by the authors follows:

(i) Subdivide the process under consideration into as many sections as desired, depending on the depth of detail and understanding desired from the analysis.
(ii) Perform conventional mass and energy balances on the process, and determine all basic quantities (e.g., work, heat) and properties (e.g., temperature, pressure).
(iii) Based on the nature of the process, the acceptable degree of analysis complexity and accuracy, and the questions for which answers are sought, select a reference-environment model (see details in section 2.6).
(iv) Evaluate energy and exergy values, relative to the selected reference-environment model (see details in section 2.5).
(v) Perform exergy balances, including the determination of exergy consumptions (see details in section 2.5).
(vi) Select efficiency definitions, depending on the measures of merit desired, and evaluate values for the efficiencies (see details in section 2.7).
(vii) Interpret the results, and draw appropriate conclusions and recommendations, relating to issues such as design changes and retrofit plant modifications.

2.11 Illustrative example

The seven steps involved in energy and exergy analyses are illustrated, based on previous analyses by the authors [11,12], for a subsection of a pressurized fluid-bed combustion (PFBC) plant for electricity generation (based on a design prepared for the U.S. Department of Energy by the General Electric Corporation [68]).

2.11.1 Description and subdivision for analysis of system considered

Step (i): We describe the overall PFBC power plant and the subsection of the plant considered in this illustrative example.

2.11.1.1 Description of overall PFBC power plant

The plant considered has an electrical efficiency of 40 per cent, and a net electrical output of 683 MW. The plant is shown schematically in Figure 2.10.

During operation of the plant, coal and dolomite enter the fluid-bed combustor at 10 atm with air. The gas exiting the combustor is cleaned using cyclones, and then expanded through a gas-turbine generator. The gas turbine also drives the primary air compressor.

The gas-turbine exhaust passes through a heat recovery section where FW from a conventional steam power cycle is heated. The FW is convened in the combustor to steam, which is expanded through steam-turbine generators.

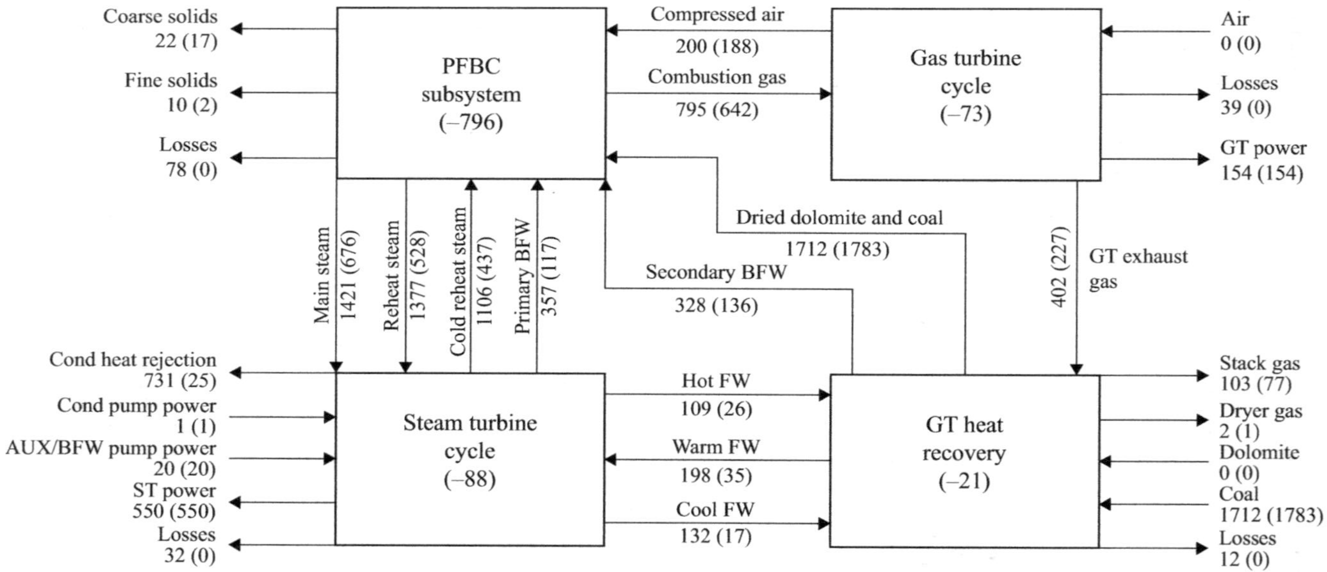

Figure 2.10 *Flow diagram for a PFBC power plant. Shown are flow rates of energy (values not in parentheses) and exergy (positive values in parentheses) associated with material and energy flows. Also shown are consumption rates of exergy (negative values in parentheses) associated with process sections. All values are in megawatts. (Aux: auxiliary, BFW: boiler feed water, Cond: condenser, FW: feed water, GT: gas turbine, ST: steam turbine.)*

Coal is combusted and cleaned in the "PFBC Subsystem," converting chemical energy to hot combustion gas and steam.

2.11.1.2 Description of PFBC subsystem

The PFBC subsystem is considered in this illustrative example. The subsystem is subject to interactions involving various materials, including coal, air, combustion gas, liquid water and steam.

The PFBC subsystem being considered, with all input and output flows identified, is shown in Figure 2.10.

During operation of the PFBC subsystem, coal is combusted and then cleaned. Combustion converts the chemical energy within the fuel to heat, which then manifests itself in the form of both hot combustion gas and steam.

2.11.2 Performance of conventional mass and energy balances

Step (ii): The performance of conventional mass and energy balances is straightforward. So, it is not described here. Rather, the focus of this illustrative example is on exergy analysis. Nonetheless, elements of energy balances are used in subsequent energy-analysis discussions.

2.11.3 Selection of reference-environment model

Step (iii): Applying exergy analysis requires that a reference-environment model be specified, and preferably it is based on the local natural environment.

The reference-environment model used in this example has the following property values:

- a reference-environment temperature of $T_o = 15°C$,
- a reference-environment pressure of $P_o = 101.325$ kPa (1 atm) and
- a reference-environment chemical composition identical to that specified in Table 2.2.

2.11.4 Evaluation of energy and exergy flow rates

Step (iv): The evaluation of the energy and exergy flow rates associated with each material and energy flow entering or exiting the subsystem is described below, by class of flow.

2.11.4.1 Water flows

For H_2O flows, steam and water tables [66] are utilized for property determination and evaluation. This approach is common, as data for water and steam are well understood and widely tabulated for all commonly applied states.

First, the reference-state enthalpy h_o and entropy s_o, which are applicable to all H_2O flows, are determined:

$$h_o = h(T_o, P_o) = 63.035 \text{ kJ/kg}, \; s_o = s(T_o, P_o) = 0.2241 \text{ kJ/kg K}$$

Table 2.4 Thermodynamic quantity values for water flows in PFBC subsection

	$\dot{m}$ (kg/s)	T (K)	P (MPa)	$\dot{E}$ (MW)	$\dot{\in}$ (MW)
Main steam	438	811	24.20	1,421	676
Reheat steam	397	811	4.19	1,377	582
Cold reheat steam	386	569	4.63	1,106	437
Primary boiler feed water (BFW)	281	573	27.57	356	117
Secondary BFW	157	670	28.96	328	136

Then, the flows are considered individually. For the "main steam" flow, for example, it is known from given information, and the results of mass and energy balances, that:

$$\dot{m} = 438.13 \text{ kg/s}, T = 811 \text{ K}, P = 24.2 \text{ MPa}$$

The required properties are then evaluated for the flow:

$$h = h(T, P) = 3{,}306.87 \text{ kJ/kg}, s = s(T, P) = 6.1302 \text{ kJ/kg K}$$

Finally, the energy flow rate, $\dot{E}$, is evaluated as

$$\dot{E} = \dot{m}(h - h_o) = 438.13 \times (3306.87 - 63.035) = 1421 \text{ MW}$$

and, using (2.12) and (2.16) (and noting that all exergy components are zero except for physical exergy), the exergy flow rate, $\dot{\in}$, is evaluated as

$$\begin{aligned}\dot{\in} &= \dot{m}[h - h_o - T_o(s - s_o)] \\ &= 438.13 \times [(3306.87 - 63.035) - 288 \times (6.1302 - 0.2241)] = 676 \text{ MW}\end{aligned}$$

This procedure is similarly applied to the other H_2O flows as follows. The same units as in the above paragraph apply, and the results for all H_2O flows are listed in Table 2.4.

2.11.4.2 Air flows

For flows of air, air tables [60] are utilized for property evaluation. This is a common approach, given air data are well understood and widely tabulated for a wide range of conditions.

First, the reference-state enthalpy h_o and entropy s_o are determined:

$$h_o = h(T_o, P_o) = 288.38 \text{ kJ/kg}, s_o = s(T_o, P_o) = 5.6606 \text{ kJ/kg K}$$

For the "compressed air" flow, it is known from given information, and the results of mass and energy balances, that:

$$\dot{m} = 666.54 \text{ kg/s}, T = 581.3 \text{ K}, P = 988.0 \text{ kPa}$$

The required properties are then evaluated for the flow:

$$h = h(T, P) = 587.66 \text{ kJ/kg}, s = s(T, P) = 5.7218 \text{ kJ/kg K}$$

Finally, the energy flow rate, $\dot{E}$, is evaluated as

$$\dot{E} = \dot{m}(h - h_o) = 666.54 \times (587.66 - 288.38) = 199 \text{ MW}$$

and, using (2.12) and (2.16) (and noting that all exergy components are zero except for physical exergy), the exergy flow rate, $\dot{\epsilon}$, is evaluated as

$$\begin{aligned}\dot{\epsilon} &= \dot{m}[h - h_o - T_o(s - s_o)] \\ &= 666.54 \times [(587.66 - 288.38) - 288 \times (5.7218 - 5.6606)] = 188 \text{ MW}\end{aligned}$$

2.11.4.3 Other gas flows

Here, the other gases are combustion gases. For evaluating physical exergy, combustion gas tables [65] for fuels with 120 per cent theoretical air (as is the case here) are utilized, and the calculation procedure is as for air flows.

It is noted that, alternatively, property values can be evaluated with slightly greater accuracy using the exact gas composition, but the simpler approach is adopted here for convenience. Thus, with (2.16):

$$\dot{E}_{ph} = 785 \text{ MW}, \dot{\epsilon}_{ph} = 580.7 \text{ MW}$$

For evaluating the chemical exergy flow rate $\dot{\epsilon}_o$ of the combustion gas, its composition (on a mass basis is used):

- 20.5 per cent CO_2
- 0.1 per cent CO
- 5.0 per cent H_2
- 3.6 per cent O_2
- 70.8 per cent N_2.

This composition is employed in with (2.15) and in the constituent chemical exergy expressions in Table 2.3. The chemical energy flow rate $\dot{E}_o$ of the combustion gas is attributable to the remaining unburned CO it contains. Thus, it can be shown that:

$$\dot{E}_o = 10 \text{ MW}, \dot{\epsilon}_o = 62.7 \text{ MW}$$

The total energy and exergy flow rates are the sums of all non-zero individual exergy components. Thus, the total energy and exergy flow rate is

$$\dot{E} = \dot{E}_{ph} = \dot{E}_o = 785 \text{ MW} + 10 \text{ MW} = 795 \text{ MW}$$

and the total exergy flow rate is

$$\dot{\epsilon} = \dot{\epsilon}_{ph} + \dot{\epsilon}_o = 580.7 \text{ MW} + 62.7 \text{ MW} = 642 \text{ MW}$$

2.11.4.4 Solid flows

For non-environmental solid substances such as coal ($\dot{m} = 68.23$ kg/s, HHV = 25.070 MJ/kg), the chemical energy for the reference-environment model utilized is evaluated as

$$\dot{E}_o = \dot{m}\,(\text{HHV}) = 68.23 \times 25{,}070 = 1{,}711\ \text{MW}$$

The chemical exergy is normally evaluated using an empirical energy-to-exergy-ratio expression which is based on the composition of the coal. The details of the calculation procedure are given in an appendix of Reference [4]. Here, knowing an ultimate analysis of the coal used:

- 59.6 per cent C
- 5.9 per cent H
- 3.9 per cent S
- 1.0 per cent N
- 20.0 per cent O
- 9.6 per cent ash.

With these values, the ratio is determined to be 1.042. Hence, the chemical exergy flow rate is

$$\dot{\epsilon}_o = \dot{E}_o \times 1.042 = 1{,}711 \times 1.042 = 1{,}783\ \text{MW}$$

Since the coal is input at T_o and P_o, the physical energy and physical exergy components are zero, and the chemical components represent the total energy and exergy flow rates. For dolomite, and the coarse and fine solids, the chemical exergy and chemical energy are taken to be zero (as they are reference environment substances), and only the physical energy and exergy terms are of significance. As the dolomite is input at T_o, its physical energy and exergy are zero. For the coarse solids:

$$\dot{m} = 11.34\ \text{kg/s}, T = 1{,}211.9\ \text{K}, P = 101.3\ \text{kPa}$$

and the energy flow rate is evaluated (using estimated properties) as 22 MW. The exergy flow rate is approximated, since entropy data are difficult to obtain for the coarse solids, as the exergy of the heat deliverable (using a modified form of (2.17)), as follows:

$$\dot{\epsilon} = (1 - T_o/T)\dot{E} = (1 - 288/1{,}211.9) \times 22 = 17\ \text{MW}$$

Similarly, for the fine solids, $\dot{E} = 10$ MW and $\dot{\epsilon} = 2$ MW.

2.11.4.5 Miscellaneous heat losses

An energy balance (2.8) is applied to determine the energy flow rate of the miscellaneous heat losses as follows:

$$\begin{aligned}\dot{Q} &= (1{,}106 + 357 + 328 + 1{,}712 + 200)\\ &\quad - (22 + 10 + 1{,}421 + 1{,}377 + 795)\\ &= 78\ \text{MW}\end{aligned}$$

Heat flows are assumed to cross the system boundary at T_o so that the associated exergy (see (2.17)) is zero, that is:

$$\dot{\epsilon}^Q = \left(1 - \frac{T_o}{T_o}\right)\dot{Q} = 0$$

2.11.5 Performance of exergy balances and determination of exergy consumptions

Step (v): We now perform exergy balances, noting that exergy is not a conserved quantity. To achieve a "balance" for exergy, we must account for exergy consumptions, which are also determined now.

The energy balance was used in Step (iv) to evaluate the miscellaneous heat loss. Clearly, energy is seen in that step to be a conserved quantity.

The exergy balance (2.11) can be used with data from Figure 2.10 to evaluate the subsystem exergy consumption rate, $\dot{I}$, as follows:

$$\begin{aligned} \dot{I} &= \sum_{in}(\textit{exergy flows}) - \sum_{out}(\textit{exergy flows}) \\ &= (188 + 1783 + 136 + 117 + 437) - (642 + 528 + 676 + 0 + 2 + 17) \\ &= 796 \text{ MW} \end{aligned}$$

2.11.6 Selection and evaluation of efficiencies

Step (vi): The efficiencies are defined here as the ratio of product outputs (combustion gas, net steam/water) to inputs (coal, compressed air) (see (2.19) and (2.20)).

Hence, for the subsystem, the energy efficiency η is

$$\eta = \frac{795 + [(1421 + 1377) - (1106 + 357 + 328)]}{1712 + 200} \times 100\% = 94\%$$

and the exergy efficiency ψ is

$$\psi = \frac{642 + [(676 + 528) - (437 + 117 + 136)]}{1783 + 188} \times 100\% = 59\%$$

2.11.7 Interpretation of results

Step (vii): The results show that the PFBC subsystem is very efficient on an energy basis, converting over 90 per cent of the input energy in the coal to steam.

This subsystem is much less efficient on an exergy basis, and, based on investigations of other electricity generating stations, the main exergy losses are due to the highly irreversible processes of combustion and heat transfer across large temperature differences.

The implications of the results, in the context of a full analysis of the PFBC plant, are presented and discussed elsewhere [11,12].

2.12 Exergy values for typical commodities encountered in cogeneration and district energy

When allocating carbon dioxide emissions based on the exergy contents of the products in a cogeneration process, as is discussed subsequently in Chapter 10, it is necessary to know the exergy values associated with electrical and thermal energy. The situation for electrical energy is straightforward, as the energy and exergy contents of electricity are equivalent. For thermal energy, however, the energy and exergy contents generally differ, and the differences in some cases can be quite significant.

2.12.1 Exergy values for thermal quantities

Values of the energy and exergy associated with thermal energy, when it is treated purely as heat, are presented in Tables 2.5 and 2.6. These values are illustrated in Figures 2.11 and 2.12, for large and small temperature ranges, respectively. This table and these figures consider heat (i.e., thermal energy transferred at temperatures above the environment temperature) and cold (i.e., thermal energy transferred at temperatures above the environment temperature), for various temperature categories.

The ratio of exergy to energy is also shown in Table 2.5 for heat flows and in Table 2.6 for cold flows. The ratio is illustrated in Figures 2.13 and 2.14, for large and small temperature ranges, respectively. Some interesting observations can be made:

- For heat, the ratio of exergy to energy varies from zero when the thermal energy is transferred at the environment temperature to unity as the temperature of heat transfer approaches infinity.

Table 2.5 Quality of heat (thermal energy above the reference-environment temperature) for a heat transfer rate of 1 MW at various temperatures

Temperature of heat flow[1]		**Exergy rate**[1] **(kW)**	**Exergy-to-energy ratio**
(K)	**(°C)**		
293	20	0	0.000
323	50	93	0.093
373	100	215	0.215
473	200	381	0.381
573	300	489	0.489
773	500	621	0.621
1,273	1,000	770	0.770
1,773	1,500	835	0.835
2,273	2,000	871	0.871

[1]The reference-environment temperature is $T_o = 20°C = 293$ K and the exergy rate is zero for heat transfer at this temperature (see top row).

Table 2.6 Quality of cold[1] (thermal energy below the reference-environment temperature) for a heat transfer rate of 1 MW at various temperatures

Temperature of heat flow[2]		Exergy rate[2] (kW)	Exergy-to-energy ratio
(K)	(°C)		
293	20	0	0.000
283	10	−35	−0.035
273	0	−73	−0.073
263	−10	−114	−0.114
243	−30	−206	−0.206
223	−50	−314	−0.314
173	−100	−694	−0.694
123	−150	−1,382	−1.382
73	−200	−3,014	−3.014
23	−250	−11,740	−11.74

[1]Cold is taken to be a transfer of thermal energy at below reference-environment temperatures.
[2]The reference-environment temperature is $T_o = 20°C = 293$ K and the exergy rate is zero for heat transfer at this temperature (see top row).

- For cold, the values of exergy rate are negative, implying that although heat is taken out of a system to make it colder, the exergy associated with the thermal energy is input to the system to make it colder. That is the flows of energy and exergy in such instances are in opposite directions. This observation implies what is intuitively understood when dealing with systems at below-environment temperatures: cold is the useful commodity.
- Also for cold, the magnitude of the ratio of exergy to energy varies from zero when the thermal energy is transferred at the environment temperature to greater than unity as the temperature of heat transfer approaches absolute zero. For very cold systems, therefore, the exergy transfer can be larger than the energy transfer.

Thermal energy is often transferred via a medium, and in cogeneration systems the medium of choice is often water. Values of the energy and exergy of water in various forms are presented in Tables 2.7 and 2.8. Table 2.7 considers water conveying heating capacity (e.g., superheated steam, dry saturated steam, hot water), while Table 2.8 considers cooling capacity (e.g., cold water). The ratio of exergy to energy is also shown in Table 2.7 and Table 2.8. Observations as similar as for Tables 2.5 and 2.6 can be made, in that the magnitudes of energy and exergy flows differ and, for cold commodities, the flows of energy and exergy are in opposite directions.

2.12.2 Exergy values for other thermodynamic quantities

Qualitative comparisons of the energy quality of a range of energy forms, where exergy is used as the measure of quality, are shown in Figures 2.15–2.18. The quality of energy based on exergy for thermal energy forms is compared qualitatively in Figure 2.15 for thermal energy above the temperature of the reference environment and in Figure 2.16 for thermal energy below the temperature of the

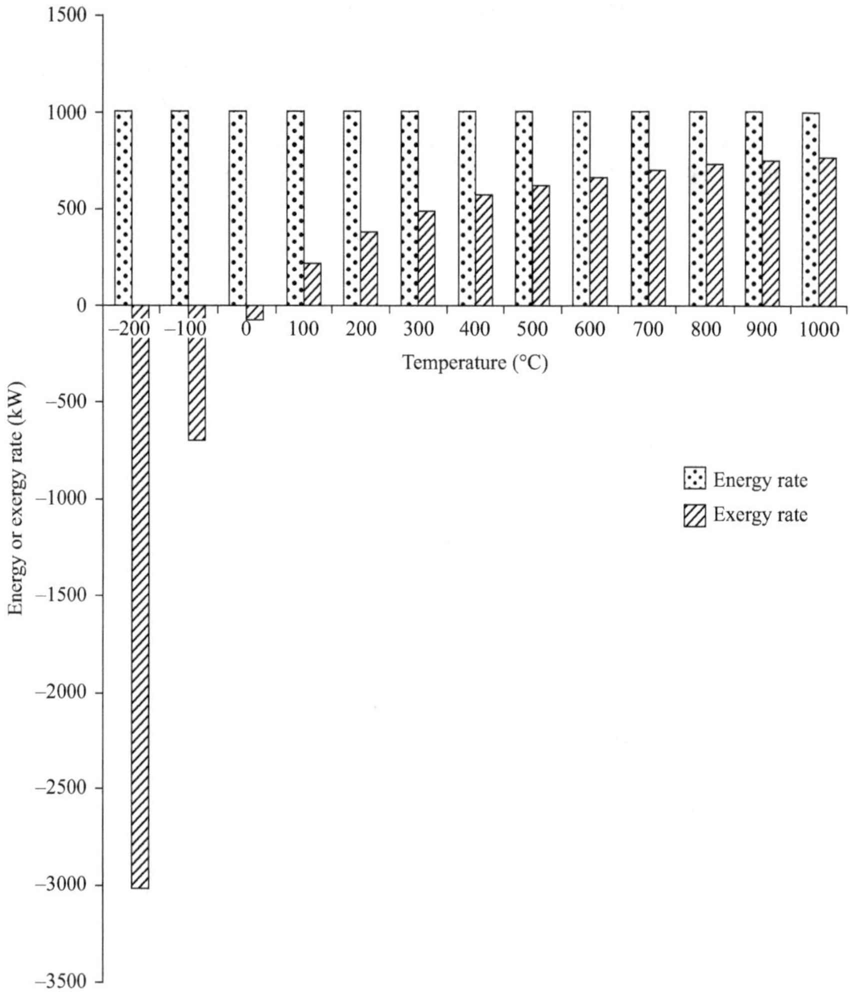

Figure 2.11 *Comparison of thermal energy and thermal exergy for a wide range of temperatures, including the reference-environment temperature (20°C here) (based on data in Tables 2.5 and 2.6). It is observed that thermal exergy above the reference-environment temperature is less than or equal to the corresponding thermal energy, and asymptotically approaches the corresponding thermal energy as the temperature associated with the thermal energy approaches infinity. Also, it is observed that thermal exergy below the reference-environment temperature is negative, indicating that the thermal exergy moves in the opposite direction to the thermal energy, or that the application of cooling is equivalent to the application of exergy. It is further observed that the magnitude of thermal exergy below the reference-environment temperature can be less than or equal to or greater than the magnitude of the corresponding thermal energy, and that it becomes much greater than the corresponding thermal energy as the temperature associated with the thermal energy approaches absolute zero*

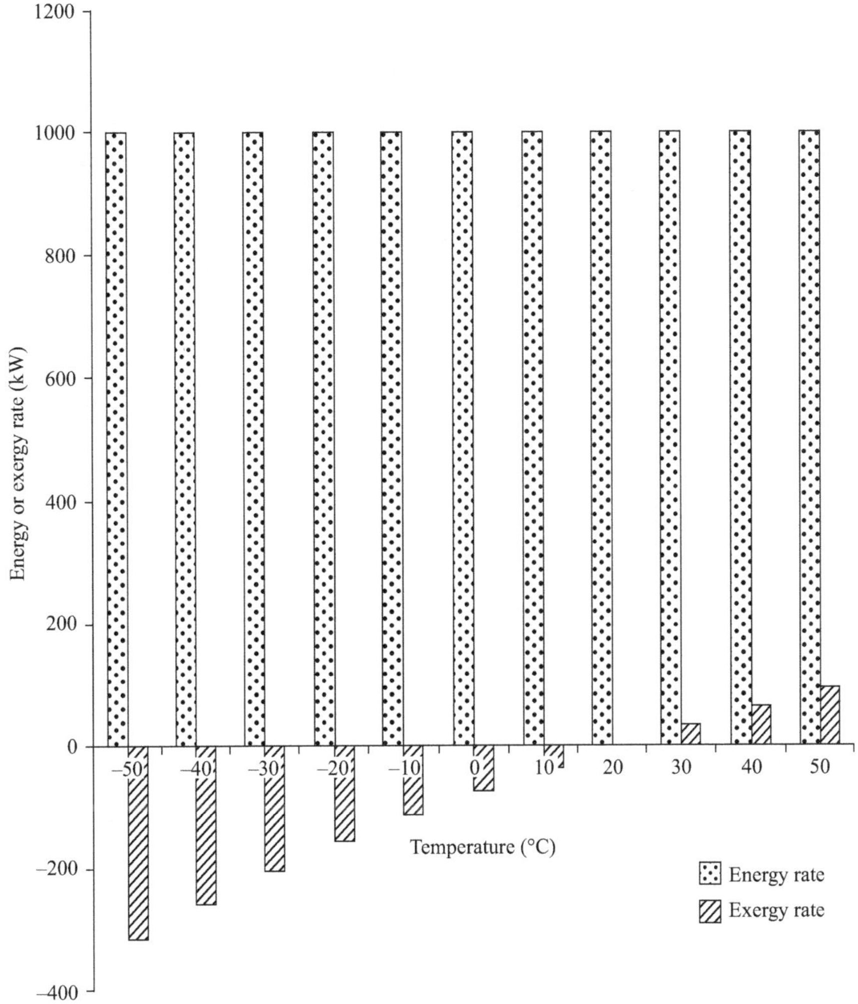

Figure 2.12 Comparison of thermal energy and thermal exergy for a narrow range of temperatures above and below the reference-environment temperature (20°C here). Such temperatures are reflective of conditions encountered in building heating and cooling. This figure is in essence a subsection of the figure in Figure 2.11, expanded to show finer details

reference environment. It is seen that the highest quality, which is reflected in the highest magnitude of the ratio of thermal exergy to thermal energy, occurs for ultra-low- and ultra-high-temperature processes. The quality of energy based on exergy for various non-thermal energy forms are compared qualitatively in Figure 2.17. An overall qualitative comparison of the energy quality of a range of energy forms,

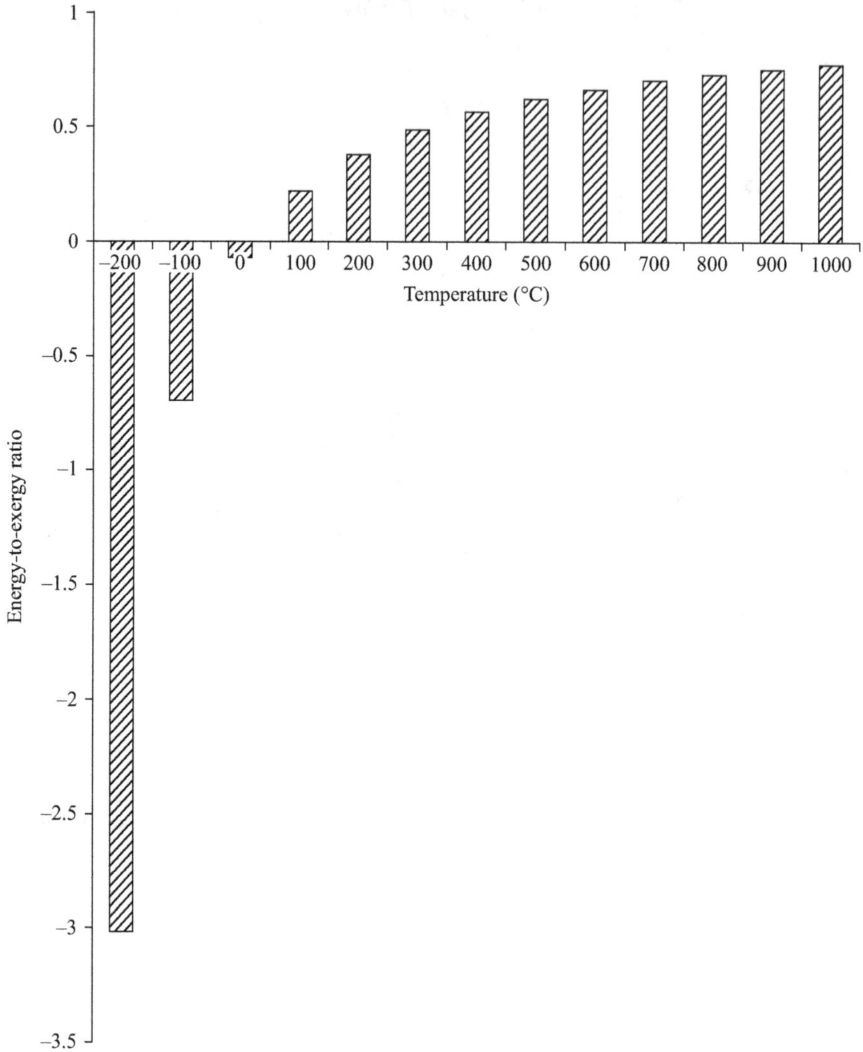

Figure 2.13 *Ratio of thermal exergy to thermal energy for a wide range of temperatures, including the reference-environment temperature (20°C here) (based on data in Tables 2.5 and 2.6). It is observed that the thermal exergy-to-thermal energy ratio above the reference-environment temperature is less than or equal to unity, and asymptotically approaches unity as the temperature associated with the thermal energy approaches infinity. Also, it is observed that the thermal exergy-to-thermal energy ratio below the reference-environment temperature is negative, indicating the same insights as described in Figure 2.11. It is further observed that the magnitude of the thermal exergy-to-thermal energy ratio below the reference-environment temperature can be less or greater than unity, and becomes many times greater than unity as the temperature associated with the thermal energy approaches absolute zero*

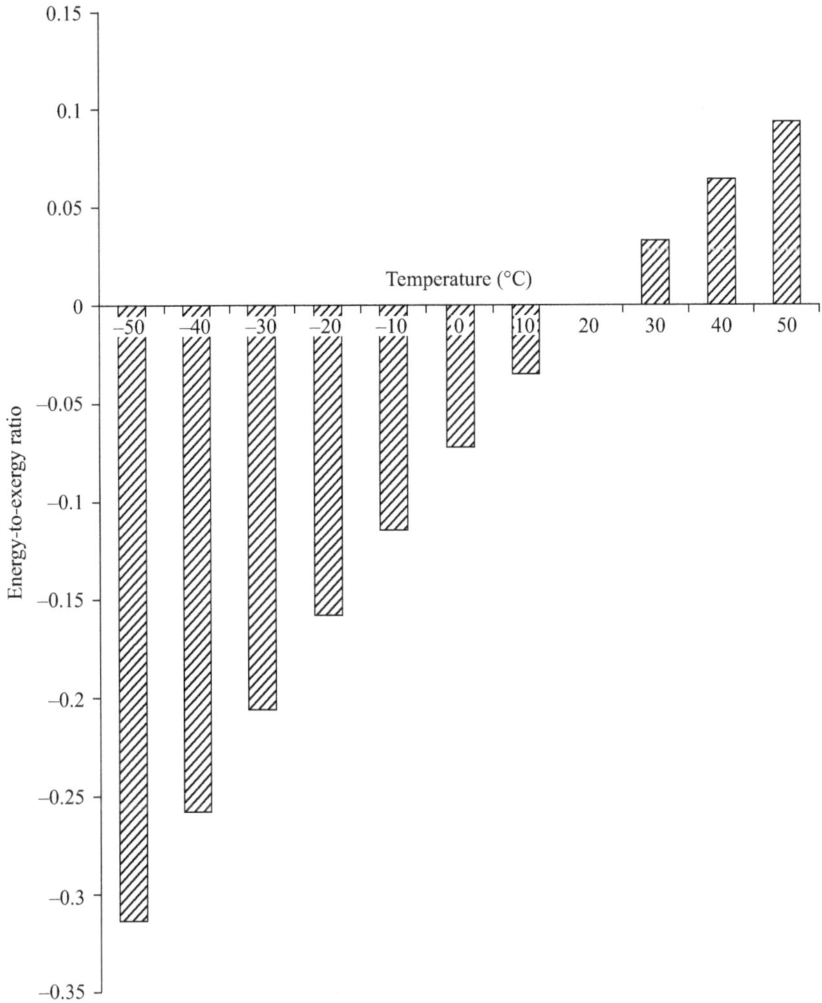

Figure 2.14 Ratio of thermal exergy to thermal energy for a narrow range of temperatures surrounding the reference-environment temperature (20°C here), as are encountered in building heating and cooling. This diagram is in essence subsection of the diagram in Figure 2.13, expanded to show finer details

where exergy is again used as the measure of quality, is shown in Figure 2.18. The analyses of energy quality for different energy commodities (e.g., electricity, work, heated substances, cooled substances, chemicals) are made using exergy methods. That information is useful as it permits, for instance, the relative quality and enthalpy of medium- and low-pressure industrial steam, and commercial hot water are assessed and compared to electricity or shaft power.

Table 2.7 Comparison of quality of H_2O at various conditions above the reference-environment temperature[1]

Condition of H_2O			Specific energy (kJ/kg)	Specific exergy (kJ/kg)	Ratio of exergy to energy
Phase	**Temperature (°C)**	**Pressure (bar)**			
Superheated steam	700	40	3,822	1,677	0.439
	500	40	3,361	1,372	0.408
	300	40	2,919	1,146	0.393
Dry saturated steam	200	15.54	2,709	912	0.337
	150	4.758	2,663	747	0.281
	100	1.014	2,592	525	0.203
Liquid hot water	100	1.014	335	39.9	0.119
	50	0.126	125	6.93	0.0553
	30	0.0425	41.8	0.78	0.0187

[1]The reference-environment temperature and pressure are $T_o = 20°C = 293$ K and $p_o = 1$ bar, respectively.

Table 2.8 Comparison of quality of H_2O at various conditions below the reference-environment temperature[1]

Condition of H_2O			Specific energy (kJ/kg)	Specific exergy (kJ/kg)	Ratio of exergy to energy
Phase	**Temperature (°C)**	**Pressure (bar)**			
Liquid cold water	10	0.0123	−42	0.778	−0.0185
	5	0.00872	−63	1.524	−0.0274
	0	0.00611	−84	3.021	−0.0360

[1]The reference-environment temperature and pressure are $T_o = 20°C = 293$ K and $p_o = 1$ bar, respectively.

2.13 Extensions of exergy methods

Exergy concepts extend and can also be applied beyond thermodynamics. Three important areas in which this is the case are environmental assessment, economics and sustainability.

The applications of exergy methods to environmental and ecological impact assessment and mitigation represent an important and increasingly used extension of exergy. Many suggest that the impact of energy resource utilization on the environment and the achievement of increased resource-utilization efficiency should be addressed by considering exergy, among other factors. Although the exergy of an energy form or a substance is a measure of its usefulness, exergy is also a measure of its potential to cause change. The latter point suggests that exergy may be, or provide the basis for, an effective measure of the potential of a substance or energy form to impact the environment. An additional important factor is the fact

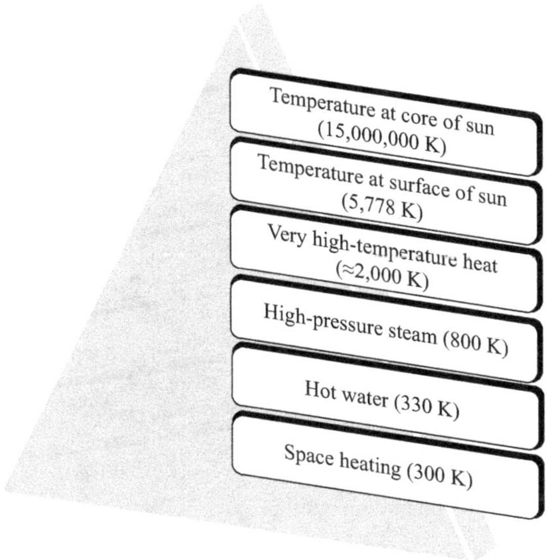

Figure 2.15 Quality of energy based on exergy (in descending order) for thermal energy above the temperature of the reference environment (taken to be a typical mean ambient temperature). It is seen that the highest quality, which is reflected in the highest ratio of thermal exergy to thermal energy, occurs for ultra-high-temperature processes and the lowest for near ambient condition heat transfers

that, since exergy efficiencies always measure approach to ideality or reversibility, exergy losses always correlate with an increased need for external resource inputs, which also contribute to environmental and ecological impact. Methods of performing exergy-based environmental analyses have been developed recently, such as environomics.

Another area in which applications of exergy are increasing is that of economics. In the analysis and design of energy systems, techniques are often used which combine technical disciplines (especially thermodynamics) with economic disciplines (such as cost and financial accounting) to achieve optimum designs. For energy systems, costs are conventionally based on energy. Many researchers, however, have recommended that costs are better distributed among outputs, based on exergy – largely because exergy reflects the quality or usefulness or value of an energy or other commodity. Methods of performing exergy-based economic analyses have evolved recently and are referred to by names such as thermoeconomics, second-law costing and exergoeconomics. These analysis techniques recognize that exergy, more than energy, is the commodity of value in a system, and assign costs and/or prices to exergy-related variables. These techniques usually help determine the appropriate allocation of economic resources so as to optimize the design and operation of a system, and/or the economic feasibility and profitability of a system. This is often accomplished by obtaining actual costs of products and their appropriate prices.

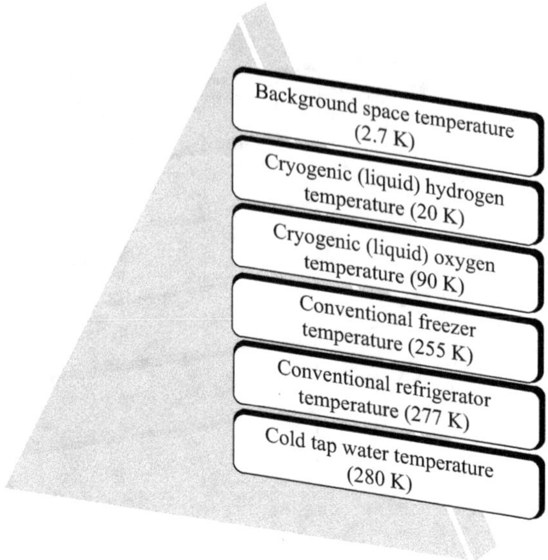

Figure 2.16 Quality of energy based on exergy (in descending order) for thermal energy below the temperature of the reference environment (taken to be a typical mean ambient temperature). It is seen that the highest quality, which is reflected in the highest magnitude of the ratio of thermal exergy to thermal energy, occurs for ultra-low-temperature processes and the lowest for near ambient condition heat transfers

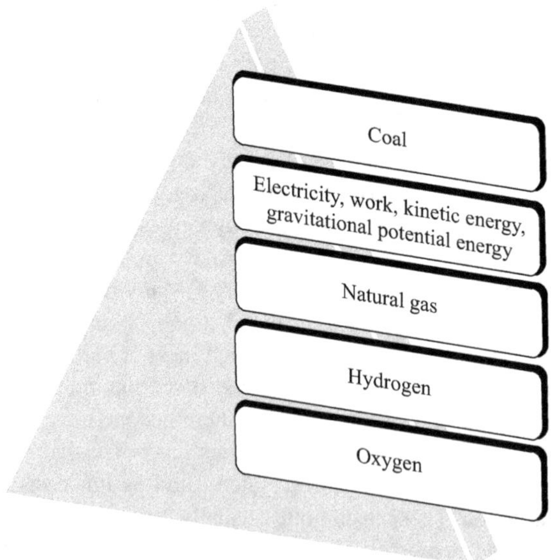

Figure 2.17 Quality of energy based on exergy (in descending order) for non-thermal forms of energy. The highest quality values are unity or slightly greater or lower, and occur for all of the items in the top four blocks

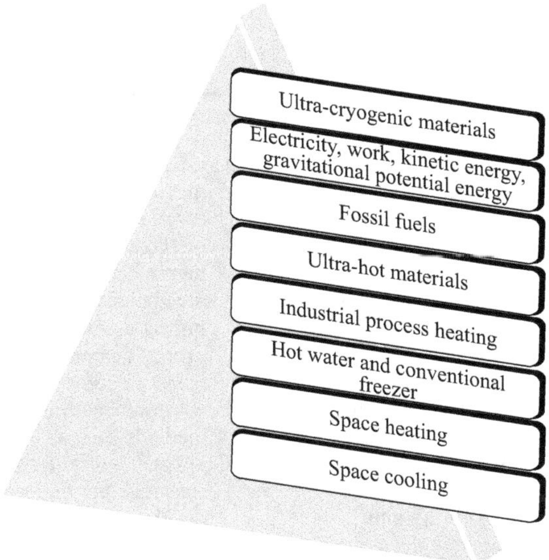

Figure 2.18 Quality of energy based on exergy (in descending order) for various thermal and non-thermal forms of energy. The rank order is somewhat approximate, as there are variations possible in the values of some energy quantities (e.g., the exergy of thermal energy forms is dependent on the specific associated temperature)

An additional important discipline where exergy concepts and methods are proving useful is sustainability or sustainable development. Since exergy efficiencies always measure approach to ideality or reversibility, which reflects a true state of sustainable activity in that interactions can proceed back and forth endlessly without losses when a process is reversible, exergy naturally lends itself to be used as the basis for measures of sustainability. This complements the beneficial use of exergy analysis in the analysis, design and improvement of energy systems and processes. Exergy analysis can reveal whether or not, and by how much, it is possible to design more efficient energy systems by reducing the inefficiencies, permitting the exergy method to assist great efforts to achieve sustainability. But the contributions that exergy can make to energy sustainability extend beyond efficiency improvements, and exploit the relations of exergy with economics and environmental factors, as discussed in the previous two paragraphs.

The rationale for exergy analysis over energy analysis in areas extending beyond thermodynamics is summarized in Table 2.9. Several points are emphasized there, including the fact that exergy methods – unlike energy methods – correlate naturally with economic methods since exergy correlates with value. It is also highlighted that, while energy does not necessarily correlate with environmental impact, exergy methods link to environmental methods since exergy quantities and exergy losses correlate loosely with the potential for environmental impact. More broadly, it is stressed that

Table 2.9 Rationale for exergy analysis over energy analysis in areas extending beyond thermodynamics

Area	Energy analysis shortcoming	Exergy analysis advantage
Economics	Does not correlate naturally with economic methods since energy quantities do not necessarily correlate with value	Correlates naturally with economic methods since exergy quantities correlate with value, leading to methods such as exergoeconomics and thermoeconomics
Environment	Does not correlate naturally with environmental impact assessment methods since energy quantities do not necessarily correlate with environmental impact	Correlates naturally with environmental impact assessment methods since exergy quantities and exergy losses correlate somewhat with potential for environmental impact, leading to methods such as environomics
Sustainability	Does not correlate naturally with the concepts of sustainability and sustainable development	Correlates with the concepts of sustainability and sustainable development since exergy inefficiencies measure deviations from reversibility, which represents a sustainable operation condition, and since exergy correlates with potential for environmental impact and economics, two important pillars of sustainability

exergy approaches correlate with the concepts of sustainability and sustainable development since exergy inefficiencies measure deviations from reversibility, which represents a sustainable condition, and since exergy correlates with two important pillars of sustainability: environmental impact and economic. Conversely, energy methods do not exhibit innate ties with sustainability and sustainable development.

2.14 Closure

Fundamentals of thermodynamics and thermodynamic analysis are provided to permit the behavior, performance and efficiency of energy systems to be described, especially with energy and exergy analyses. The implications of energy and exergy analyses are explained, and a procedure is provided for energy and exergy analyses with an example. The relevance of the material to cogeneration and district energy is explained where necessary, as are extensions of exergy methods to economics and the environment.

Chapter 3
Cogeneration systems

Overview

Cogeneration – and systems for it – are described, starting with fundamentals and definitions and extending to benefits, uses, operation and applications. Although the term cogeneration can in general refer to any process that generates two (and sometimes more than two) products, we typically in this book restrict our use of the term cogeneration to its most common meaning: the simultaneous production in a single process of electricity and heat (typically in the form of steam and/or hot water). In such instances, cogeneration is often referred to as CHP. For clarity, cogeneration is contrasted with thermal electrical generation, pointing out similarities and differences, and highlighting how cogeneration is a logical extension of many systems for thermal electricity generation. The trade-off between thermal and electrical outputs of cogeneration is discussed, demonstrating how electrical output decreases as thermal output (and the quality of the thermal output as measured by its temperature) increases. The interface between cogeneration and energy storage is also discussed, where the storage can apply to the thermal and/or electrical product of cogeneration. A general cogeneration system model, suitable for thermodynamic and engineering assessments, is provided that is sufficiently flexible and robust to model most cogeneration system options. Energy and exergy balances for the general cogeneration system model are given and illustrated. The advantages and applications of cogeneration are given, and the main categories of heat demands that can normally be satisfied with cogeneration are described: residential, commercial and institutional processes (e.g., heating of air and water) and industrial processes (e.g., drying, heating, boiling). The selection of the size and type of a cogeneration system to match as optimally as possible the thermal and electrical demands is discussed, as are matching schemes that can be used. Systems for thermal electricity generation are described, including steam-turbine, gas-turbine, combined-cycle, reciprocating engine and renewable energy-based thermal electricity-generation technologies. Cogeneration systems are also detailed, and it is shown how the types of cogeneration systems available parallel to the types of systems for thermal electricity generation.

3.1 Introduction

A cogeneration process involves the simultaneous production of electricity and heat (usually in the form of steam and/or hot water). The main advantage of cogeneration is that less input energy (usually ranging from 10 to 30 per cent [92]) is consumed than would be required to produce the same thermal and electrical products in separate processes. Additional benefits of cogeneration often are reduced environmental emissions (due to reduced energy consumption and the use of modern technologies in large, central installations), and more economic [91], safe and reliable operation. Cogeneration has been used, particularly by industry, for more than a century [93].

3.2 Fundamentals

Many general descriptions and studies of cogeneration systems have been reported [7–9,92–118]. For example, over 4,000 cogeneration projects are listed in one cogeneration database [95–97], and numerous selected cogeneration systems are described in Table 3.1. Although research and development on cogeneration is ongoing, the basic technology is mature, well understood and proven [92–118].

In general, when an electrical generating station is modified for cogeneration, there is a trade-off between thermal and electrical power outputs. That is, electrical output decreases as thermal output (and the quality of the thermal output as measured by its temperature) increases.

Two main categories of heat demands can normally be satisfied with cogeneration: (i) residential–commercial processes (e.g., heating of air and water) and (ii) industrial processes (e.g., drying, heating, boiling). The use of a central heat supply to meet residential–commercial heat demands is often referred to as district heating, and has been applied and studied extensively [98,99,107,108]. Cogeneration-derived heat can also be used to drive absorption chillers for applications such as district cooling, which is similar to district heating except that it utilizes natural cooling water sources or cool water from central chillers.

The size and type of a cogeneration system is normally selected to match as optimally as possible the thermal and electrical demands. Many matching schemes can be used [93]. Systems can be designed to satisfy the electrical or thermal base-loads, or to follow the electrical or thermal loads. Electricity or heat can be imported from or exported to the electrical or thermal utilities. Storage systems for electricity (e.g., batteries) or heat (e.g., hot water or steam tanks) are often used to overcome periods when demands and supplies for either electricity or heat are not coincident. Cogeneration systems are sometimes used to supply only the peak portions of the electrical or thermal demands.

3.3 Cogeneration and related energy systems

3.3.1 Cogeneration definition

Cogeneration usually refers to the simultaneous production of two energy forms (electricity and heat in the form of steam and/or hot water) from one energy source (normally a fossil fuel).

A cogenerator can be a utility, an industry, a government or any other party.

Table 3.1a Data for selected cogeneration facilities (less than 75 MW electrical)[1]

Cogeneration facility	Type of industry	Main equipment notes	Fuel	Energy rates (MW)		Process steam data (lb/hr)
				Electrical	Thermal	
NRG, Dover, DE	Food[2]	One heat-recovery steam generator[2] One condensing steam turbine[2,3] One gas turbine[2,4]	Natural gas[2] Oil[2]	62[2,4]	38	48,000 at 175 psig 66,100 at 35 psig
Drexel Inc., Oildale, CA	Oil refining	One heat-recovery steam generator and gas turbine[3] One condensing steam turbine	Natural gas	31	9	15,000 at 50 psig (80% quality)
Power Systems Eng., Corona, CA	Milk processing	One heat-recovery steam generator One gas turbine with steam injection	Natural gas	35	19	53,000 at 270 psig (600°F) and 112 psig (saturated)
Packaging Corp. of America, Counce, TN	Pulp and paper[2]	One steam generator One back-pressure steam turbine	Black liquor[2] Natural gas[2] Wood waste[2] Coal[2] Waste oil[2]	35	51	75,000 at 610 psig 75,000 at 160 psig
Midlands Elec. Board, Hereford, UK	Utility	Two waste-heat boilers Diesel engines[3]	Diesel fuel	15	13	10,000 at 300 psig (saturated)
O'Brian Energy systems, Newark, NJ	Boxboard manufacturing	One heat-recovery steam generator and gas turbine with steam injection One condensing steam turbine Selective catalytic reduction	Natural gas	52	14	40,000 at 265 psig
Mt. Carmel Cogen Northumberland, PA	Mining[2]	Circulating fluidized-bed boiler Steam turbine	Culm[2]	47		Hot water
United Airlines, San Francisco, CA	Aircraft maintenance	One heat-recovery steam generator One gas turbine One steam turbine	Aircraft fuel	27		Steam at 100 psig (saturated)
Rohr Industries, Riverside, CA	Aerospace	One heat recovery steam generator One gas turbine with NO_x reduction by water injection	Natural gas	4		20,000 at 150 psig (450°F)

[1]Omitted data were not available in referenced data source.

[2]Data updated in December 2015.

[3]All turbines and diesel engines included in this table are coupled to an electrical generator.

[4]An additional 44-MW gas turbine operates at this plant but is not coupled to a heat recovery system and, therefore, is not part of the cogeneration system. The total electrical capacity of the plant is 106 MW if this turbine is included.

Table 3.1b Data for selected cogeneration facilities (between 75 and 125 MW electrical)[1]

Cogeneration facility	Type of industry	Main equipment notes	Fuel	Energy rates (MW)		Process steam data (lb/hr)
				Electrical	Thermal	
Southeast Paper Mfg. Co., Dublin, GA	Pulp and paper[2]	One circulating fluidized-bed boiler One condensing steam turbine[3]	Coal[2]	78		400,000 at 185 psig and 85 psig
Southport Power Plant, Southport, NC	Utility[2]	Eight coal-fired boilers Two condensing steam turbines	Coal[2] Wood[2] Residuals[2] Tire-derived fuel[2]	110		110,000 at 500 psig
Basic American Foods – American I, King City, CA	Food[2]	One heat recovery steam generator One gas turbine[3] One high-pressure steam turbine One low-pressure steam turbine	Natural gas[2]	120		200,000 at 160 psig
Tesoro, Martinez, CA	Oil refinery[2] Sludge drying	Two heat recovery steam generators and gas turbines One condensing steam turbine NO˙ reduction by steam injection	Natural gas[2] Refinery gas	90	68	160,000 at 600 psig (600°F) 27,000 at 170 psig (400°F)
Argus Cogen Project, Trona, CA	Chemical[2]	One circulating fluidized-bed boiler One back-pressure steam turbine One condensing steam turbine	Coal[2]	96	253	≤650,000 at 450 psig (710°F)
Gilroy Energy Co., Gilroy, CA	Food[2]	One heat recovery steam generator One gas turbine with steam injection One condensing steam turbine	Natural gas[2] Propane[2]	120	64	≤190,000 at 150 psig (saturated)
Indian Petrochem. Corp., Baroda, India	Chemical	Two heat recovery steam generators Two gas turbines with steam injection Two steam turbines	Residual oil	75	75	155,000 at 600 psig (780°F) 44,500 at 80 psig (380°F)
Vulcan Chemicals, Wichita, CA	Chemical	Two heat recovery steam generators Two gas turbines One condensing steam turbine NO_x control by steam injection	Natural gas	110	160	381,000 at 900 psig (830°F) 13,000 at 265 psig (saturated)
Contra Costa, San Francisco, CA	Refinery	Circulating fluidized bed	Petroleum Coke	82		166,000 at 1,500 psig (1,000°F) from boiler

[1]Omitted data were not available in referenced data source.
[2]Data updated in December 2015.
[3]All turbines and diesel engines included in this table are coupled to an electrical generator.

Table 3.1c Data for selected cogeneration facilities (more than 125 MW electrical)[1]

Cogeneration facility	Type of industry	Main equipment notes	Fuel	Energy rates (MW)		Process steam data (lb/hr)
				Electrical	Thermal	
Midland Cogen Venture, Midland, MI	Chemical[2]	One heat recovery steam generator and gas turbine[3]	Natural gas[2]	1,400		629,000
Gulf states Util's, Boston, MA	Utility	Three gas-fired boilers and back-pressure steam turbines[3] One heat recovery steam generator and gas turbine	Natural gas Refinery gas	210		1,800,000 at 600 psig and 135 psig
Midway Sunset, Kern County, CA	Oil refinery[2]	Three heat recovery steam generator and gas turbines	Natural gas[2]	225[2]		1,215,000 at 1,000 psig (80% quality)
Power Systems Eng. Inc., Houston, TX	Chemical	Five heat recovery steam generator and gas turbines with steam injection One condensing steam turbine	Natural gas	471	383	963,000 at 750 psig (625°F) 87,000 at 70 psig (320°F)
TransAlta Sarnia, ON	Chemical[2]	Heat recovery steam generators Three gas turbines[2] Two steam turbines	Natural gas[2]	506[2]		150,000 at 450 psig 474,000 at 334 psig 375,000 at 44 psig
Florida Stone & Cement, Lakeland, FL	Stone and cement[2]	Fluidized bed Uses process air Limestone product	Coal[2]	125		
Dow Chemical, Plaquemine, LA	Chemical[2]	Four gas turbines[2] One steam turbine[2]	Natural gas[2] Hydrogen-gas stream[2]	880[2]		5,000,000[2]
Fluor Daniel, Irvine, CA		Two heat recovery steam generators and gas turbines One steam turbine	Natural gas	250		Variable
Deepwater, TX	Refinery	Air-fired furnace Steam turbine Electrostatic precipitation Flue gas desulfurization Gypsum byproduct	Petroleum Coke	135		600,000
Louisiana Power/Gulf States Utilities	Refinery	Three gas boilers One heat recovery steam generator and gas turbine Two steam turbines	Refinery gas	228		1,800,000 at 135 psig and 600 psig

[1]Omitted data were not available in referenced data source.
[2]Data updated in December 2015.
[3]All turbines and diesel engines included in this table are coupled to an electrical generator.

3.3.2 Cogeneration vs. thermal electrical generation

Cogeneration systems are often extensions of thermal electricity generation systems. In thermal electrical generating stations, the energy content of a resource (normally a fossil fuel) is converted to heat (in the form of steam or hot gases) which is then converted to mechanical energy (in the form of a rotating shaft), which in turn is converted to electricity. A portion (normally 20–45 per cent) of the heat is converted to electricity, and the remainder is rejected to the environment as waste.

Cogeneration systems are similar to thermal electricity generation systems, except that a percentage of the generated heat is delivered as a product, normally as steam or hot water, and the quantities of electricity and waste heat produced are reduced. Overall, cogeneration efficiencies based on both the electrical and thermal energy products of over 80 per cent are achievable.

Most thermal systems for large-scale electricity generation are based on steam and/or gas turbine cycles, and can be modified relatively straightforwardly for cogeneration.

3.3.3 Cogeneration benefits

The advantages generally reported from cogenerating thermal and electrical energy rather than generating the same products in separate processes include the following:

- **Increased efficiency**. By simultaneously producing electricity and useful heat in an advantageous manner, cogeneration systems achieve higher efficiencies – often over 80 per cent on an energy basis – than separate processes for electricity generation and heat production.
- **Reduced energy consumption**. By producing electricity and useful heat more efficiently via cogeneration, relative to separate processes for electricity generation and heat production, less energy is consumed to drive the process.
- **Reduced environmental emissions**. The environmental emissions associated with cogeneration are reduced because of their reduced energy consumption compared to separate processes for electricity and heat, and also because cogeneration typically involves the use of modern technologies in large, central installations with advanced environmental controls and technology.
- **Enhanced economics**. In many instances, the production of electricity and useful heat via cogeneration is less expensive, on a lifecycle basis, than using separate processes for electricity generation and heat production.
- **Enhanced safety**. Cogeneration is often carried out in large plants outside of buildings where electricity and heat are used. By effectively removing the heat generation devices, which often consume fossil fuels, from those buildings, they have lower risks of accidents due to events such as fires, explosions and suffocations.
- **Enhanced reliability**. Cogeneration systems are often incorporated in district energy systems, so as to allow the thermal energy product to be distributed

where needed. There are often several thermal energy suppliers linked to a given district energy system, providing redundancy and the ability to keep the system operating when one plant becomes inoperable (due to a breakdown or regular servicing).

The benefits and potential of cogeneration has been discussed extensively elsewhere [119–124].

3.3.4 Cogeneration uses and operation

Three main categories of heat demands can normally be satisfied through cogeneration:

- **Residential, commercial and institutional heating processes**. These require large quantities of heat at relatively low temperatures (e.g., for air and water heating). The use of a central heat supply to meet residential, commercial and institutional heat demands is often referred to as district heating.
- **Industrial heating processes**. These require heat at a wide range of temperatures (e.g., for drying, heating and boiling in, for instance, chemical processing, manufacturing, metal processing, mining and agriculture).
- **Residential, commercial and institutional cooling processes**. Cogenerated heat can drive chillers, as well as provide heating services. This category requires large quantities of heat at relatively low temperatures, and can use of a central heat supply through district heating, or can transport cooling capacity via district cooling. This application can be particularly beneficial in locations where the annual peak electrical demand is associated with the summer cooling load.

The size and type of a cogeneration system are normally selected to match as optimally as possible the thermal and electrical demands. Many matching schemes can be used. Systems can be designed to satisfy the electrical or thermal base-loads, or to follow the electrical or thermal loads. Cogeneration systems are sometimes used to supply only the peak portions of the electrical or thermal demands.

3.3.5 Cogeneration applications

Cogeneration has been used, particularly by industry, for approximately a century. Many general descriptions and studies of cogeneration systems have been reported [125–139]. Cogeneration systems are in use throughout the world (e.g., thousands are listed by the Association of Energy Engineers), and the basic technology is proven. Numerous examples exist of large cogeneration systems:

- A steam-turbine plant in Switzerland generates 465 MW of thermal power and 135 MW of electrical power, with an overall efficiency of 75 per cent.
- A nuclear power plant in Michigan left incomplete due to lack of funding was eventually completed as a gas-fired combined-cycle cogeneration plant having 12 heat recovery steam generators and gas turbines and two steam turbines, producing 1,400 MW of electrical power and 285,000 kg/hr of steam.

- Approximately ten plants are used to generate 240 MW of electrical power and to supply 90 per cent of the 1,500 MW thermal demand for the city of Malmo, Sweden (population 250,000). Fuel drives two of the plants (an extraction steam-turbine plant generating 110 MW of electrical power and 240 MW of thermal power, and a back-pressure steam-turbine plant generating 130 MW of electrical power and 300 MW of thermal power), while the remaining plants operate on waste heat from neighboring industries (e.g., smelting, carbon-black production, sewage treatment and refuse incineration).

3.3.6 Cogeneration and energy storage

Storage systems for electricity (e.g., batteries) or heat (e.g., hot water or steam tanks) are often used to overcome periods when demands and supplies for either electricity or heat are not coincident.

3.4 General cogeneration system model

Several cogeneration plant models exist that fulfill various purposes such as economic, performance and optimization analyses [112]. The basic principle is the same in all cogeneration systems. Fuel and air enter at the temperature and pressure of the environment. Electrical- and thermal energy products and wastes exit.

A model for a general cogeneration system is shown in Figure 3.1. It is shown that a cogeneration (CHP) plant takes in fuel ($\dot{m}_f$) and air ($\dot{m}_a$) and transforms them into useful products that appear in the form of useful work ($\dot{W}_p$) and heat ($\dot{Q}_p$).

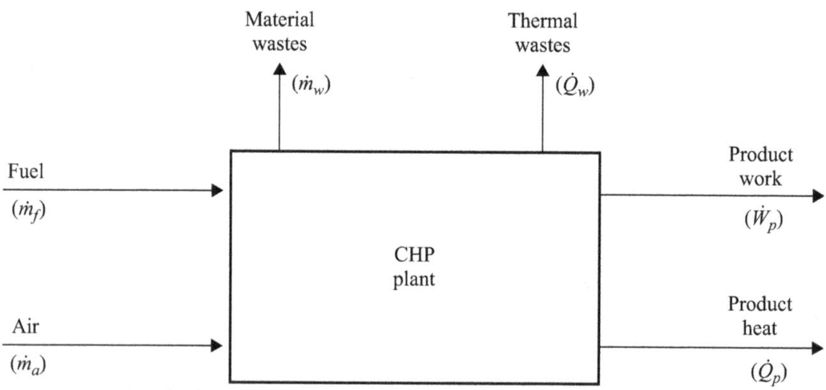

Figure 3.1 A model for a general cogeneration (CHP) system. The system simplifies to a general heat production system if the product work term is zero, and to a general electricity generation system if the product heat term is zero

Wastes generated throughout the process include waste materials ($\dot{m}_w$) and waste thermal energy ($\dot{Q}_w$). A variety of specific cogeneration systems are shown in Figures 3.2 and 3.3, and are discussed later. For simplicity, only one heat product is considered. Wastes can include materials and heat.

The general cogeneration system represents an electrical generation system if there is no thermal energy product, and a heat production system if there is no electrical product.

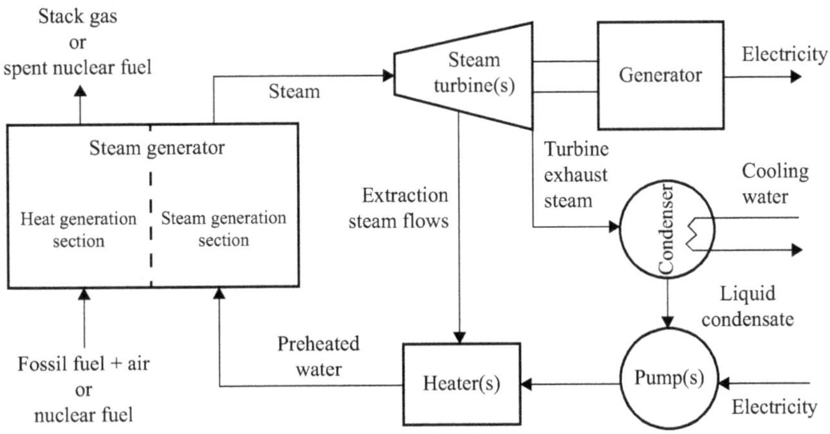

Figure 3.2a Simplified diagram of a steam-turbine electricity-generation system

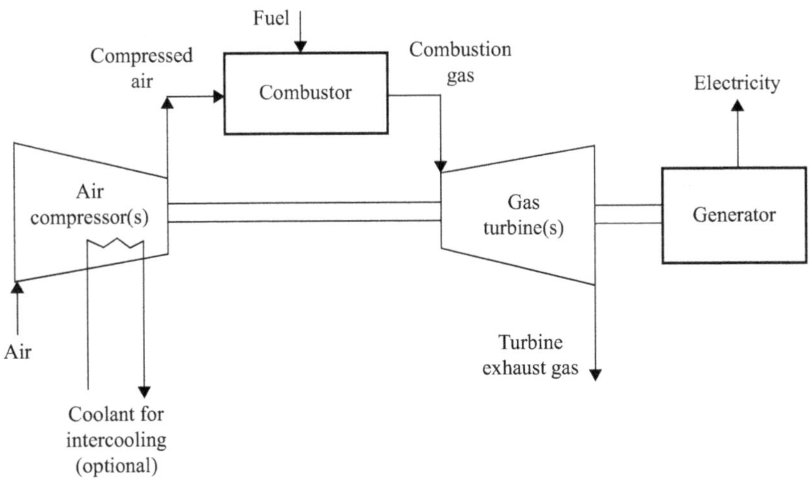

Figure 3.2b Simplified diagram of a gas-turbine electricity-generation system

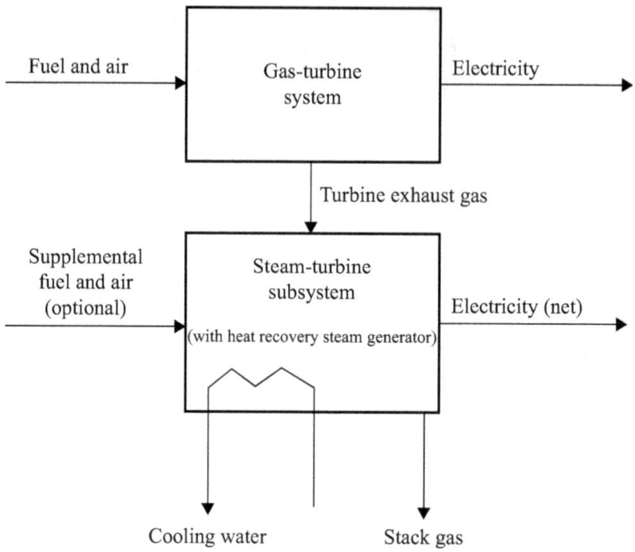

Figure 3.2c Simplified diagram of a combined-cycle electricity-generation system

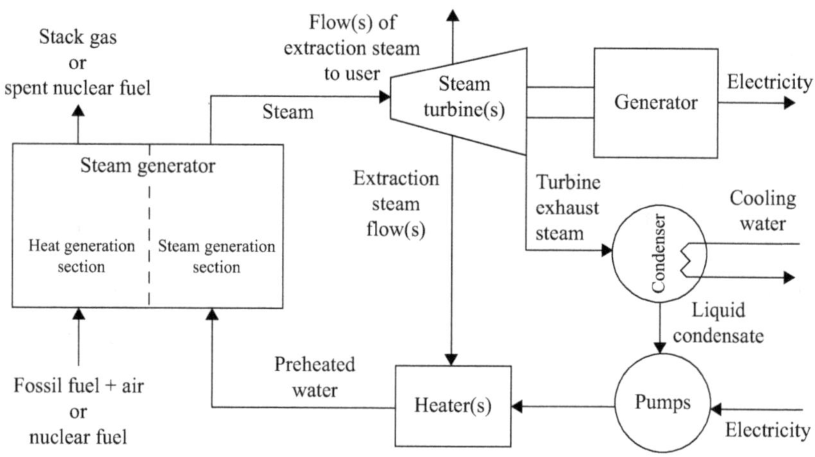

Figure 3.3a Simplified diagram of an extraction steam-turbine cogeneration system

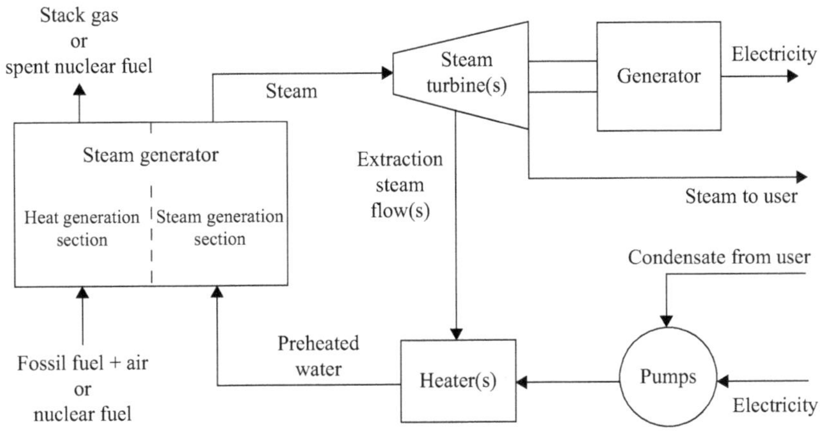

Figure 3.3b Simplified diagram of a back-pressure steam-turbine cogeneration system

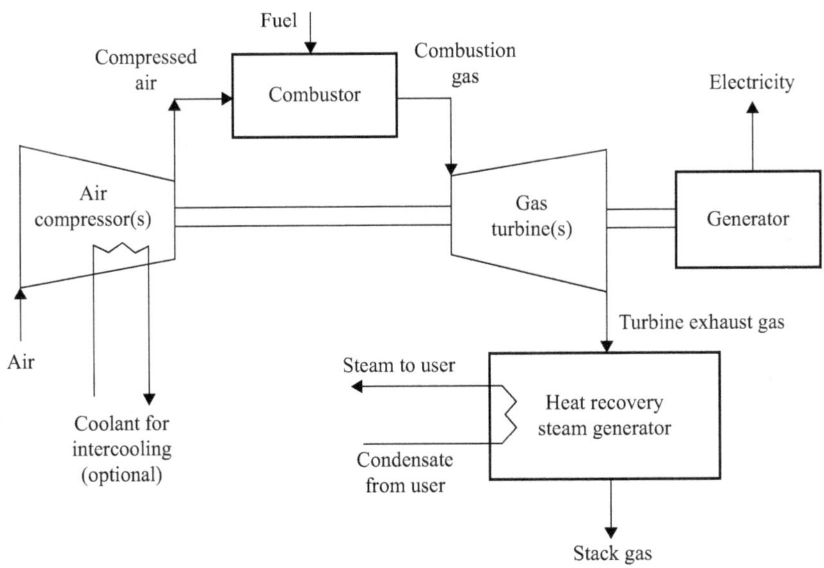

Figure 3.3c Simplified diagram of a gas-turbine cogeneration system

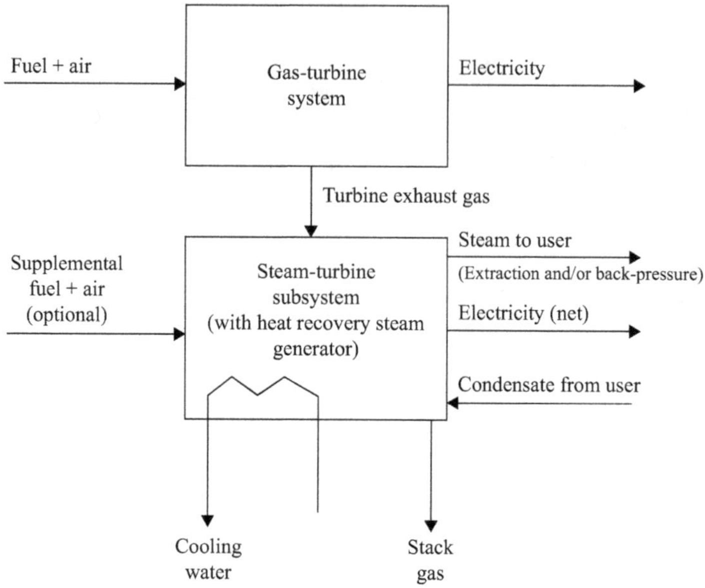

Figure 3.3d Simplified diagram of a combined-cycle cogeneration system

3.5 Description of a general cogeneration system

This section reviews the aspects of thermodynamics that are more relevant to energy and exergy analyses specific to a general model for cogeneration systems. General efficiency definitions based on energy and exergy flows for cogeneration systems are included and the trade-off between variation of the electrical and thermal outputs of a cogeneration plant is discussed.

3.5.1 Energy and exergy balances

Assuming steady-state operation and neglecting kinetic and potential energies, an energy balance for the system in Figure 3.1 can be written, following (2.8), as

$$\dot{m}_f h_f + \dot{m}_a h_a = \dot{W}_p + \dot{Q}_p + \sum_w \dot{Q}_w + \sum_w \dot{m}_w h_w \tag{3.1}$$

where h denotes specific enthalpy; $\dot{m}$, $\dot{Q}$ and $\dot{W}$, respectively, denote rates of mass flow, heat flow and work; the subscripts f, a, p and w denote, respectively, fuel, air, product and waste; and the summations are over all material and thermal waste streams. The terms on the left-hand side of (3.1) represent the total energy input rate and are entirely associated with flows of matter. The first two terms on the right-hand side represent the energy output rate associated with products (electrical and thermal), and the remaining two terms represent the energy output rate associated with wastes (thermal and material). The ratio $\dot{Q}_p/(\dot{Q}_p + \sum \dot{Q}_w)$ denotes the fraction of the total thermal energy output rate which is product, and ranges from as low as zero for electricity generation systems to unity for systems in which all heat is utilized.

A corresponding exergy balance can be written, following (2.11), as

$$\dot{m}_f\varepsilon_f + \dot{m}_a\varepsilon_a = \dot{W}_p + \dot{Q}_p\tau_p + \sum_w \dot{Q}_w\tau_w + \sum_w \dot{m}_w\varepsilon_w + \dot{I} \quad (3.2)$$

where ε denotes the specific exergy of a material flow; τ_p and τ_w are the exergetic temperature factors which are functions of the relevant temperatures T_p, T_w and T_o; T_p, T_w and T_o denote, respectively, the temperatures associated with the product heat flow, the waste heat flow(s) and the environment and other terms are as previously defined. The terms on the left-hand side of (3.2) represent the exergy input rate, while, on the right-hand side, the first two terms represent the exergy output rates associated with electrical and thermal energy products, the second two terms represent the exergy output rates associated with thermal and material energy wastes and the final term represents the exergy consumption rate. The terms τ_p and τ_w range from zero to unity for all heat flows in normal cogeneration systems where heat is output at above-environment temperatures (i.e., where $T_p \geq T_o$ and $T_w \geq T_o$).

It is noted that nuclear-based cogeneration systems can be handled with the model in Figure 3.1. In such systems, no air enters (i.e., $\dot{m}_a = 0$), and the fuel represents nuclear fuel. Further, it is stipulated that $\dot{m}_f h_f = \dot{Q}_n$, where $\dot{Q}_n$ is the heat transferred to the system (at a temperature T_n) due to fissioning of the nuclear fuel. Then:

$$\dot{m}_f\varepsilon_f = \dot{Q}_n\tau_n \quad (3.3)$$

Here, it is assumed that T_n is potentially infinite in value (as is often done [4–6]), so that the energy and exergy quantities of the thermal energy delivered from the fissioning of the nuclear fuel are identical. Consequently:

$$\dot{m}_f\varepsilon_f = \dot{Q}_n = \dot{m}_f h_f \quad (3.4)$$

for nuclear systems.

3.5.2 *Efficiencies*

Following (2.19) and (2.20), the energy efficiency for the cogeneration system in Figure 3.1 can be written as

$$\eta_{CHP} = \frac{\dot{W}_p + \dot{Q}_p}{\dot{m}_f h_f} \quad (3.5)$$

and the corresponding exergy efficiency as

$$\psi_{CHP} = \frac{\dot{W}_p + \dot{Q}_p\tau_p}{\dot{m}_f\varepsilon_f} \quad (3.6)$$

While energy and exergy efficiencies are both sensitive to $\dot{Q}_p$, only exergy efficiencies are sensitive to the temperature at which the product heat is delivered.

When $\dot{Q}_p = 0$, the energy and exergy efficiencies reduce to electrical energy and exergy efficiencies, respectively. Since for many reference-environment models (including the one described in Table 2.2), $h_a = \varepsilon_a = 0$, h_f is often the HHV of a fuel evaluated at T_o and P_o, and $\varepsilon_f = h_f$ (within approximately 15 per cent), the electrical energy and electrical exergy efficiencies are approximately equal for a given system, although the general energy and exergy efficiencies can differ significantly. As $\dot{Q}_p$ increases, energy and exergy efficiencies both increase, but energy efficiencies increase more rapidly than exergy efficiencies.

Since $0 \leq \tau_p \leq 1$ for $T_p \geq T_o$, it can be shown with (3.5) and (3.6) that $\psi_{CHP} \leq \eta_{CHP}$ when $\varepsilon_f = h_f$, and that the amount by which the energy efficiency exceeds the exergy efficiency is

$$\eta_{CHP} - \psi_{CHP} = \frac{\dot{Q}_p}{\dot{m}_f h_f} \frac{T_o}{T_p} \tag{3.7}$$

For a fixed reference-environment and set of inputs, the difference in efficiencies (i) decreases as the product-heat delivery temperature T_p increases and output rate $\dot{Q}_p$ decreases and (ii) is independent of the product-work rate, $\dot{W}_p$. The largest deviations between ψ_{CHP} and η_{CHP} normally occur for low-temperature systems, and the smallest deviations for high-temperature systems. On the one extreme, if $T_p = T_o$, the largest deviation between ψ_{CHP} and η_{CHP} occurs (with $\psi_{CHP} = \eta_{CHP}$ only if no heat is produced, and $\psi_{CHP} = 0$, regardless of the value of η_{CHP}, if no electricity is produced). On the other extreme, if $T_p = \infty$, $\psi_{CHP} = \eta_{CHP}$ for all cases.

Exergy efficiencies can be argued to give more illuminating insights into cogeneration process performance than energy efficiencies, because exergy efficiencies provide a measure of how nearly the performance of the system approaches ideality, and of the magnitude of the potential for improvement. Energy efficiencies appear overly optimistic, particularly when large fractions of the heat output from the system are utilized for cogeneration. These observations stem from the fact that exergy efficiencies weigh heat and electricity according to the amount of work they can be used to theoretically produce, and imply that cogeneration systems are better compared to using exergy efficiencies. Thus, the use of energy efficiencies in most practical situations, where thermal energy is normally produced at low-to-moderate temperatures, can lead to erroneous interpretations and conclusions.

3.5.3 *Trade-off between cogeneration electrical and thermal outputs*

To illustrate more clearly the trade-off for a cogeneration plant between electrical output and product heat output (in terms of quantity and quality, i.e., temperature), the following COP is introduced:

$$\text{COP}_{\text{CHP}} = \frac{\text{Product heat output rate}}{\text{Electricity generation rate curtailed}} = \frac{\dot{Q}_p}{\dot{W}_p} \tag{3.8}$$

The denominator represents the difference between the electrical power generated by the plant when it operates in an all-electric mode ($(\dot{W}_p)_{elec}$) and that when it operates in a cogeneration mode ($\dot{W}_p$). That is,

$$\dot{W}_c = (\dot{W}_p)_{elec} - \dot{W}_p \tag{3.9}$$

This measure of performance treats the cogeneration plant as a heat pump in that a cogeneration plant in general foregoes electrical output to produce useful heat, while a heat pump uses electricity to produce useful heat. The term COP_{CHP} is in principle analogous to a heat pump COP. This comparison is important in that it quantifies the ability to generate electricity which is transferred to producing heat, when heat-driven absorption chillers are used in integrated cogeneration/chiller systems (as discussed later).

In some instances, such as for diesel and other reciprocating engines, little electrical output is lost when they are converted to cogeneration and produce useful heat. Hence, values of COP_{CHP} for these do not vary extensively and are high (approaching infinity as the electrical output curtailed in producing useful heat approaches zero). In most instances, however, COP_{CHP} values vary significantly with variations in the product heat parameters.

The behavior of COP_{CHP} is illustrated for a simplified scenario, following Figure 3.1, in which the outputs from the cogeneration plant are electrical ($\dot{W}_p$) and thermal ($\dot{Q}_p$) power, and a fixed amount of wastes. The total "product" energy output rate is, therefore, $\dot{W}_p + \dot{Q}_p$. It is assumed that, as electrical output is curtailed (i.e., $\dot{W}_p$ decreases), $\dot{Q}_p$ increases and approaches the value of $\dot{W}_p + \dot{Q}_p$ in the electric mode of operation, i.e., $(\dot{W}_p + \dot{Q}_p)_{elec}$. For simplicity, $\dot{Q}_p$ is entirely regarded as useful (product). Then, when electrical power output is curtailed by $\dot{W}_c$, the thermal power output $\dot{Q}_p$ is, from an energy balance, $(\dot{Q}_p)_{elec} + \dot{W}_c$. Consequently, from (3.8):

$$\text{COP}_{\text{CHP}} = \frac{\dot{Q}_p}{\dot{W}_c} = \frac{(\dot{Q}_p)_{elec} + \dot{W}_c}{\dot{W}_c} \tag{3.10}$$

or, in an alternative form:

$$\text{COP}_{\text{CHP}} = \frac{(\dot{Q}_p/\dot{W}_p)_{elec} + \dot{W}_c/(\dot{W}_p)_{elec}}{\dot{W}_c/(\dot{W}_p)_{elec}} \tag{3.11}$$

The behavior of COP_{CHP} is illustrated in Figure 3.4 as a function of the fraction of the electrical output rate in the all-electric mode of operation which is curtailed in the cogeneration mode of operation (i.e., $\dot{W}_c/(\dot{W}_p)_{elec}$). When all electrical power generation is foregone (i.e., $\dot{W}_c/(\dot{W}_p)_{elec} = 1$), (3.11) simplifies to:

$$\text{COP}_{\text{CHP}} = (\dot{Q}_p/\dot{W}_p)_{elec} + 1 \tag{3.12}$$

which would, for example, equal 3 for a cogeneration plant having a ratio of 2 for its thermal-to-electric output in the electric mode of operation. As the curtailed fraction decreases, COP_{CHP} monotonically increases and asymptotically approaches infinity as the curtailed fraction approaches zero.

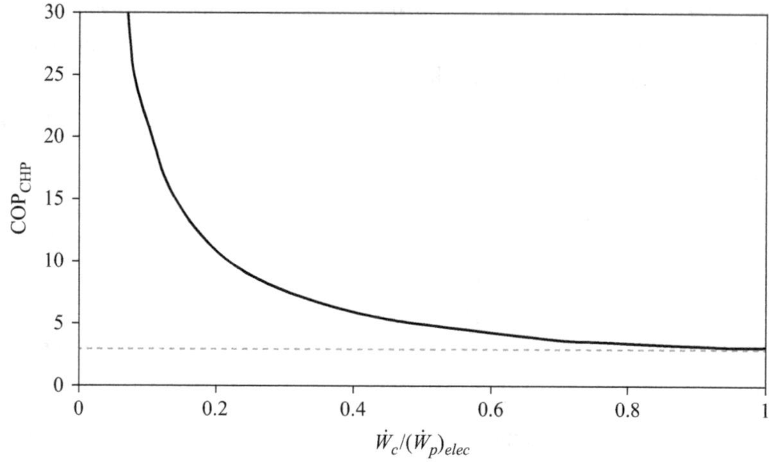

Figure 3.4 Variation of the COP of a cogeneration plant with the fraction of the electrical output rate in the all-electric mode of operation that is curtailed in the general cogeneration mode of operation. Here, the ratio of thermal to electrical output is 2 in the electric mode of operation, and it can be seen that $COP_{CHP} = 3$ (dashed horizontal line) when $\dot{W}_c/(\dot{W}_p)_{elec} = 1$, and that COP_{CHP} asymptotically approaches infinity as $\dot{W}_c/(\dot{W}_p)_{elec}$ approaches zero

The above discussion omits the issue of temperature. In actuality, the temperature associated with the thermal power output decreases from the maximum electric-mode cycle temperature (when $\dot{W}_c/(\dot{W}_p)_{elec} = 1$) to the minimum electric-mode cycle temperature (when $\dot{W}_c/(\dot{W}_p)_{elec} = 0$). An alternative way of viewing the information in Figure 3.4 so as to highlight the effect of temperature is to envision a steam-turbine generator within the cogeneration system. In the all-electric mode of operation, steam expands fully through the turbine and (i) a maximum amount of electrical power is generated and (ii) the steam exits at a minimum temperature (or pressure). To curtail electrical output, the steam is extracted at higher temperatures (or pressures) from the turbine. The quantity COP_{CHP} is plotted as a function of extraction-steam temperature and pressure in Figure 3.5 for an idealized case, and in Figure 3.6 for a realistic case. The steam is assumed to enter the turbine at 500°C and 160 bar, and to exit, in the all-electric mode, at 0.045 bar. The isentropic-turbine, shaft and generator efficiencies are assumed to be 80, 98 and 99 per cent, respectively, for the realistic case, and 100 per cent for all in the idealized case.

It can be seen that similar behavior is exhibited in the plots in Figures 3.5 and 3.6 for both the idealized and realistic cases. When steam is extracted at the highest possible temperature or pressure (i.e., at the turbine inlet), all electrical output is curtailed and COP_{CHP} takes on a value between 2 and 3. As the temperature or pressure of the extraction steam decreases toward the minimum

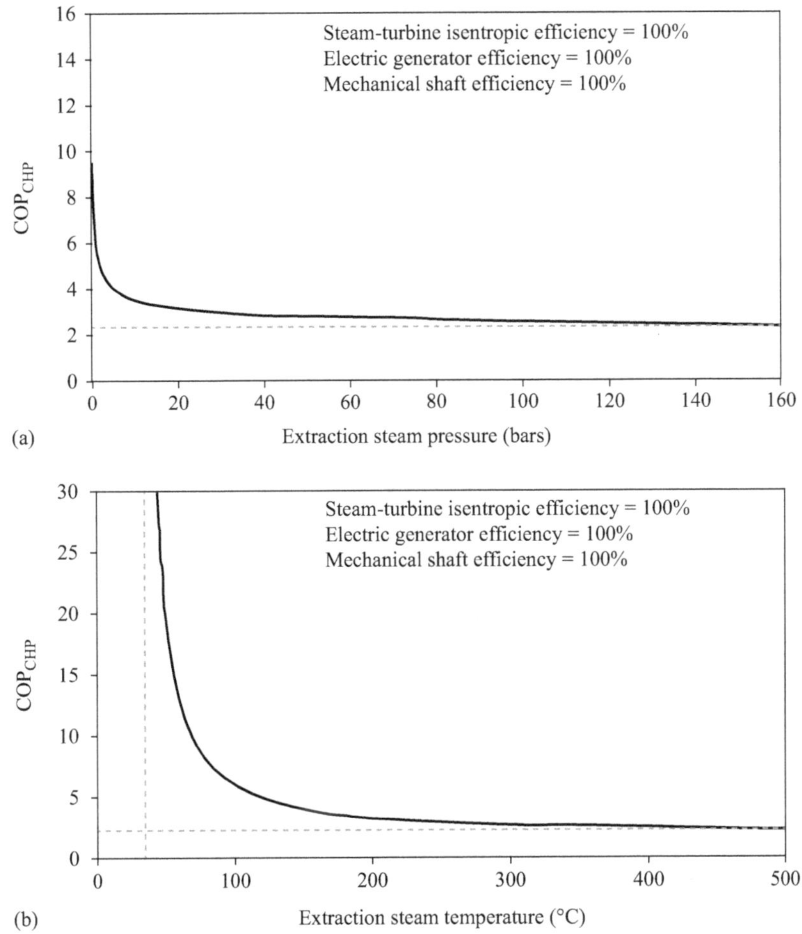

Figure 3.5 *Variation of cogeneration COP with extraction-steam pressure (a) and extraction-steam temperature (b) for an idealized steam turbine-based cogeneration plant. Steam enters the turbine at 50°C and 160 bar and exits in the all-electric mode of operation, at 0.045 bar. The dashed horizontal lines represent the minimum COP_{CHP} values, which apply when steam is extracted at the input condition. The vertical dashed line identifies the minimum possible extraction steam temperature (i.e., the temperature corresponding to the minimum extraction steam pressure of 0.045 bar)*

possible value (i.e., the condition at the turbine outlet after full expansion), the electrical output curtailed reduces to zero and COP_{CHP} increases and asymptotically approaches infinity.

From the above discussions, it can be clearly inferred that, thermodynamically, cogeneration is a very efficient method of producing useful heat, compared to

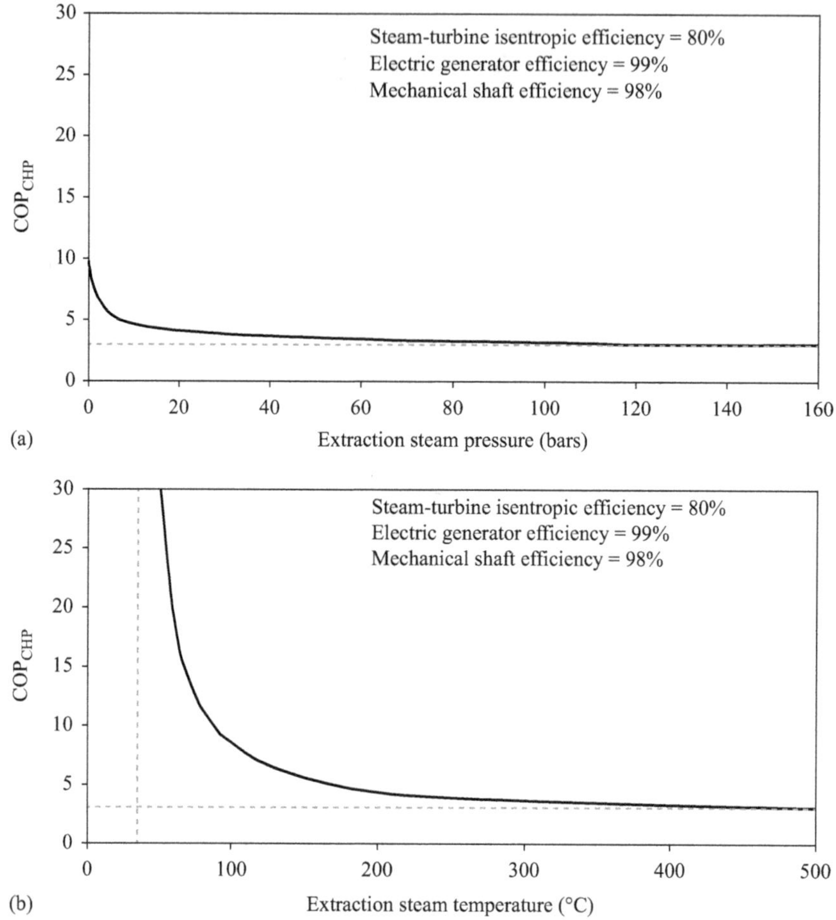

Figure 3.6 Variation of cogeneration COP with extraction-steam pressure (a) and extraction-steam temperature (b) for a realistic case for a steam turbine-based cogeneration plant. Other details are as in Figure 3.5

typical heat pumps (for which COP values range approximately from 2 to 5) and electrical resistance heating (for which COP values usually lie between 0.95 and 1). Particularly in instances in which heat is required at low temperatures, the trade-off inherent to cogeneration systems between electrical and thermal output greatly favors cogeneration as a method of producing the required heat from work. Energy and exergy efficiencies of cogeneration systems with variation of several system characteristics have been investigated in several studies [13–21].

Relations on energy and exergy bases between the electrical and thermal outputs for cogeneration systems are illustrated in Figures 3.7 and 3.8. The relation between the outputs for a cogeneration system for electricity and heat is shown in Figure 3.7 on energy and exergy bases, emphasizing how energy analysis overweights the value of

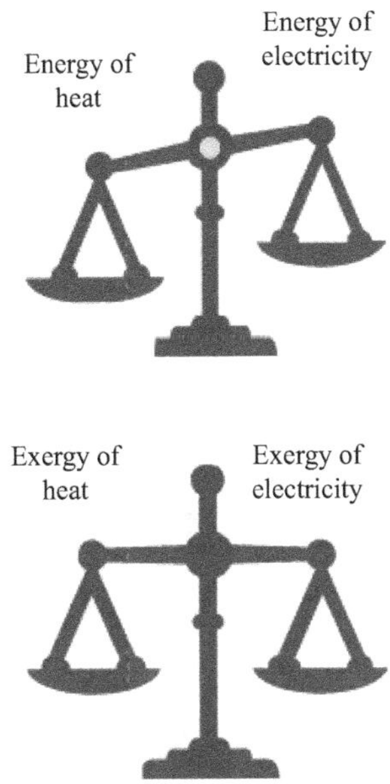

Figure 3.7 Illustration of the relation between the outputs for a cogeneration system for electricity and heat, on an energy basis (top) and an exergy basis (bottom). Energy analysis is seen to overweight the value of heat relative to the value of electricity, while exergy analysis weights the value of these commodities equitably

heat relative to the value of electricity, while using exergy analysis weights the value of those commodities fairly. Similarly, the relation between the outputs for a cogeneration system for electricity and cool is shown in Figure 3.8, emphasizing how energy analysis overweights the value of cool relative to the value of electricity, while using exergy analysis weights the value of these commodities fairly. Although the thermal product can be either heat or cool, the phenomenon of overweighting the thermal product is similar in both instances.

3.6 Systems for electricity generation and cogeneration

Cogeneration systems and thermal electricity-generation systems, which form the basis of most cogeneration systems, are discussed in this section. In most thermal electricity-generation systems [1,2], an energy resource (normally a fossil or nuclear fuel) is converted to heat (in the form of steam or hot gases) which is then converted to mechanical energy (in the form of a rotating shaft), which in turn is converted to electricity. A portion (normally 20–45 per cent) of the heat is converted to electricity, and the remainder rejected to the environment as waste. Cogeneration systems are similar to thermal electricity-generation systems, except that a percentage of the generated heat is delivered as a product (normally as steam

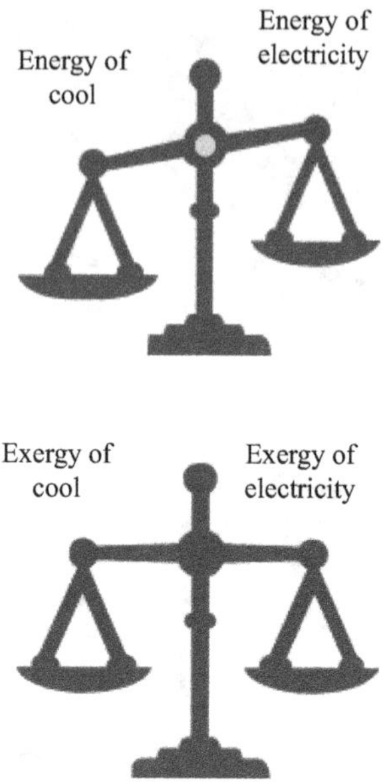

Figure 3.8 Illustration of the relation between the outputs for a cogeneration system for electricity and cool, on an energy basis (top) and an exergy basis (bottom). Energy analysis is seen to overweight the value of cool relative to the value of electricity, while exergy analysis weights the value of these commodities fairly

or hot water), while electrical production is held constant or reduced and waste-heat output reduced. Cogeneration energy efficiencies (based on both electricity and heat) of over 80 per cent are achievable.

3.6.1 Thermal electricity-generation systems

Most thermal systems for large-scale electricity generation are based on steam and/or gas turbine cycles, and reciprocating engines. These and other systems are illustrated in Figure 3.9 and are described below.

3.6.1.1 Steam-turbine electricity-generation systems

In steam-turbine electricity-generation systems (Figure 3.2a), heat is normally generated either by the combustion of fuels (e.g., coal, oil, natural gas, municipal solid waste [MSW], biomass) in air or by the nuclear reaction of a nuclear fuel (normally uranium). The heat vaporizes pressurized water in a steam generator. The steam passes through one or more turbines, causing the turbine shaft to rotate. The shaft is coupled to an electrical generator which converts the mechanical energy in the rotating shaft to electricity. The steam exiting the turbines is condensed, and the condensate is pressurized in one or more pumps, and preheated (using small quantities of steam extracted from the turbines), before it returns to the steam generator.

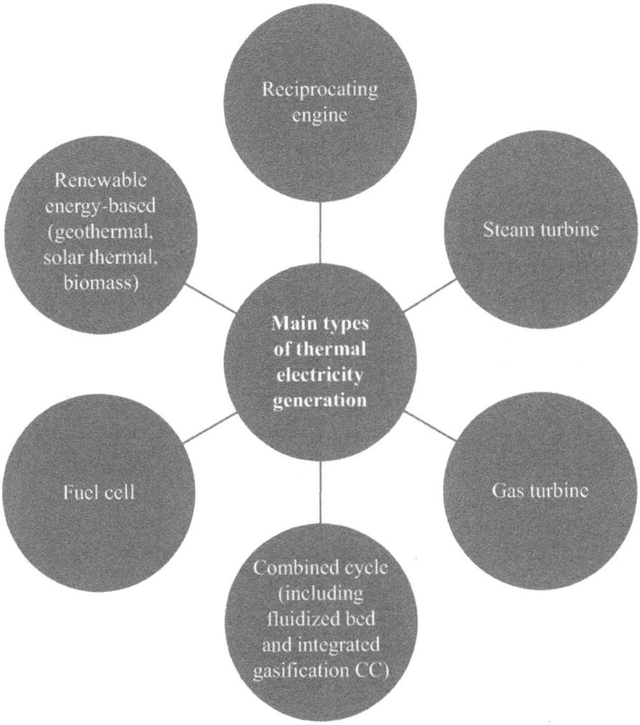

Figure 3.9 The main types of thermal electricity generation, including traditional methods using fossil fuels, as well as new methods based on renewable energy and other fuels

Electrical efficiencies are approximately 35–40 per cent for most modern fossil fuel steam turbine plants, and 30 per cent for most modem nuclear plants. Steam turbine units with electrical outputs of 300–600 MW exist, and are being developed for outputs up to 1,000 MW [93].

Solar thermal technologies can also produce high-pressure vapor that can expand through a steam turbine to produce work. In order to achieve sufficiently high temperatures and pressures for operation of a steam turbine, solar technologies that are able to produce heat at high temperatures, such as concentrating solar collectors, are used. Low-temperature solar thermal technologies such as flat plate solar thermal collectors can also be used to provide power for small-scale applications when coupled to an organic Rankine cycle (ORC) [94]. ORC-based cogeneration systems have lower operation pressures and temperatures compared to conventional steam-Rankine cycle cogeneration systems.

3.6.1.2 Gas-turbine electricity-generation systems

In gas-turbine electricity-generation systems (Figure 3.2b), heat is generated by the combustion of gaseous fuels (e.g., natural gas, landfill gas, coal-derived fuel gas) in air that has been pressurized in one or more compressors. The air is sometimes cooled during compression. The resulting high-temperature gas passes through one or more turbines, causing the turbine shaft to rotate and drive a coupled electrical generator, and then exits the system.

Electrical efficiencies are approximately 30–45 per cent for most modern gas turbine plants. Gas turbine units with electrical outputs up to approximately 100 MW exist, and are being developed for outputs up to 200 MW [93].

3.6.1.3 Combined-cycle electricity-generation systems

In combined-cycle electricity-generation systems (Figure 3.2c), steam- and gas-turbine cycles are combined by using the moderately hot exhaust gas from the gas turbine to generate steam in a "heat recovery steam generator" for use in the steam-turbine cycle. Such combined-cycle plants can achieve efficiencies of 45–55 per cent. Heat recovery steam generators, with or without capabilities for additional combustion, are available in a variety of types and sizes [1,2,93].

3.6.1.4 Reciprocating engines

Reciprocating engines are heat engines that use a heat source, such as hot fossil fuel combustion gases, in a piston-cylinder device to provide rotating motion, i.e., work output that can be converted to electricity when connected to an electrical generator. Devices such as gas and diesel reciprocating engines can use a variety of fuels, and are often applied for small electricity-generation applications [69,93]. Among the various types of reciprocating engines, two are described in this section: Stirling engines and internal combustion engines.

Internal combustion engines use the heat produced from combustion of a fuel, often a fossil fuel, with air to produce a reciprocating motion of a piston, which can be converted to electricity when coupled to an electrical generator. Depending on how the combustion starts (manual or automatic ignition), such engines are classified as spark-ignition engines or compression-ignition engines. Spark-ignition engines involve compression of the air–fuel mixture followed by the combustion of the mixture via spark ignition. The combustion gases at high pressure and temperature expand and force the piston to move providing output work, which can be converted to electricity when coupled to an electrical generator. The piston then pushes out the exhaust combustion gases and allows fuel and air to enter the cylinder for the repetition of the cycle. Compression-ignition engines are basically similar to spark-ignition engines but compress the air to a higher temperature than for the air-fuel mixture in spark-ignition engines. Thus, combustion starts spontaneously on contact after the fuel is injected into the hot air. Such engines are mainly used to provide power for automobiles, trucks, aircrafts and ships, but they are also used as electric power generators for other applications.

In Stirling engines, a heat source, such as hot gases from an external combustion, is used to heat a gas, such as helium or air, and to expand it isothermally to produce output work. The gas then passes through a regeneration process where some of the heat that it contains is transferred to a regenerator in a constant-volume process. The low-pressure gas then loses additional heat to a low-temperature environment in an isothermal compression process before it is heated by the regenerator back to its initial state in a second constant-volume process. Note that the Stirling engine contains a fixed amount of working fluid throughout the process and all energy flows in and out of the engine are through heat exchangers. Such engines are generally more efficient than internal combustion engines but are still not as efficient as some other power generation technologies.

3.6.1.5 Fuel cells

Fuel cells are devices used for conversion of chemical energy to electrical energy. They operate based on the electron exchange between a hydrogen-rich fuel, such as renewable biogas or natural gas, and oxygen. Hydrogen is ionized at the surface of an anode and the hydrogen ions flow through an electrolyte to the cathode. The free electrons from ionizing hydrogen flow from the anode to the cathode through an external circuit, i.e., through an electrical generator, to produce electricity. The free electrons and oxygen are combined with hydrogen ions at the surface of the cathode and form water. The overall process uses a hydrogen-rich fuel and oxygen as input products and produces water and electricity. Thus, the process does not create pollutants as opposed to some other power generation plants that use such fuels.

Similar to batteries, fuel cells can be grouped together to generate more power in the form of a fuel cell stack. While a typical battery has a fixed supply of energy, fuel cells continuously generate electricity as long as fuel is supplied. Therefore, they are good candidates for sustainable power generation in applications with power usage profiles that require baseload power. They are also a secure and reliable option for on-site power generation.

3.6.1.6 Other systems

Other thermal electricity-generation systems exist, and are in various stages of development, including fluidized-bed combustion systems, integrated coal-gasification combined-cycle systems.

3.6.2 Cogeneration systems

The main cogeneration systems are described below. The operating cycles used by these systems are often separated into two categories:

- topping cycles, in which electricity is generated from the fuel-derived heat before product heat is delivered and
- bottoming cycles, in which the process order is reversed.

Topping cycles are the more common of the two [93].

Like thermal systems for large-scale electricity generation, most cogeneration systems are based on the types of systems shown in Figure 3.9, for example, steam turbine cycles, gas turbine cycles and reciprocating engines. The principal types of cogeneration systems are illustrated in Figure 3.10.

3.6.2.1 Steam-turbine cogeneration systems

Steam-turbine cogeneration systems are extensions of steam-turbine electricity-generation systems, in which, at the expense of reduced electricity output, heat is

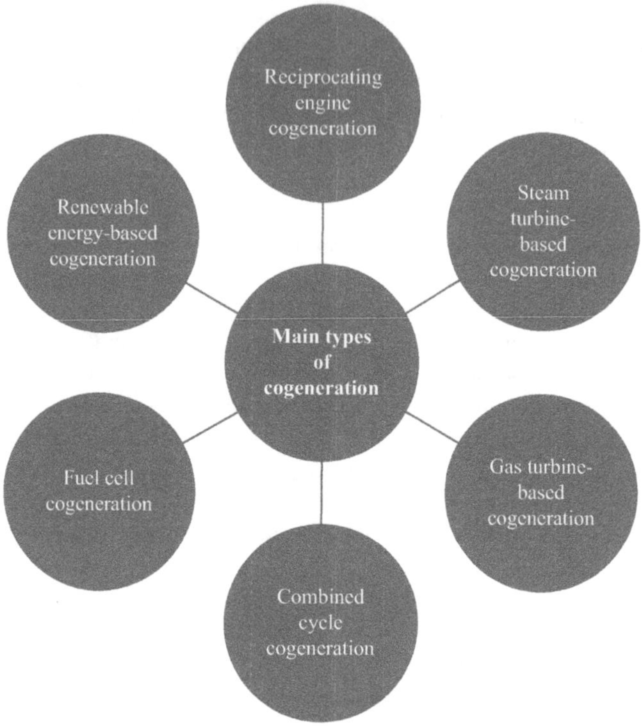

Figure 3.10 The main types of cogeneration, including traditional methods using fossil fuels, as well as new methods based on renewable energy and other fuels

delivered as a product via steam or hot water at useful temperatures. There are two main types of steam-turbine cogeneration systems: extraction and back-pressure. In extraction steam-turbine cycles (Figure 3.3a), extraction steam from one or more points on the turbines, at the desired pressure(s) and temperature(s), is delivered as a product. In back-pressure steam-turbine cycles (Figure 3.3b), the steam produced in the steam generators is expanded only to the pressure and temperature desired for the product heat. The condenser and some of the turbines, pumps and preheaters are replaced by the heat user. Back-pressure systems have high energy efficiencies (over 80 per cent), and lower costs than steam-turbine electricity-generation systems, since some equipment is eliminated in the cogeneration mode and back-pressure turbines are less expensive than condensing turbines [100–102].

Steam turbines (extraction and back pressure) are available in a wide range of sizes (up to several hundred megawatts electrical) and types [93]. They also operate on a number of fuels. Numerous examples of cogeneration systems operating on back-pressure and extraction steam-turbine cycles are listed in Table 3.1, including a back-pressure plant in Switzerland that generates 465 MW of thermal power and 135 MW of electrical power, with an overall efficiency of 75 per cent [103], and an extraction plant that was installed in the former Soviet Union that generates 250 MW of thermal power and 385 MW of electrical power, with an overall efficiency of 87 per cent [103].

3.6.2.2 Gas-turbine cogeneration systems

In gas-turbine cogeneration systems (Figure 3.3c), the relatively hot turbine-exhaust gas from a gas-turbine electricity-generation system is used directly for heating, or to heat a heat-transfer medium such as water, steam or air in a heat exchanger. Overall, energy efficiency is increased significantly (to as high as over 80 per cent), with little loss of electrical output. Gas-turbine cogeneration systems normally have small costs and installation times, and relatively high electricity-to-heat ratios [93]. Newer systems have achieved emission reductions through the use of technologies such as steam injection, selective catalytic reduction and dry combustion. These cogeneration systems are most often used in smaller industrial applications. Numerous examples of gas-turbine-based cogeneration systems are listed in Table 3.1, including a DuPont cogeneration system which has two 100-MW gas turbines.

3.6.2.3 Combined-cycle cogeneration systems

Combined-cycle cogeneration systems (Figure 3.3d) are similar to combined-cycle electricity-generation systems, except that product heat is delivered from the steam-turbine section of the combined cycle (via back-pressure or extraction techniques). These cogeneration systems can be assembled in many possible configurations, and have overall efficiencies as high as 90 per cent. In some combined-cycle cogeneration systems, steam from the steam-turbine section, when it is available in excess, is injected into the gas turbine to increase electrical output and reduce emissions.

Combined-cycle cogeneration systems are used in industry today. Some examples follow (more are listed in Table 3.1). DOW Chemical Canada's Sarnia plant has two gas turbines and two steam turbines, and generates 235 MW of thermal power and 165 MW of electrical power with an overall efficiency of 83 per cent [93]. A nuclear power plant left incomplete due to lack of funding was eventually completed as a gas-fired combined-cycle cogeneration plant having 12 heat recovery steam generators and gas turbines and two steam turbines, and producing 1,400 MW of electrical power and 285,000 kg/hr (629,000 lb/hr) of steam.

3.6.2.4 Reciprocating engine-based cogeneration systems

Reciprocating engines often exhaust most of their thermal input (e.g., 60.75 per cent thermal losses in internal combustion engines). Waste heat from reciprocating engines can be recovered from jacket cooling water, exhaust gas and lubricating oil, and used for small-scale cogeneration applications that require hot water or low-pressure steam for uses such as space heating [69,93], allowing them to produce heat and electricity. The ratio of heat to power production of internal combustion engines is about 1–2. Therefore, they are best suited for smaller sites where the demand for electricity is reduced and the requirement for heat and hot water is high. An example of heat and power generation by use of spark-ignition engines in a wastewater treatment plant is described in section 3.7.

3.6.2.5 Fuel cells

As described in section 3.6.1.5, fuel cells utilize electrochemical power generation. Production of water from hydrogen ions, oxygen and free electrons at the cathode surface is an exothermic chemical reaction. Since fuel cells are often used mainly to produce electricity, the heat that is released as a product of the chemical reaction is sometimes discharged. Fuel cells have thermal efficiencies of about 40.60 per cent, losing about 25.40 per cent of their chemical energy potential in the form of heat. If such energy is used in heating applications (as a CHP generation process), efficiencies of up to 85 per cent can be achieved. In cogeneration of heat and power using fuel cells, the waste heat from water production is captured and used in local heating and cooling and hot water production, district heating and cooling, and industrial operations such as heating ovens for industrial processes or heating anaerobic digesters for renewable biogas applications. Since they are mainly used in small-scale applications, fuel cells are suitable for microcogeneration applications [113,114].

Example application: fuel cell cogeneration for a university campus. A 1.4-MW molten carbonate fuel cell power plant is installed at the University of Bridgeport in Connecticut to generate electricity on campus, using natural gas, to heat and power campus buildings. Exhaust heat from the fuel cell is used for hot water production, reducing usage of combustion-based boilers. The power produced at the plant satisfies 80 per cent of the campus power needs.

As with any fuel cell, power is generated without combustion and the operation of the plant involves no pollution. Compared to the local electric grid, the fuel cell installation is estimated to avoid annually the emission of 28 tons of smog-producing nitrogen oxide (NO_x), 64 tons of sulfur dioxide (SO_x), 1.5 tons of particulate matter (PM10) and more than 7,000 tons of CO_2.

3.6.2.6 Solar cogeneration

Similar to all thermal electricity generation systems, solar thermal systems used for power generation have a component that removes excess heat once the steam leaves the turbine. Heat is often removed by circulating cold water or air that is discharged from the system, thereby discharging the excess heat. Use of cold water to remove the excess heat from the system may be a challenge in water-deprived regions.

Concentrating solar power generation is the most common solar thermal technology used for power generation in larger scales and has been used for more than a decade in countries such as the USA. To improve the efficiency of power production by such technologies, the waste heat can be recovered and used. Figure 3.11 illustrates the schematic of a solar cogeneration plant. A challenge with using excess heat from solar concentrating power generation plants is that they are sometimes located in isolated areas, such as deserts, where heating applications are not easy to find. Moving heat to demand regions incurs energy losses and costs related to transmission of heat and power [104].

In some thermal applications such as desalination, the use of solar cogeneration heat not only improves the overall efficiency of the system in use of solar energy but also reduces the cooling water requirements of a solar thermal power

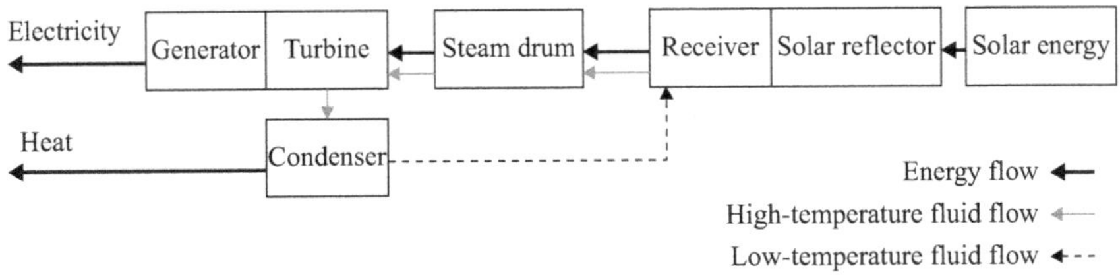

Figure 3.11 Solar cogeneration plant process diagram

generation, which can be a challenge. Thus, cogeneration in such cases makes the use of solar thermal power generation more viable.

3.6.2.7 Micro-cogeneration

Cogeneration (or CHP) systems are often used for large-scale industrial applications. Their large-scale use in the residential sector is limited to the deployment of district energy systems. Although cogeneration systems are associated with higher energy efficiencies in fuel use than thermal electricity generation plants, they still incur some losses related to the heat and electricity production at the central plant as well as losses related to distribution of heat and electricity to the consumer. Small-scale domestic use of cogeneration using "micro-cogeneration" technologies is a relatively new method based on the same principles as cogeneration. Generally, a cogeneration system is called a micro-cogeneration system when it generates less than 15 kW of power. Another term, "small-scale cogeneration," is sometimes used for cogeneration systems providing power up to 100 kW [115]. Micro-cogeneration technologies can provide heating for space and hot water production as well as electricity for residential use. They are widely used with ORCs, which are similar to the conventional steam-based Rankine cycle in principle but use an organic working fluid with a lower boiling temperature, instead of water. They sometimes can prove more efficient than cogeneration systems as the distribution losses are avoided [115]. However, systems with low cost and simple operation in the residential sector are still being developed. Some factors hindering the development of micro-cogeneration are a lack of suitable system components in the market, challenges in matching electrical and thermal loads to consumer demands, and trends in gas and electricity prices [116,117].

3.6.2.8 Other systems

Other technologies such as fluidized-bed combustion and Stirling engines [118] can be used for cogeneration applications.

3.7 Case study: reciprocating engine heat and power generation for wastewater treatment

The Rochester Wastewater Reclamation Plant in Rochester, MN, treats up to 23.8 million gallons of wastewater per day, and produces organic sludge that requires treatment before its disposal. The treatment can be anaerobic (in the presence of oxygen) or anaerobic (in the absence of oxygen). The plant includes a 2-MW cogeneration system and two anaerobic digesters. The main components of the plant and its benefits are described in the following subsections.

3.7.1 Cogeneration system

Two 1-MW spark-ignition engines that can burn biogas (anaerobic digester gas) or natural gas generate power at the Rochester plant. The plant's full generation capacity is utilized when backup power is needed for operating critical units, during utility grid power outages. The digesters produce enough biogas per day to generate 550–750 kW of continuous electric power [119,141]. At this capacity, a heat rate of up to 1.5 MW is captured from the engine water jacket and up to 1.2 MW from the exhaust gases in the form of hot water (82–88°C) that is utilized for keeping the anaerobic digesters at the optimum operating temperature (37°C) and for providing space heating throughout the plant.

3.7.2 Anaerobic digesters

The Rochester plant uses two anaerobic digesters (1.85 million gallons each) to break down the organic waste contained in the sludge from wastewater treatment. The process produces bio-solids that are ready for farmland application, a liquid high in nutrient content (mainly nitrogen) that must be further treated, and a biogas that contains approximately 66 per cent methane. Biogas can be used to replace natural gas in boilers for heating and in engines for distributed generation (including cogeneration), or it can be cleaned to utility-grade gas and used for fueling compressed natural gas vehicles or injected into natural gas pipelines (which can be a costly option). The digesters produce approximately 9,570 m^3 of 66 per cent methane biogas per day, enough to generate 550–850 kW of continuous electric power [119,141] (depending on the time of the year) and heat. In winter, the biogas is sometimes utilized in boilers for space heating, thus reducing the amount of wasted biogas.

3.7.3 Biogas conditioning

The biogas produced at the digesters meet the H_2S requirements of the engine (smaller than 74 ppmv) and, therefore, there is no conditioning system at the plant.

3.7.4 Biogas storage tank

A 15-m diameter spherical biogas storage tank is installed for storing compressed biogas (at 317 kPa) to provide the cogeneration system with the needed fuel during utility grid power outages.

3.7.5 System benefits

Originally, the cogeneration system was installed in 1982 as two 400-kW engines, mainly to reduce the annual energy costs related to the electricity and natural gas needed for wastewater treatment as well as heating the facility. The engines were upgraded in 2002 and 2008 to their current size. The new cogeneration system has several benefits including the following:

- **Efficiency**. The new 2-MW engines are 20 per cent more efficient than the original engines.
- **Economics**. The installation of the engines cost $4 million and the annual cost for maintaining it was about $169,700 in 2014 [141]. The cogeneration system has achieved annual energy savings of more than $650,000 since its operation. In 2014, the cogeneration system saved $505,700 in energy cost for the city of Rochester: nearly $348,000 in electric energy and $157,700 in natural gas costs. Economic savings are made from electricity generation as well as replacing natural gas with biogas.
- **Emissions**. The use of biogas for power generation in the cogeneration system reduces emissions that would be present if the gases were flared as a product of anaerobic digestion.
- **Sustainability**. The engines are capable of using biogas, a renewable energy source, for power generation which supports their sustainability in the long run.
- **Reliability**. The cogeneration plant provides adequate emergency power for the wastewater treatment plant.

3.8 Closure

Cogeneration systems and their benefits, uses, operation and applications are described. The trade-off between thermal and electrical outputs of cogeneration is examined, demonstrating how electrical output decreases as thermal product quantity and quality increase. The interface between cogeneration and energy storage is also discussed. A general cogeneration system model is provided and assessed. The main categories of heat demands that can normally be satisfied with cogeneration are described, as is the selection of a cogeneration system to match thermal and electrical demands.

Chapter 4

Heating and district heating systems

Overview

Systems for heating and district heating are described. This material is of great importance to the book for at least two reasons: heating systems are the foundation of most district heating systems and heating systems constitute one of the main technologies against which cogeneration must compete. Heating plants typically convert energy in the form of a fossil fuel or electricity to direct thermal heat or a heated medium such as hot gases, steam or hot water. A general model for heating and district heating systems is provided and illustrated, along with analyses of it using energy and exergy analyses. The resultant energy and exergy balances and efficiencies are also given. The various types of heating technologies are listed along with their characteristics. The types of heating technologies include fuel-based heating (e.g., heating using furnaces), electricity-based heating, heating via waste heat recovery, ground-based heating, solar-based heating and heating using heat pumps. The coverage for these systems includes basics, applications, types, classifications and operation factors. Extending this material, the many types of technologies for district heating are listed and described. Characteristics of district heating systems are provided, including advantages, operation, applications and energy sources. The types of district heating systems covered include fossil fuel, geothermal, solar, biomass and waste-to-energy-based district heating. Typical energy conversion efficiencies are given for heating and district heating, and it is demonstrated how the thermal product can be used in the residential, commercial and institutional sectors for processes such as space and water heating, and in the industrial sector for a wide range of activities. Attention is paid to the use of a central heat supply to meet residential–commercial heat demands via district heating given its prevalence (sometimes in tandem with the use of cogeneration-derived heat to drive absorption chillers for applications such as district cooling). Many of the uses for heat from heating systems are seen to be the same as the uses for heat from cogeneration systems.

4.1 Introduction

Heating processes are pervasive in society and industry, and are a necessity for life in many countries. Heating can be performed using a variety of energy resources. Heating operations are one of the underlying foundations of cogeneration systems, and are the primary component of district heating systems.

In heating plants, energy usually in the form of a fossil fuel or electricity is converted to heat (in the form of hot gases or another heated medium such as steam or hot water), often with an energy conversion efficiency of over 80 per cent.

Heat is used in the residential, commercial and institutional sectors for processes such as space and water heating, and in the industrial sector for a wide range of activities. Many of the uses for heat from heating systems are the same as the uses for heat from cogeneration systems, some of which were described in the last chapter (see section 3.1).

4.2 General heating system model

The model for a general cogeneration system shown in Figure 3.1 serves as a model for a general heating system when the product work term ($\dot{W}_p$) is zero. Then, fuel and air enter at the temperature and pressure of the environment, and product heat and wastes exit (see Figure 4.1).

4.3 Analyses for general heating system

Commodity balances as well as efficiencies for the general model of a heat production system are presented in this section.

4.3.1 Energy and exergy balances

Energy and exergy balances for the general model of a heat production system are presented in this section. The energy and exergy balances from section 3.5.1, which apply to the system shown in Figure 3.1, are applicable for a general heating system when the product work term is zero. For consistency, the format of the treatment in section 3.5.1 is followed here.

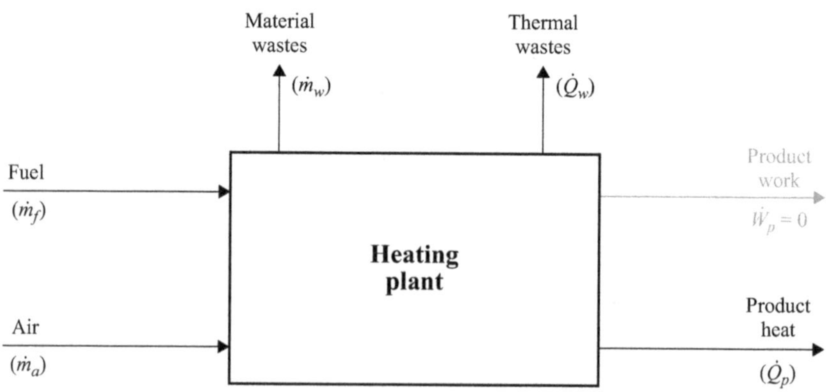

Figure 4.1 A model for a general heat production system

Assuming steady-state operation and neglecting kinetic and potential energies, an energy balance for the system in Figure 4.1 can be written, following (2.8), as:

$$\dot{m}_f h_f + \dot{m}_a h_a = \dot{Q}_p + \sum_w \dot{Q}_w + \sum_w \dot{m}_w h_w \tag{4.1}$$

As before, h denotes specific enthalpy; $\dot{m}$ and $\dot{Q}$, respectively, denote rates of mass flow and heat flow; the subscripts f, a, p and w denote, respectively, fuel, air, product and waste; and the summations are over all material and thermal waste streams. The terms on the left-hand side of (4.1) represent the total energy input rate and are entirely associated with flows of matter. The first term on the right-hand side represents the energy output rate associated with the thermal product, and the remaining two terms represent the energy output rate associated with wastes (thermal and material). The ratio $\dot{Q}_p/(\dot{Q}_p + \sum \dot{Q}_w)$ again denotes the fraction of the total thermal energy output rate which is product, and can at most be unity. That occurs for systems in which all output heat is utilized.

A corresponding exergy balance can be written, following (2.11), as:

$$\dot{m}_f \varepsilon_f + \dot{m}_a \varepsilon_a = \dot{Q}_p \tau_p + \sum_w \dot{Q}_w \tau_w + \sum_w \dot{m}_w \varepsilon_w + \dot{I} \tag{4.2}$$

As before, ε denotes the specific exergy of a material flow; τ_p and τ_w are the exergetic temperature factors which are functions of the relevant temperatures T_p, T_w and T_o; T_p, T_w and T_o denote, respectively, the temperatures associated with the product heat flow, the waste heat flow(s) and the environment; and other terms are as previously defined. The terms on the left-hand side of (4.2) represent the exergy input rate, while, on the right-hand side, the first term represents the exergy output rate associated with the thermal energy product, while the second two terms represent the exergy output rates associated with thermal and material energy wastes, and the final term represents the exergy consumption rate. The terms τ_p and τ_w range from zero to unity for all heat flows in conventional heating systems where heat is output at above-environment temperatures (i.e., where $T_p \geq T_o$ and $T_w \geq T_o$).

Nuclear-based heating systems can be modeled following Figure 4.1, but with no air entering (i.e., $\dot{m}_a = 0$) and the fuel representing nuclear fuel. That is, the energy delivered as heat from the nuclear fuel can be expressed as

$$\dot{m}_f h_f = \dot{Q}_n \tag{4.3}$$

and the corresponding thermal exergy delivered from the nuclear fuel can be expressed as

$$\dot{m}_f \varepsilon_f = \dot{Q}_n \tau_n \tag{4.4}$$

As before, $\dot{Q}_n$ is the heat transferred to the system (at a temperature T_n) due to fissioning of the nuclear fuel. The temperature T_n is potentially infinite in value (as is often done [4–6]), so that the energy and exergy quantities of the thermal energy delivered from the fissioning of the nuclear fuel are identical. Consequently, for nuclear systems:

$$\dot{m}_f \varepsilon_f = \dot{Q}_n = \dot{m}_f h_f \tag{4.5}$$

4.3.2 Efficiencies

Energy and exergy efficiencies for the general model of a heat production system are presented in this section. The energy and exergy efficiencies from (3.5) and (3.6) of section 3.5.2, which apply to the system shown in Figure 3.1, are applicable for a general heating system when the product work term is zero. For consistency, the format of the treatment in section 3.5.2 is followed here.

Following (2.19) and (2.20), the energy efficiency for the heat production system in Figure 4.1 can be written as

$$\eta_H = \frac{\dot{Q}_p}{\dot{m}_f h_f} \tag{4.6}$$

and the corresponding exergy efficiency as

$$\psi_H = \frac{\dot{Q}_p \tau_p}{\dot{m}_f \varepsilon_f} \tag{4.7}$$

While energy and exergy efficiencies are both directly a function of $\dot{Q}_p$, only exergy efficiencies are also a function of the temperature at which the product heat is delivered. When $\dot{Q}_p = 0$, the energy and exergy efficiencies both reduce to zero. Since for many reference-environment models (including the one described in Table 2.2), $h_a = \varepsilon_a = 0$, h_f is often the HHV of a fuel evaluated at T_o and P_o, and $\varepsilon_f = h_f$ (within approximately 15 per cent), the electrical energy and electrical exergy efficiencies are approximately equal for a given system, although the general energy and exergy efficiencies can differ significantly. As $\dot{Q}_p$ increases, energy and exergy efficiencies both increase, but energy efficiencies increase more rapidly than exergy efficiencies.

Since $0 \leq \tau_p \leq 1$ for $T_p \geq T_o$, it can be shown with (4.6) and (4.7) that $\psi_H \leq \eta_H$ when $\varepsilon_f = h_f$, and that the amount by which the energy efficiency exceeds the exergy efficiency is

$$\eta_H - \psi_H = \frac{\dot{Q}_p}{\dot{m}_f h_f} \frac{T_o}{T_p} \tag{4.8}$$

For a fixed reference-environment and set of inputs, the difference in efficiencies decreases as the product-heat delivery temperature T_p increases and output rate $\dot{Q}_p$ decreases. The largest deviations between ψ_H and η_H normally occur for low-temperature systems, and the smallest deviations for high-temperature systems. On the one extreme, if $T_p = T_o$, the largest deviation between ψ_H and η_H occurs and $\psi_H = \eta_H = 0$ if no heat is produced. On the other extreme, if $T_p = \infty$, $\psi_H = \eta_H$ for all cases.

Exergy efficiencies can be argued to give more illuminating insights into heating process performance than energy efficiencies, because exergy efficiencies provide a measure of how nearly the performance of the system approaches ideality, and of the magnitude of the potential for improvement. Energy efficiencies appear particularly overly optimistic for heating systems. These observations imply that cogeneration systems are better compared using exergy efficiencies. Thus, the use of energy efficiencies in most practical heating systems, where thermal energy is normally produced at low-to-moderate temperatures, can lead to erroneous interpretations and conclusions.

4.4 Systems for heating

Heat produced in heating plants in the form of hot gases, steam or hot water. It is used in the residential, commercial and institutional sectors for two main tasks:

- space heating
- domestic hotwater heating

There are also many other heating activities in residential, commercial and institutional sectors that are less consuming of energy. These include cooking as well as clothes washing and drying.

Heating is utilized extensively in the industrial sector, for a wide range of processes such as the following:

- process heating
- sterilization
- drying
- curing
- phase change (e.g., raising steam)
- distillation
- refining metals and chemicals
- petrochemical processing
- manufacturing

Examples of industries that require large amounts of heat for their operations include the following [119,147]:

- food and beverage
- chemicals
- petrochemicals
- pulp and paper
- metal surface treatment
- cement
- bricks and concrete blocks
- textile

Many of the uses for heat from heating systems are the same as the uses for heat from cogeneration systems, some of which were described in the last chapter (see section 3.1).

4.4.1 Heating technologies and their characteristics

Numerous types of heating systems exist, and are widely used. These systems can produce heat from the secondary energy form electricity, from fuels such as fossil fuels, biomass and solid waste, as well as from energy resources such as nuclear fuel, solar radiation and geothermal energy. The choice of particular heating methods and technologies depends on several factors such as their availability at affordable cost and the temperature at which heating is required. For example, space heating in newly designed buildings can be performed with heat at temperatures as low as 45°C while heat requirements for industrial process applications range from 60°C to over 1,600°C (see Table 4.1). The quantities of energy use for the various industrial

Table 4.1 Breakdown and characterization of common industrial heating processes[1]

Industrial operation category	Sample industrial application in category	Heating temperature range (°C)	Energy use, in the USA in 2010	
			(TWh)	(per cent of total)
Non-metal melting	Plastics and rubber manufacturing Food preparation Softening and warming materials	930–1,650	77	3.7
Smelting and metal melting	Casting Steelmaking Other metal production Glass production	720–1,650	377	17.8
Calcining	Lime calcining	620–1,170	154	7.3
Metal heat treating and reheating	Hardening Annealing Tempering Forging Rolling	500–1,180	79	3.7
Coking	Iron making Other metal production	380–1,100	35	1.7
Drying	Water removal Organic compound removal	160–550	457	21.7
Curing and forming	Coating Polymer production Enameling Molding Extrusion	140–650	42	2.0
Fluid heating	Food preparation Chemical production Reforming Distillation Cracking Hydrotreating	110–460	620	29.4
Other	Preheating Catalysis Thermal oxidation Incineration Other heating	100–1,650	271	12.8

[1]Source for some data: References 119–151.

process application categories are also given, to provide some context on the importance and degree of commonality of the different industrial heating operations.

A variety of heating systems that are used for space heating and hot water production as well as those used in industrial processes are described in this section. The systems are used for various applications according to their temperature requirements and energy source availability. Sometimes, a combination of heating systems based on different energy sources and/or different heating methods of a single energy source is used to optimize their energy use and increase overall thermal efficiency. As most heating technologies are well understood, the discussions included here are somewhat limited to avoid excessive technical detail, so as to leave the focus of the book on cogeneration-based district energy.

The main types of heating systems are illustrated in Figure 4.2 and in the below subsections.

4.4.2 Fuel-based heating

Fuel-based heating occurs when a mixture of air and a fuel – in solid, liquid or gaseous form – are burnt to provide heat. The fuels used for such heating systems

Figure 4.2 The main types of heating systems, including traditional methods using fossil fuels, as well as new methods based on renewable energy and other fuels

include fossil fuels such as natural gas, propane, fuel oil, coal and some types of biomass such as wood.

Heat is captured from the combustion gases either directly via convection or by passing the combustion gases through radiant burner tubes or panels.

Furnaces and boilers are examples of fuel-based space heating equipment. Examples of heating systems used for industrial purposes include ovens, fired heaters, kilns and melters.

A key challenge facing fuel-based heating systems is to cost-effectively and simultaneously reduce their emissions and increase thermal efficiencies. This depends on factors such as control of air–fuel ratios during all stages of heating, fuel-mix variability, completeness of combustion and performance of the burner over the range of its operation. In the subsections below, some general characteristics of furnaces and boilers are described [147].

4.4.2.1 Furnaces

A furnace is a central heating system in which a mixture of air and a hydrocarbon fuel, such as natural gas, propane, fuel oil and coal, are burnt to provide heat. In space heating, the heat is often transferred to air via metal heat exchangers. Air is then distributed in the building via a fan for heating spaces. Such space heating systems are common and popular in North America.

There are several types of such furnaces that vary according to their fuel and efficiency in capturing the heat in the combustion gases. Recent advances in furnace design have led to energy efficiencies higher than 90 per cent, as is the case with condensing furnaces. Furnace losses often account for start-up, cool-down and other operating losses that occur while the system is operating.

4.4.2.2 Boilers

Similar to furnaces, boilers use fuels to provide heat for spaces. Boilers, in fact, can use a variety of hydrocarbon fuels to provide space heating. However, instead of using air as a heat transport fluid, the flames and the combustion gas are used to heat and circulate water through pipes in the building to provide space heating. Sometimes the water remains a liquid, while in other cases, steam is circulated in the space and condenses as it heats spaces in the building. Other systems involve use of boilers as a central heating system and air for distribution of heat to spaces where heating is required.

4.4.3 Electricity-based heating

Electricity-based heating systems convert electricity to heat. The conversion is often done either directly via resistance heating by passing electric current through a resistant material or indirectly via induction heating. Electricity-based heating technologies such as electric boilers and baseboard heaters are used in space heating and hot water production.

Electricity-based systems used for industrial heating include electric arc furnaces, infrared emitters, induction heating, radio frequency drying, laser heating and microwave processing. Electricity-based systems are sometimes combined with a fuel-based boiler to avoid the high cost of electricity use at peak times.

Another electricity-based technology that can be used for heating for various purposes is a heat pump. Heat pumps differ from the aforementioned systems in that they primarily use electricity to move or "pump" heat from one medium to another, rather than simply converting it to heat. Heat pumps are described in more detail in section 4.4.7.

The source of the electricity used in electricity-based heating systems does not affect the type of heating technology employed. However, it greatly affects the initial selection of electricity-based heating systems over other heating systems. Where there is a surplus in electricity production or a high proportion of the electricity mix in the grid is from renewable energy sources such as solar, hydropower and wind energy, or from nuclear energy, the provision of heating through use of electricity-based heating systems can be a competitive choice compared to other heating technologies from both economic and environmental perspectives.

4.4.4 Waste heat recovery

A significant portion of the total energy used in heating systems of various types is often wasted. But heat that would otherwise be wasted after its main use can be recovered and used in lower temperature heat applications. Such waste heat is available from various mediums such as exhaust gases from combustion chambers and spent cooling water.

When the temperature of waste heat is not sufficiently high to provide the required heating for a given application, the waste heat can be used for preheating. Heat pumps are also commonly used to recover waste heat and provide heating at higher temperatures.

Significant energy cost savings can be realized through the utilization of advanced heat recovery systems. Systems to collect, recover and use waste heat are often available in large scale and are currently used in some industrial heating applications. Their use in residential, institutional and commercial applications often requires a central facility to collect and recover waste heat and to distribute it via a district heating network.

4.4.5 Ground-based heating

Some of the processes that occur within the earth result in geothermal heat that is conducted or convected to the upper layers of the earth's crust resulting in high-temperature regions within the earth crust. Ambient conditions, which are mainly due to the effects of solar radiation, also provide heat at lower temperatures in shallow levels of the crust of the earth.

Geothermal heat is harvested through use of ground heat exchangers which are installed in the ground from depths of 1 m below the ground surface up to hundreds of meters below the surface.

Shallow systems have heat exchangers placed within a few meters of the surface. In shallow systems, the heat exchangers are installed by excavating the ground and then covering the heat exchangers with the ground material after installation is complete. Heat exchangers are typically laid out horizontally in the ground in shallow systems.

Deeper systems have heat exchangers that are oriented vertically and extend between about 20 m and hundreds of meters below the surface of the earth. In deeper systems, the heat exchangers are installed by drilling holes to depths in the ground of 250 m or more. The vertical ground heat exchangers are then installed in these "boreholes." A fluid is circulated through the heat exchangers to collect heat from the ground.

Ground-based energy can provide heating directly in winter, since the ground below the surface is often warmer than the air above. Such applications include space heating, greenhouse heating, aquaculture pond heating, agricultural drying, industrial heating uses, heating of bathing and swimming facilities, and snow melting. Sometimes the ground temperature is only adequate to provide preheating. The ground temperature can also be boosted via devices such as heat pumps, allowing ground-based energy to provide heating at higher temperatures (see section 4.4.7 for description of heat pumps). Due to their high efficiencies, their ability to utilize groundwater or ground temperatures anywhere in the world as well as progress in drilling technology, the use of geothermal energy using ground-source heat pumps has grown considerably in recent decades compared to the other heating methods.

Geothermal energy is a renewable source of energy. The availability of non-ambient geothermal energy depends on the geological conditions in the earth's crust. Where it can be harvested, this source of energy does not depend on climate or weather conditions and its use may prove advantageous relative to other renewable energy sources such as solar heat.

4.4.6 Solar-based heating

Solar-based technologies collect the solar energy and convert it to heat or electricity. In this section, we are primarily concerned with solar thermal collectors which produce useful heat from solar radiation.

4.4.6.1 Basics and applications

Solar thermal collectors are able to provide heat at a wide variety of temperatures to provide heating for various applications such as pool, space and hot water heating as well as heating for industrial purposes such as sterilizing, pasteurizing, drying, hydrolyzing, distillation and evaporation, washing and cleaning, and polymerization. Heat requirements for industrial heating applications that use solar-based heating systems commonly range from 60°C to 260°C [145].

Solar-based heating systems often consist of an absorption mechanism where solar energy is absorbed and a transport medium that carries the heat to its application or to a storage. In order to increase the solar energy absorbed, solar-tracking collectors are designed to be oriented toward the solar rays over longer periods of the day and year.

4.4.6.2 Types and classifications

Various types of solar thermal collectors exist. Many of these are listed with their primary design parameters in Table 4.2. Also, shown are key performance parameters for solar thermal collectors such as typical temperature ranges for which they can provide heat and typical ranges for their concentration ratios. The concentration ratio is defined as the heat absorber aperture area divided by the receiver/absorber area of the collector, and represents how much heat a given solar thermal collector provides relative to a flat plate solar thermal collector. Among the collectors listed in Table 4.2, stationary and single-axis sun-tracking parabolic through collectors, because of the temperatures they can provide, are able to satisfy the heating needs of industrial heating applications [145]. Solar thermal collectors are commonly categorized and this can be done according to any of the parameters listed in Table 4.2 (e.g., collector type, absorber type, ability to move to orient surface, temperature and concentration).

The most common type of solar collector used for space heating as well as some industrial processes is flat-plate collectors. They do not track the sun and are oriented directly toward the equator, facing south in the northern hemisphere and north in the southern hemisphere. For this type of collector, a large portion of solar rays are absorbed as heat by a flat black surface with high radiation absorptivity. The heat is then transferred to a fluid circulating in tubes that are attached to the absorber and carried away for storage or use. How the tube is attached to the

Table 4.2 Characteristics of various types of solar thermal energy collectors[1]

Type of solar thermal collector	Design parameters		Typical performance parameters	
	Geometry of heat absorber	Ability to orient collector surface	Output heating temperature range (°C)	Concentration ratio range
Flat plate	Flat (plane)	Stationary	30–80	1
Evacuated tube	Flat (plane)	Stationary	50–200	1
Compound parabolic	Tubular (line)	Stationary	60–240	1–5
Fresnel lens	Tubular (line)	Single-axis tracking	60–250	10–40
Parabolic trough	Tubular (line)	Single-axis tracking	60–300	15–45
Cylindrical trough	Tubular (line)	Single-axis tracking	60–300	10–50
Parabolic dish	Small shape at focal position (point)	Two-axis tracking	100–500	100–1,000
Heliostat field	Small shape at focal position (point)	Two-axis tracking	150–2,000	100–1,500

[1]Data drawn from various sources [142–146].

absorber plate greatly affects the system efficiency. High-efficiency flat-plate collectors can reach temperatures of up to 100°C.

Evacuated tube collectors are more efficient than flat plate collectors and are able to provide heat at higher temperatures. The heat transport fluid in this case is placed inside an evacuated tube absorber. By surrounding the absorber with a vacuum space, thermal losses due to convection and conduction are greatly reduced.

While flat plate and evacuated tube collectors have absorber and aperture areas of the same size, other types of solar collectors concentrate the light rays on the absorber (also called the receiver in this case). For example, parabolic reflectors are shaped based on the optical principle of a reflecting parabola; all rays of light parallel to its axis are reflected to a point. Parabolic through and parabolic dish reflectors concentrate the light rays to a line and a point, respectively.

4.4.6.3 Operation factors

Since solar thermal energy is available only during periods of time when the sun is shining and solar irradiation is available, they are often integrated with energy storage units, such as water tanks, to increase the use of collected heat.

Solar-based heating systems consist of a mechanism to collect the heat and one to transport the heat via a heat exchanger. The main type of solar collector that is applied for space and hot water-heating purposes is the non-tracking (stationary) collector, while the main type used for industrial heating is the single-axis sun-tracking parabolic trough collector.

4.4.7 Heat pumps

The heat pump is a device that operates on a cycle to transfer heat from a low- to high-temperature medium. A heat pump can therefore be used for either heating a building (using heat drawn from the environment) or cooling a building (by transferring heat from it to the environment). A heat pump cycle for cooling is conceptually similar to the cycle that drives a common refrigerator.

4.4.7.1 Description

A heat pump typically has four main components: condenser, expansion device, evaporator and compressor. The condenser and evaporator are heat exchangers. These are combined as shown in Figure 4.3. A working fluid, commonly a refrigerant, flows through the components comprising the cycle.

As heat cannot of itself flow to a higher temperature, external energy is needed to drive the process. A heat pump typically uses electricity to force heat to move against its natural direction of flow.

During operation of a heat pump, the following steps occur:

- The working fluid at low-pressure and low-temperature flows through the evaporator and receives heat from a region constituting the low-temperature heat source, causing it to become a vapor.
- The warmer low-pressure working fluid in gaseous form enters through a compressor where its pressure is boosted, simultaneously causing its temperature to rise.

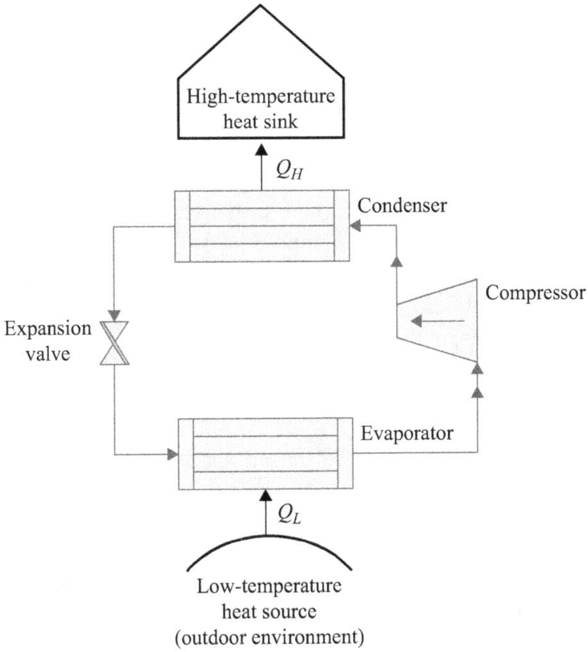

Figure 4.3 Heating a house using a heat pump, showing heat extracted from the low-temperature heat source (the environment) Q_L and heat delivered to the high-temperature heat sink Q_H

- The hot, high-pressure vapor flows through a condenser where heat is removed to a region constituting the high-temperature heat sink. As heat is transferred from the working fluid, it is converted back to a liquid.
- The cool high-pressure liquid refrigerant from the condenser flows through an expander, which reduces the pressure of the refrigerant, simultaneously causing its temperature to drop. The working fluid is then conveyed to the evaporator, continuing the cycle.

4.4.7.2 Thermal reservoirs

Like heat engines, heat pumps operate between two thermal reservoirs at different temperatures. But heat pumps act in the reverse direction of heat engines.

Heat pumps that are used for heating consider the space as the high-temperature medium, whereas heat pumps that are used for cooling consider the space as the low-temperature medium. A variety of options exist for the other medium from/to which heat is transferred. These include natural mediums such as outdoor air, local bodies of water and the ground. Artificial mediums can also be utilized. For instance, waste heat from industrial processes, cooling equipment or ventilation air from buildings can be used as heat sources for heat pumps for heating. The choice of such media depends on numerous factors, including availability and temperature.

4.4.7.3 Coefficient of performance

The efficiency of a heat pump is typically measured in terms of a coefficient of performance (COP_{HP}). The higher the COP of a heat pump, the more efficient it is. When a heat pump is used for space or water heating, the aim is to transfer the required heat from the low-temperature environment and it is normally desired that be done with little electricity. Heat pumps can reduce the energy use for heating significantly since they can transfer more thermal energy than the electrical energy they utilize. Similarly, when a heat pump is used for space cooling, the aim is to transfer the required heat from the low-temperature space being cooled to the environment, with as little electricity as is reasonable accounting for other system constraints and objectives.

The efficiency of a heat pump used for heating is conventionally expressed in terms of the following COP:

$$COP_{HP,heating} = \frac{\text{Heat provided to heated space}}{\text{Electricity consumption}} \tag{4.9}$$

Conversely, the efficiency of a heat pump coefficient used for cooling (e.g., used as a conventional building air conditioner) is conventionally expressed in terms of the following COP:

$$COP_{HP,\ cooling} = \frac{\text{Heat removed from cooled space}}{\text{Electricity consumption}} \tag{4.10}$$

4.4.7.4 Ideal coefficients of performance

The maximum efficiency that may be achieved by a thermodynamically perfect or ideal heat pump can be determined by considering a reversed theoretical Carnot cycle. In such a cycle, all the processes are reversible. Hence, the COP of an ideal or reversible heat pump (COP_{rev}) used for heating can be expressed as follows:

$$COP_{rev} = \frac{1}{1 - \frac{T_L}{T_H}} \tag{4.11}$$

Similarly, the maximum efficiency that may be achieved by a thermodynamically perfect or ideal heat pump used for cooling can be expressed as follows:

$$COP_{rev} = \frac{1}{\frac{T_H}{T_L} - 1} \tag{4.12}$$

It is observed here that the COP of an ideal or reversible heat pump is a function of only the high and low temperatures between which it operates. It can also be shown with (4.11) and (4.12) that the COP of a reversible heat pump used for heating always exceeds that of a reversible heat pump used for cooling by unity.

Table 4.3 Coefficient of performance of an ideal heat pump for operation between selected high and low temperatures

Temperature of cooler medium, T_L		Temperature of warmer medium, T_H		Temperature ratio, T_L/T_H	COP of reversible heat pump for heating	COP of reversible heat pump for cooling
(°C)	(K)	(°C)	(K)			
Heating to a fixed temperature for various cold outdoor environmental temperatures						
–40	233	40	313	0.74	3.9	2.9
–30	243	40	313	0.78	4.5	3.5
–20	253	40	313	0.81	5.2	4.2
–10	263	40	313	0.84	6.3	5.3
0	273	40	313	0.87	7.8	6.8
10	283	40	313	0.90	10.4	9.4
Cooling to a fixed temperature for various hot outdoor environmental temperatures						
10	293	25	298	0.98	59.6	58.6
10	293	30	303	0.97	30.3	29.3
10	293	35	308	0.95	20.5	19.5
10	293	40	313	0.94	15.7	14.7
10	293	45	318	0.92	12.7	11.7

Values of the COP of an ideal heat pump for operation between various sets of high and low temperatures are shown in Table 4.3. The values confirm that the COP of a reversible heat pump used for heating exceeds that of a reversible heat pump used for cooling by unity in all cases. It is seen that in both heating and cooling modes, as the temperature lift across the heat pump increases, the COP decreases. When heating a 40°C medium by moving heat from a –40°C medium compared with a 10°C medium, the COP of the heat pump decreases from 10.4 to 3.9. The same pattern is observed in cooling a medium by delivering heat to a warmer environment. When removing heat from a 10°C medium and transferring it to a 35°C environment compared to a 45°C environment, the heat pump COP decreases from 19.5 to 11.7.

4.4.7.5 Actual coefficients of performance

For a real heat pump for heating, the value of COP_{HP} is typically about three to four, i.e., for every one unit of energy used to power the system, about three to four units of heat are delivered to the space. This value represents about ten per cent of the theoretical maximum. Nonetheless, compared to efficiencies of 0.92 (i.e., 92 per cent) for a high-efficiency natural gas furnace, or 0.99 (i.e., 99 per cent) for an electric resistance heater, heat pumps are very efficient for heating.

For real heat pumps, the value of the heat pump COP decreases the temperature difference between the high- and low-temperature mediums increases. This is because, as for ideal heat pumps, the COP of an actual heat pump, COP_{HP}, is dependent on temperatures of the high- and low-temperature mediums between which it operates.

For a real heat pump for cooling, the value of COP_{HP} is typically about two to three, i.e., for every one unit of energy used to power the system, about two to three units of heat are extracted from the space. This again represents about ten per cent of the theoretical maximum. Again, for real heat pumps for cooling, the value of the heat pump COP decreases the temperature difference between the high- and low-temperature mediums increases. This remains attributable to the fact that the COP of an actual heat pump is dependent on temperatures of the high- and low-temperature mediums between which it operates, as is the case for ideal heat pumps.

The COP of a heat pump declines notably as the large temperature difference between the heat sink and the heat source grows. Therefore, it is often prudent to seek reasonable temperature levels for the heat source and heat sink.

Various measures can be taken to improve heat pump performance. Since the space is often kept at predefined temperatures, the temperature of the other medium greatly affects the value of the coefficient of performance COP_{HP} and capacity of the heat pump. When heating a space, the lower the temperature of the medium from which the heat is extracted, the higher is the rate of heat pump electricity consumption. Developments in compressors, motors and controls are very effective in improving heat pump efficiencies.

4.5 Systems for district heating

District heating is a method for supplying heat to a space heating and domestic hot water heating, and is used in residential, institutional and commercial buildings as well as in businesses and industry.

4.5.1 Description

In essence, district heating requires a central heat production plant and a distribution system consisting of insulated pipes in which a heat transport medium, almost always steam or hot water, is transported from the central plant to the heat users. The system usually incorporates return pipes in which spent (or cooled) water is transported back to the heating plant for reheating. Figure 4.4 illustrates the concept of district heating compared to individual heating systems in a community. In some systems, the distribution system consists of a primary network for water distribution to a secondary network that distributes heat to consumers. This central heat production replaces individual heating systems, conventionally furnaces or electric heaters, used by each user.

A well-designed district heating system provides safe connections between heat users, expansion capabilities, backup for increased energy demand and allows the connection of additional heating plants.

The use of district heating for the provision of heating to a community is illustrated in Figure 4.5. In the heat transport loop, the heated fluid is transported to the community after exiting the heating plant and then returned after use in the community for reheating. Once-through systems exist in which the fluid is water and it is discarded after use, but these are increasingly uncommon due to the cost of losing of the transport

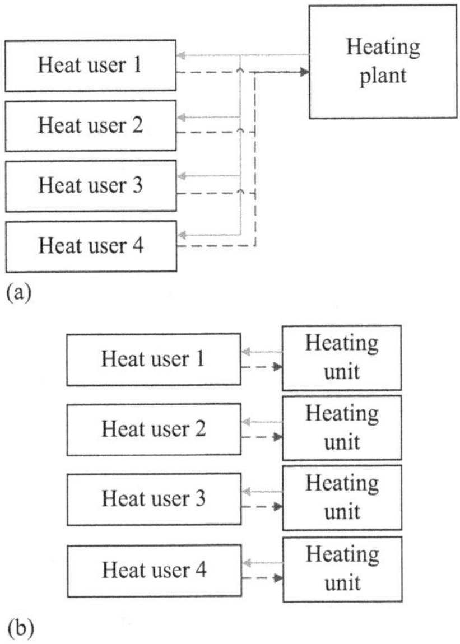

Figure 4.4 Comparison of (a) district heating and (b) individual heating, for four heat users in a community. Solid lines denote the outgoing heated heat transport fluid, and dashed lines the return spent heat transport fluid

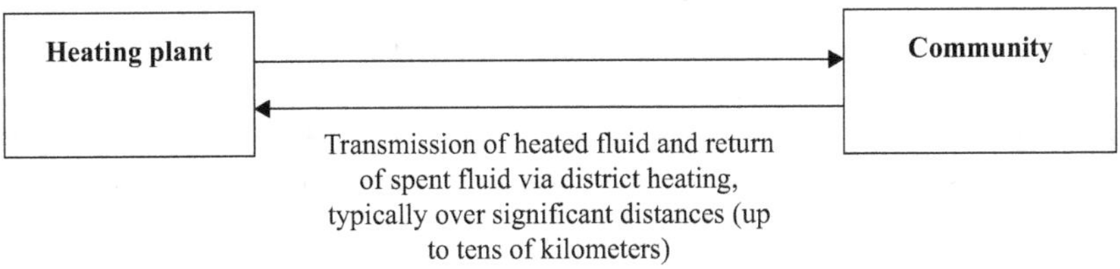

Figure 4.5 The use of district heating for the provision of heating to a community. The fluid is transported to the community after it is heated in the heating plant and the spent fluid is returned after use in the community for reheating. Less commonly, once-through systems exist in which the fluid is water and it is discarded after use

fluid and the associated environmental impact of discarding it and extracting new water. The transmission of the heat transport fluid in district heating is typically over significant distances, from tens of meters to up to tens of kilometers.

4.5.2 Advantages of district heating

District heating systems have numerous advantages over conventional heating systems. But there are two main advantages over conventional heating systems that are the predominant reasons for the growing use of district heating.

The first advantage relates to economy of scale. Despite the distribution losses related to district energy systems, providing heat for buildings and industries via large central heating plants is usually on balance more efficient than using individual small systems based on the same heating concept. The higher efficiency of central heating plants that consume fossil fuels directly or indirectly during operation typically lead to lower environmental impacts.

The second main advantage of using district heating systems is that they permit a range of heat source options to be exploited that are not available for individual use. These often include use of geothermal heat, industrial waste heat, bioheat including heat from municipal wastes and waste heat from power generation plants in the form of cogeneration. These cases are often economically competitive with conventional heat sources as they are available at very low or no cost when the technologies to harness them are installed. Such systems are often associated with almost no environmental impacts (e.g., solar and geothermal) and their environmental impacts are typically far lower than those associated with fossil fuels.

Boilers in heating plants are a conventional source of heat in district heating, but have been rarely used. The main advantage of boilers using a hydrocarbon fuel, such as natural gas, is their low cost and simplicity in providing heat. However, they are normally not considered a sustainable source of heat particularly because they emit CO_2. Therefore, boilers are often used only for peak-load or back-up capacity in district heating systems in which the main part of the heat comes from other sources such as biomass.

4.5.3 Operation and applications

Since heat distribution incurs costs related to temperature and pressure losses, district heating is often used where the total cost of operation is lower than the cost of local heat generation alternatives. This is often the case when exploiting low value or waste heat. For example, the use of technologies to reduce the combustion of fossil fuels often requires a district heating system [152,153]. Examples of such technologies include the following:

- cogeneration to improve system efficiency,
- industrial waste heat recovery to offset fossil fuel use and
- replacement of fossil fuels with residual resources such as waste or biomass fuels.

Note that heat is usually a byproduct of such plants. The development of a district heating network is dependent on the availability of waste heat or waste from other plants. Thus, district heating networks offer improved energy utilization and management by incorporating a range of energy resources other than natural gas for heating. Other benefits of district energy systems include enhanced operations management since it can be accommodated for a large, centralized plant and lower emissions related to the provision of heat centrally.

Many district heating configurations exist. One of the parameters that is often used to categorize district heating systems is the temperature of the transport fluid that supplies heat to the user via the district network. Low-temperature systems typically use water at temperatures of approximately 50°C and 70°C in the summer and winter seasons, respectively. High-temperature district heating systems operate at temperatures as high as 120°C. Steam at higher temperatures is also used in some district heating systems as the working fluid. Low-temperature district heating systems allow a higher share of renewable energy penetration by the use of low-temperature renewable energy sources such as geothermal energy. More efficient building designs are lowering the temperature requirements for district heating networks and allow low-temperature district energy systems to become sufficient for providing space heating at the community level. Heat losses related to the heating supply network and a seasonal TES, if used, are reduced for low-temperature district heating systems.

4.5.4 History

District heating has a storied history, somewhat paralleling the growth of societies, and driven by the importance of heating and the need to transport heat to where it is needed.

The origins of district heating traces extend back as far back as ancient Rome, where hot water distribution systems supplied heat to baths and greenhouses.

In Europe, during the middle ages and the Renaissance, district heating systems became more prominent. In France, for instance, one system has been in continuous operation since the 1300s [154].

District heating became used in North America in the mid-1800s. For example, steam-based district heating was used by the U.S. Naval Academy in Annapolis, Maryland, beginning in 1853. Despite the operation of these and other systems, they were not commercially profitable. The first commercially successful district heating system began operation in 1877 in Lockport, New York. The system was designed by B. Holly, a hydraulic engineer from the U.S. who is often considered the founder of district heating. In subsequent decades, hundreds of additional district heating systems were installed globally [154].

4.5.5 Energy sources for district heating

Several cogeneration plants that can provide heat for district heating systems have already been described in section 3.6.2, and several that are integrated with district cooling systems are discussed subsequently in section 5.5. With the penetration of renewable energy sources into heating and electricity networks expected to grow in the future, it becomes necessary to identify the appropriate integration of current cogeneration units with renewable energy sources, if we hope to optimize their interactivity with the grid and the district heating network.

Many district heating systems are linked to neither cogeneration nor district cooling. Renewable sources such as geothermal and solar as well as renewable fuels such as biomass can be used to provide heat for communities through the use of a district energy network. In many cases, more than one of the above-mentioned energy sources for district energy are used to satisfy the heating needs. An interesting example exists in Denmark, where a large network of pipes is installed under towns and cities. The system collects waste heat from factories, incinerators and transport systems, and combines it with heat generated from solar thermal energy plants, wind turbines, and conventional gas and coal power stations, to provide a low cost and highly efficient heat supply.

4.5.6 Types of district heating systems

Various types of district heating systems exist. The differences are often related to the source of energy used to provide the heat, although other distinguishing characteristics are also used for classification purposes (e.g., type and phase of heat transport fluid).

The main types of district heating systems are illustrated in Figure 4.6, and are described in the subsequent subsections, pointing out their similarities and differences.

4.5.6.1 Fossil fuel-based district heating

Fossil fuels are commonly used as the heat source for district heating systems, usually depending on the availability of energy resources. In some jurisdictions,

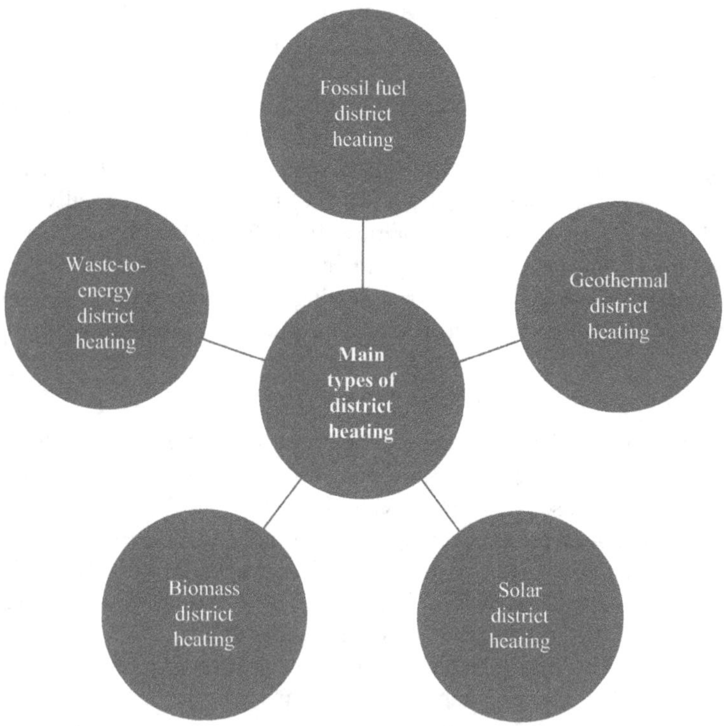

Figure 4.6 The main types of district heating, including traditional methods using fossil fuels, as well as new methods based on renewable energy and other fuels

coal is plentiful and used for district heating, while in others natural gas is the preferred fossil fuel, often for environmental reasons. Fossil fuel-based heating systems are extensively described earlier in the book.

4.5.6.2 Geothermal district heating

Apart from individual use of geothermal heating systems described in section 4.4.5, geothermal heat can be harvested in a central station to provide large-scale heating for residential, commercial, institutional and industrial heating needs. The heat from one or multiple production wells is transferred to the main network often through a geothermal heat exchanger. The heated network water circulates among users and returns to the geothermal heat exchanger for reheating. The cooled geothermal water is then pumped back to the reinjection well. For geothermal waters with relatively lower mineral content, the efficiency of the geothermal system can be improved by avoiding reinjection and discharging cooled geothermal water into high-discharge streams and rivers [154]. This eliminates risks associated with variations in geothermal wastewater reinjection parameters, reduces the electricity needed to power water reinjection pumps and increases the heating capacity of the geothermal system [154].

The temperature of geothermal sources is often close to or higher than the requirements of heat users in a low-temperature district heating network. Those requirements are about 47°C for the district heating distribution network and 45°C for the building circuits. In the latter case, a cogeneration plant is used to produce electricity from the high-temperature geothermal water or steam and the cooled water from the cogeneration plant is used to provide district heating.

To satisfy the heat demand of users via a district heating system at acceptable operating costs, geothermal heat sources are often coupled to a back-up heating plant that uses fossil fuels to cover peak user energy needs. To reduce the fossil fuel use of back-up systems, thermal storage systems can be coupled to a geothermal district heating network. This allows constant geothermal heat production.

4.5.6.3 Solar district heating

Solar thermal collectors can provide heat at adequately high temperatures to meet a community's space heating and domestic hot water needs. As pointed out in section 4.4.6, solar energy systems are often integrated with energy storage to help offset and manage the intermittent nature of solar resources. The storage is often designed to maintain the supply temperature from the solar collector field below approximately 90–95°C. Yearly dynamic simulations of solar plants are required to properly design plant components such as the solar collector field size. The storage volume is complex and requires local detailed climatic and demand data.

Many parameters influence the performance of the solar heating plant such as the district heating network return temperature, the solar collector area and the type of solar collector (low- or high-temperature collector). Solar fraction is often used as an indicator of system performance. In challenging climatic conditions for solar district heating systems, e.g., situations with high heating demands and low levels of solar irradiation in winter, high solar fractions can only be achieved through the use of auxiliary heating systems such as floor heating and passive solar building designs [156].

Solar district heating systems that are designed to heat communities normally require the installation of solar collectors over large areas. When space is limited for the installation of solar collectors on the ground, decentralized systems are sometimes employed.

An interesting successful example of a solar district heating system is provided by the Drake Landing Solar Community in the town of Okotoks, Alberta, Canada. The homes in that system are heated by a district system that is coupled to solar collectors. Underground pipes are used to distribute water that is heated by the solar collectors [157]. A solar fraction of 97 per cent has been achieved in this community [158–160].

In Europe, the use of solar thermal district heating is increasing in cases where cogeneration is not used for the base load. This can be due to the high shares of wind and solar photovoltaic (PV) energy in the electricity grid [161]. The Marstal District Heating System in Denmark provides heat for 1,500 consumers via solar production (55 per cent) and locally produced biomass (willow) (45 per cent) [162]. A water reservoir is used in Marstal to store the heat in summer for use in winter.

4.5.6.4 Biomass district heating

Biofuels and wastes are currently the most exploited renewable energy source worldwide [163]. An important reason for this is the renewable nature of using biomass for energy, provided its use is in balance with its production, as can be seen in the potentially renewable cycle biomass production cycle illustrated in Figure 4.7. The net effect of the cycle is work production from solar radiation.

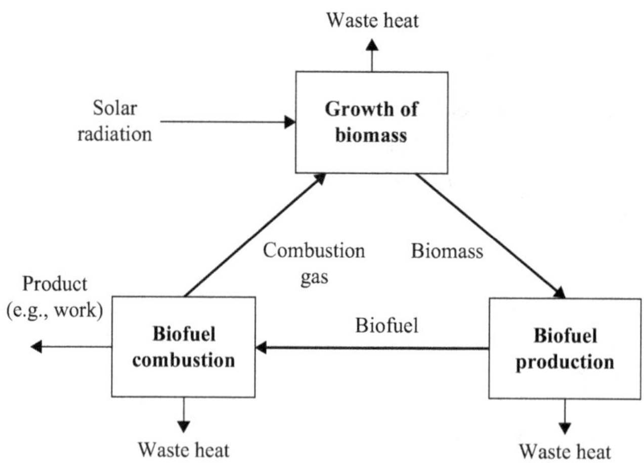

Figure 4.7 Renewable cycle for biofuel production from biomass and use, where the main driving energy form is solar energy. Heavy lines denote material flows, and light lines energy flows

Nonrenewable resources are not utilized in the cycle and, although waste heat exits, no material wastes are produced.

Biofuels that are used for district heating systems include agricultural, forest, and manure residues and some urban and industrial wastes. Historically, biomass in the form of wood has been used for heating purposes long before use of fossil fuels became common. Compared to solar and geothermal energy sources, for which emissions are typically small, biomass sources of energy produce CO_2 that could potentially have negative environmental impacts. However, biomass can be used to generate heat and electricity with limited environmental impacts if harvested in a sustainable fashion and burned under controlled conditions.

Since biofuel heating systems share some characteristics with conventional heating plants that use fossil fuels, their integration to current heating and cogeneration plants is possible by making design changes to the conventional systems according to the requirements of the biomass heating systems. However, due to their lower energy density and mass density than some fossil fuels, they require larger storage spaces and their transportation to remote distances (over about 80 km) may not be economically viable. Some types of compact biomass such as wood pellets are being increasingly used to overcome such limitations and are frequently distributed on the global scale.

In areas where wood chips are harvested locally and their transportation costs are eliminated, they can be used as a cheaper option for district heating. Furthermore, despite low fuel costs, biomass boilers typically have high capital costs compared to fuel oil boilers. These can be lowered when they are used in district heating in densely populated areas. Use of biomass district heating in more dispersed communities still involves economic challenges and is sensitive to variations in fuel oil heating prices.

Distribution pipes integrated to biomass heating stations carry superheated water (approximately 115°C) or steam or thermal oil.

4.5.6.5 Waste-to-energy heating

Waste management consists of various treatment options, such as reuse, recycling, composting, conversion to energy and landfilling. The selection of processes depends on the characteristics and properties of various waste constituents. Waste-to-energy processes recover energy from waste that cannot otherwise be reused, recycled or composted. Waste-to-energy is often a preferred option to landfilling in terms of sustainability.

Processes involved in the conversion of waste to energy, also called waste-to-energy processes, include incineration, gasification, pyrolysis and anaerobic digestion. Waste-to-energy processes can be categorized in terms of those which convert heat to electricity and/or heat directly (e.g., incineration) and those that produce a fuel that can be used for heat and/or power generation (e.g., gasification and pyrolysis). When wastes are converted to heat, they are often transferred to hot water or steam that can be transported to homes, businesses and institutions through a district heating network. Facilities for waste-to-energy have been installed in several countries worldwide. They are usually closely monitored to ensure safety

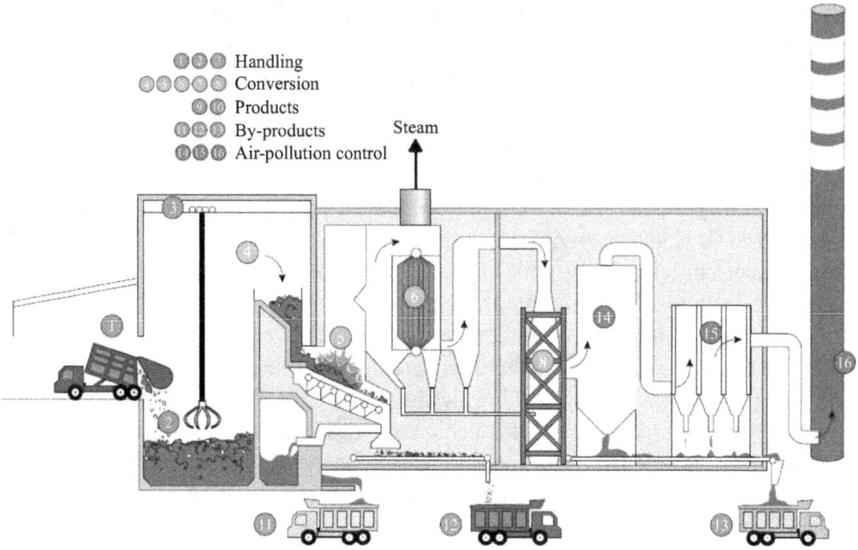

Figure 4.8 Schematic of a waste incineration plant providing steam that can be used in power generation and/or district heating (adapted from Reference 164 with permission)

and no risks human health or the environment. Although waste-to-energy facilities lead to a degree of emissions, they are still often among the cleaner energy plants.

Waste incineration is the combustion of waste organic material at high temperatures, and is usually carried out to produce steam in a boiler (see Figure 4.8). The steam can be used in power generation through a turbine–generator and/or can be distributed with a district heating system to provide heat for buildings. Toxins present in the garbage are destroyed due to the high temperatures of the process. However, ash and recovered metals are some of the byproducts of such plants. Waste incineration is the most common and proven option among waste-to-energy processes.

Waste-to-energy plants based on incineration are common in some cities. One city with a high use of waste-to-energy incineration is Paris. Three waste incineration plants provide heating for 50 per cent of Paris including homes, offices, hospitals and other buildings. Using waste form about 3.6 million Parisian households, these plants mitigate equivalent of 300,000 tonnes of oil use each year as well as prevent the release of some 90,000 tonnes of CO_2 into the atmosphere. The hot water produced with these plants is distributed to the users using a district heating network.

While waste incineration involves combustion of organic material that directly provides energy, gasification and pyrolysis are high-temperature processes where the organic material is converted to a combustible fuel that can be used in direct heating or cogeneration. One technology usable for generating the high temperatures needed in gasification and pyrolysis is the plasma arc systems. Eco-Valley in Japan is one of the first waste-to-energy facilities in the world to utilize plasma gasification technology on a commercial basis. The facility processes up to 220 tonnes per day of MSW or up to 165 tonnes per day of a 50/50 mixture of MSW and auto-shredder residue. The technology used at Eco-Valley is a result of a successful collaboration between Westinghouse Plasma Corp. and Hitachi Metals in the early 1990s.

Similar to gasification and pyrolysis, anaerobic digestion processes are used to convert waste to a fuel. In anaerobic digestion, bacterial degradation of wastes occurs. The types of wastes employed include agricultural residues; manure; municipal wastewater solids; food waste; high strength industrial wastewater and residuals; fats, oils and grease; and various other organic waste streams. The bacterial degradation results in the production of biogas (mostly methane and carbon dioxide) which can then be used to generate electricity and/or heat. In addition to anaerobic digestion, biogas can also be generated from landfilling.

4.6 Case study: solar/biomass district heating in Marstal, Denmark

4.6.1 System description

The Sunstore plant in Marstal, Denmark, provides district heating for approximately 1,550 buildings from two renewable energy sources: solar and biomass. It currently consists of solar thermal collectors, a biomass boiler, a heat pump and energy storage [165]. Details on some of these components follow:

- **Solar collectors**. A major part of the Marstal plants is their 87,100 m^2 of installed solar collector area. The plant has been grown to its current size in various stages from a small-scale plant with 75 m^2 of solar collector area in 1994 to 33,365 m^2 solar collector area in 2012 (Sunstore 4 plant) (see section 4.6.2).
- **Cogeneration unit**. The cogeneration unit consists of a 4-MW biomass boiler coupled with an ORC unit with 750 kW electric capacity. The low-emission boiler consumes about 81 TJ of willow woodchips annually.
- **Energy storage**. The energy storage installations at the Marstal plants have been added through various stages over time (see Table 4.4) and currently have an overall volume of 87,100 m^3.
- **Heat pump**. The heat pump uses CO_2 as a refrigerant and has operated at an average COP of 3.1. The heating capacity of the heat pump is 1.5 MW and it consumes 0.962 MWh of electricity annually.

4.6.2 System installation history

The Marstal plants were installed over time, starting with the installation of the original plant and followed by two subplant additions: Sunstore 2 and Sunstore 4.

In 1994, the Marstal solar heating plant was first installed as an adjustable-flow 75-m^2 solar plant that could provide supply temperatures of 70–72°C using 100 per cent solar energy in the summer months. An 8,000-m^2 solar plant with a 2,100-m^3 water tank was later installed to cover 13–15 per cent of the heat demand, by

Table 4.4 History of solar collector and energy storage system installations at Marstal plants[1]

Year	Solar collector overall area (m^2)	Thermal energy storage	
		Type	Volume (m^3)
1994	75		
1996	8,038	Water	2,100
1999	9,043		
2003	18,365	Pit	10,000
2012	33,365	Pit	75,000

[1]Data sources: References 165–168.

covering 100 per cent of heating loads for 3–4 months in the summer. The temperature in the water tank is normally about 70–75°C and increases to 95°C in especially sunny periods. In 1999, the solar collector plant was expanded by 1,000 m^2 of collector area, to an overall collector area of 9,043 m^2. The plant consisted of 720 solar panels with a size of 12.5 m^2 each, which were placed in 36 rows covering a land area of 20,000 m^2.

Sunstore 2. The system was enlarged in 2003, by adding a solar collector area of 9,322 m^2 and a pit heat storage system of 10,340 m^3, resulting in 30 per cent coverage of the yearly energy consumption. Overall, Sunstore 2 includes 18,365 m^2 of various-sized solar plants as listed below:

- 8,019 m^2 of flat plate solar collectors. The efficiency of the new collectors was improved by 18 per cent compared to the old solar collectors from 1996.
- 881 m^2 of flat plate solar collectors from another manufacturer.
- 211 m^2 of concentrating solar collectors.
- 108 m^2 of vacuum tube solar collectors.
- 103 m^2 roof integrated solar collectors.

Sunstore 4. A biomass boiler, a heat pump and a new set of solar collectors were added to the Marstal plants in 2012, providing an annual heat production of 28,000 MWh for the district energy network (see Figure 4.9). The plant is composed of the following additions to the previous plants:

- 15,000 m^2 of solar collectors.
- A cogeneration system consisting of a wood chip boiler paired with an ORC electricity generation plant.
- A 75,000 m^3 pit storage for TES.
- 1.5 MW heat pump, supplementing Sunstore 2.

Compared to the previous stages, the solar collectors were 10 per cent more efficient due to technological improvements over this period, and cost 11.7 per cent less.

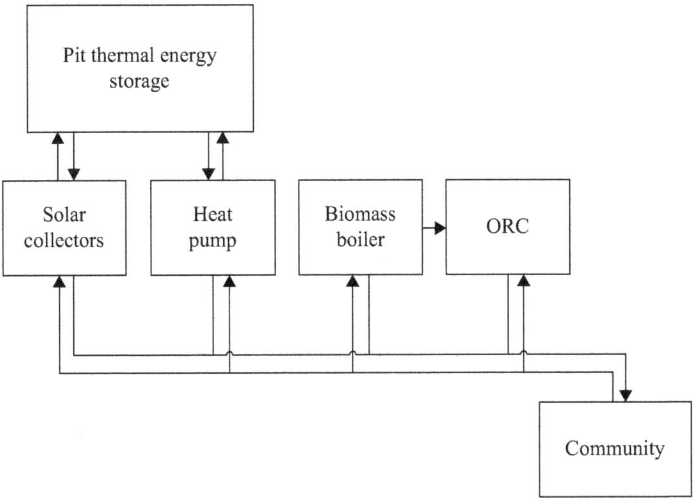

Figure 4.9 Configuration of Sunstore 4 biomass/solar plant

A temporal summary of the solar collector and energy storage system installations at the Marstal plants is provided in Table 4.4.

4.6.3 Cogeneration plant operation

Annually, the Sunstore 4 cogeneration plant produces approximately 3.2 GWh of electricity as well as 32 GWh of heat. The plant operation varies throughout the year and the various integrated components contribute to its flexibility in dealing with the intermittency of solar energy. The Sunstore 4 plant operation can be classified seasonally as follows:

- **Summer**. The solar collectors provide Marstal with district heating while also charging the storage pit with heat.
- **Fall (from the end of September)**. The solar collectors supply the district heating network while, depending on the cost of electricity, either the wood chip boiler or the heat pump supplement the heat supply in order to meet the demand. The boiler approximately has a heat output rate of 3.25 MW and the integrated ORC produces 750 kW electricity. When electricity prices are low, the heat pump operates, extracting heat from the storage and preheating the return water to a temperature of 75°C.
- **Winter**. The biomass boiler is operated regularly to supply the heat, while backup boilers (using wood chips) may be operated for a few hours on cold days to meet the total district heating demand. The heat pump is also operated during these periods, although they typically have higher electricity prices.
- **Spring (staring in February)**. The solar collectors start to supply heat to the storage pit again. The biomass boiler continues to operate until April when the solar collectors start to collect sufficient heat for district heating.

4.6.4 Energy distribution network

The Mastal solar plants supply water to the distribution network at 72°C and 76°C in the summer and winter, respectively, regulated according to the outdoor temperature. The return temperatures are approximately 33°C in winter and as high as 40°C in summer, and the plants operate more efficiently when the return temperatures are low. To reach such return temperatures, the Marstal district heating network uses local storages (typically tanks of 110–160 liters) at consumer substations to absorb heat from the higher return temperature fluid and provide the required low return temperatures to the central solar plants. The consumer substations also lower network piping to households as well as network heat losses.

4.6.5 Economics

Nearly, all inhabitants in Marstal are district heating consumers. The amount of energy sold in 2012 by Sunstore 4 was 26,500 MWh, with distribution losses of about 17 per cent. Overall, the plant provides district heating at a lower cost (average of 50–60 EUR/MWh) compared to heat production price of more than 70 EUR/MWh from bio-oil.

4.6.6 System benefits

The Sunstore plant has numerous benefits:

- The plants provide consumers with district heat from 100 per cent renewable energy sources.
- The plant model allows for economic consumption and production of electricity.
- The plant model can be expanded gradually.
- The energy storage systems and heat pumps make it possible to integrate other sources of renewable energy into the system in its future expansions.

4.7 Closure

Systems for heating and district heating are described, including basics, applications, types, classifications and operation factors, and a general model for a heating and district heating systems provided and assessed. Typical efficiencies are given, and applications in the residential, commercial, institutional and industrial sectors explained. Attention is paid to the use of a central heat supply to meet residential–commercial heat demands via district heating, with and without cogeneration.

Chapter 5
Chilling and district cooling systems

Overview

Systems for cooling and district cooling are described. This material is of great consequence to the book since cooling systems are the foundation of most district cooling systems as well as a technology against which cogeneration systems sometimes must compete. A general model for cooling and district cooling systems is provided to facilitate engineering assessments and descriptions, along with thermodynamic analyses based on energy and exergy. The resultant energy and exergy balances and efficiencies are also provided. The various types of cooling technologies are listed along with their characteristics. The types of cooling technologies include electrically driven vapor-compression chillers and heat-driven absorption chiller systems as well as systems for free cooling. The relation between chillers and heat pumps, which typically operate on the same thermodynamic cycle, is outlined. The coverage for cooling systems includes subjects such as basics, applications, types, classifications, operation, performance and efficiency. Extending the coverage of cooling or chilling, the many types of technologies for district cooling are listed and described. Characteristics of district cooling systems are provided, including advantages, operation, applications, distribution and capacity. The district cooling systems covered that operate using hydrocarbon fuels, electricity, surface waters, the ground, solar energy, biomass and industrial waste energy. It is shown how district cooling can be used to meet the space cooling requirements of buildings in the residential, commercial and institutional sectors, and how district cooling systems often allow cooling services to be provided to buildings with advantages over conventional cooling systems.

5.1 Introduction

Cooling processes are quite common in society and industry, especially for space cooling and food preservation. Cooling can be performed using a variety of energy resources. Cooling operations are one of the underlying foundations of cogeneration systems that include cooling capacity, i.e., trigeneration systems, and are the primary component of district cooling systems.

In cooling plants, energy, usually in the form of a fossil fuel or electricity, is used to drive a chiller or first converted to heat which is then used to drive a chiller.

Cooling process efficiencies are typically measured in terms of a COP. Cooling is used in the residential, commercial and institutional sectors for processes such as space cooling and food refrigeration or freezing, and in the industrial sector for a wide range of processes.

District cooling can be used to meet the space cooling requirements of buildings in the residential, commercial and institutional sectors [99]. District cooling systems often allow cooling services to be provided to buildings with advantages over conventional cooling systems, including increased efficiency, reduced environmental impact, better economics and improved reliability and safety [99,196].

In a district cooling system, a chilled fluid, normally treated water, is supplied from a central chiller plant and transported by pipe to users of the cooling capacity. The spent fluid is then returned by pipe to the central chiller plant to be cooled again. The central chiller plant can use many types of chillers, including electrically driven vapor-compression chillers and heat-driven absorption chillers.

District cooling systems using absorption chillers often complement district heating systems when both use heat supplied from a cogeneration plant because the demand for heat in a district heating system is lower in summer than in winter and heat-driven district cooling, which requires heat mainly in the summer, can help to balance the seasonal demands for cogeneration-derived heat.

5.2 General cooling system model

A model for a general chiller system is shown in Figure 5.1. Electrical power $\dot{W}_i$ and/or thermal power $\dot{Q}_i$ are input to drive the system. For simplicity, only one thermal energy input is considered. A thermal power load $\dot{Q}_{load}$ is taken into the chiller from the space being cooled. Waste heat exits. The general chiller system represents an electrical chiller if there is no heat input to drive the system, and an absorption chiller if the driving inputs are heat and a small amount of electricity.

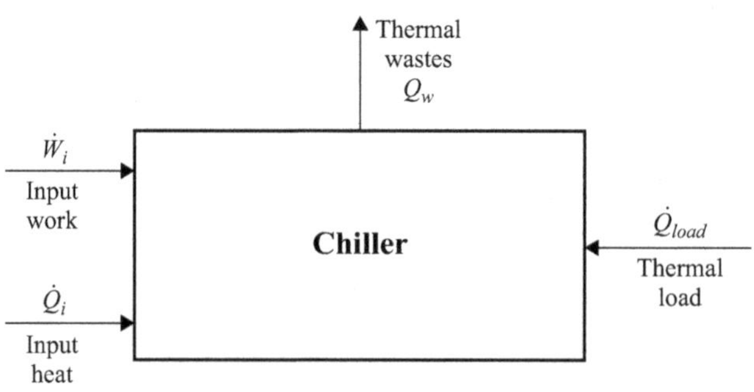

Figure 5.1 A model for a general chiller

5.3 Analyses for general cooling system

Commodity balances as well as efficiencies for the general model of a cooling system are presented in this section.

5.3.1 Energy and exergy balances

Assuming steady-state operation and neglecting kinetic and potential energies, an energy balance for the system shown in Figure 5.1 can be written as:

$$\dot{W}_i + \dot{Q}_i + \dot{Q}_{load} = \sum_w \dot{Q}_w \tag{5.1}$$

where the summation is over all thermal waste streams. The terms on the left-hand side of the balance represent the total energy input rate, while the terms on the right-hand side represent the total energy output rate (entirely associated with wastes).

An exergy balance can be written as:

$$\dot{W}_i + \dot{Q}_i\tau_i + \dot{Q}_{load}\tau_{load} = \sum_w \dot{Q}_w\tau_w + \dot{I} \tag{5.2}$$

where τ_i, τ_w and τ_{load} denote, respectively, the exergetic temperatures associated with the input driving heat flow, the waste heat flows and the cooling load and are functions of the relevant temperatures T_i, T_w and T_{load}, and T_i, T_w and T_{load} denote, respectively, the temperatures associated with the input driving heat flow, the waste heat flows and the cooling load. The two terms on the right-hand side of (5.2) represent the exergy rates associated with the waste heat outputs, and the exergy consumption rate. On the left-hand side, the first two terms represent the exergy input rate associated with the electrical and thermal driving inputs, and the third term is the exergy flow rate associated with the heat removal from the space being cooled. This third term is negative since the temperature at which heat is input to the chiller T_{load} is lower than the reference environment temperature T_o. In such instances, the following constraint applies: $-\infty \leq \tau_{load} \leq 0$. Physically, the negative value of the third term implies that a positive product (cool) is delivered from the chiller, flowing in the opposite direction as $\dot{Q}_{load}$.

5.3.2 Efficiencies

Following (2.19) and (2.20), the energy-based efficiency measure known conventionally as COP can be written for the chiller system in Figure 5.1 as:

$$COP_{ch} = \frac{\dot{Q}_{load}}{\dot{Q}_i + \dot{W}_i} \tag{5.3}$$

and the corresponding exergy efficiency as:

$$\psi_{ch} = \frac{-\dot{Q}_{load}\tau_{load}}{\dot{Q}_i\tau_i + \dot{W}_i} \tag{5.4}$$

When $\dot{Q}_i = 0$, the energy and exergy efficiencies reduce to those for an electrical chiller. While COP_{ch} can take on values between 0 and infinity, ψ_{ch} is constrained between 0 and unity.

The dependency of exergy efficiency on the temperatures T, T_{load} and T_o is particularly significant. The greater the deviation from T_o of T_i and T_{load}, the greater the exergy associated with the driving heat input and with the product cool delivered from the chiller.

Exergy efficiencies provide more illuminating insights into chiller performance than energy efficiencies, because exergy efficiencies provide a measure of how nearly the performance of the system approaches ideality, and of the magnitude of the potential for improvement. These observations stem from the fact that exergy efficiencies weight heat, cool and electricity according to their "equivalent work potentials," and imply that chiller systems are better compared using exergy efficiencies.

5.4 Systems for cooling

Various types of systems exist for the provision of cooling services. These are described, along with their characteristics, in this section.

The main types of heating systems are illustrated in Figure 5.2. The systems shown are driven by various energy sources: electricity, heat and ambient conditions. Of course, the electricity and heat can each be derived from a range of energy sources, as described in the previous two chapters.

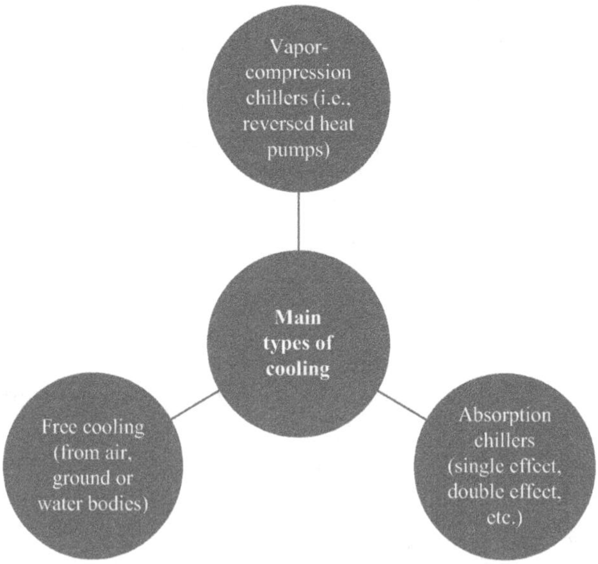

Figure 5.2 The main types of cooling systems, including traditional and newer methods. The systems shown are driven by ambient conditions or electricity or heat, and the latter two can be derived from various energy sources

5.4.1 Cooling technologies and their characteristics

Cooling capacity can be delivered by chillers in the form of chilled water, ice or ice slurries, brine solutions and cooled air. Cooling is used in the residential, commercial and institutional sectors for space cooling, and in the industrial sector for processes such as gas liquefaction, materials treatment, food chilling and various others. Examples of industries that require large amounts of cooling for their operations include food and beverage, chemical and industrial gas processors [119].

District cooling systems contain a central cooling system, a distribution network and an energy transfer station at the consumer side, also called the consumer substation, where the district water is directly or indirectly connected to the terminal building equipment such as air handling units.

Systems used for cooling, called chillers, use energy in order to move heat from a low-temperature environment to a high-temperature environment. This thermodynamic process operates on a refrigeration cycle, which is made up of various system components. Refrigeration systems can be classified based on the energy used to drive their operation: electricity or heat. There are several refrigeration cycles that operate using electricity. These include the following: vapor-compression cycle, reversed Brayton cycle, transcritical cycle and thermoelectric cycle. Refrigeration cycles that operate using heat include absorption and adsorption cooling systems, desiccant systems, and ejector refrigeration cycles.

Similar to district heating systems, the central cooling system is integrated with a district cooling network either directly or indirectly and distributes cooled water to consumers via supply and return pipes.

5.4.2 Vapor-compression chiller systems

The most common type of refrigeration system for space cooling is based on the vapor-compression cycle. Vapor-compression chillers can be driven by electricity, turbines or reciprocating engines. Electrically driven vapor-compression chillers, which use centrifugal or screw compressors, are the most common technology used in central cooling applications [1,2,170–172].

5.4.2.1 Description and operation

An electrically driven vapor-compression chiller system is shown in Figure 5.3a, where the chiller is seen to have four main components: two heat exchangers, one compressor and one expansion valve.

The operation of a vapor-compression chiller involves four main steps:

- A compressor, which can be of reciprocating, rotating or centrifugal type, driven by electrical power $\dot{W}_i$, takes in refrigerant vapor and raises its pressure, causing the refrigerant's temperature to rise.
- The high-pressure vapor flows from the compressor to a condenser where the vapor is condensed and heat is rejected to a heat sink, usually water or air.
- The high-pressure liquid is then passed through a pressure reducing device to the evaporator.

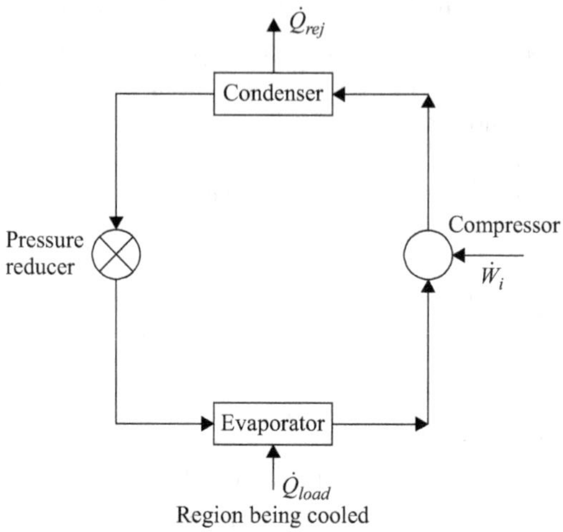

Figure 5.3a Simplified schematic diagram of a vapor-compression chiller system

- In the evaporator, the low-pressure liquid absorbs thermal power $\dot{Q}_{load}$ from the region being cooled and evaporates. The cold vapor is then returned to the compressor to begin a new cycle.

Vapor-compression chiller systems specifically designed for cooling spaces operate on the same refrigeration cycle as heat pumps and deliver the heat to the outside air.

5.4.2.2 Relation between chillers and heat pumps

Heat pumps that are used for space cooling are similar to vapor-compression chiller systems in principle. However, they can be reversed in processes and be used for heating as well. Using a reversing valve, the condenser of the cycle in the summer operates as an evaporator in the winter. To operate, heat pumps also use electrical power, a high-grade energy form. The objective of cooling systems based on vapor-compression cycles is to remove as much heat as possible from the cooled space to the high-temperature environment using a given amount of electricity, while accounting for other objectives and constraints such as economics. When cooling a space, the higher the temperature of the medium to which the heat is delivered, the higher is the rate of heat pump electricity consumption. In addition to the outside air as medium for heat delivery when cooling, heat pumps can be integrated with other options such as nearby ground or water bodies. The ground serves as a suitable heat source and sink for heating and cooling operations since it is cooler than the outside air in the summer and warmer than the outside air in the winter. The heat extracted from the central cooling station or an individual building can be delivered to the ground using a heat pump.

5.4.2.3 Modifications to enhance efficiency

Several modification can be made to a vapor-compression cycle to improve its efficiency. A common modification is to use multiple compressors to divide the compression into multiple stages and to utilize intercooling between the compressor stages. Another method is to connect multiple cycles in series to provide the refrigeration. These and other possible system modifications are described elsewhere [173–179].

5.4.3 Absorption chiller systems

A common heat-driven option for space cooling systems is based on the absorption of the refrigerant in a transport medium. Compared to vapor-compression cycles that use electricity to provide cooling, this cycle uses heat, a low-grade energy form, at temperatures as low as 100°C. There are several types of absorption chiller systems, many of which can be used when waste or surplus heat is available (e.g., the thermal co-product of cogeneration systems). The temperature requirements of the low-grade heat that can be integrated with absorption chillers as well as the required temperature of the product cooling for cooling applications both depend on the refrigerant that is used in the absorption cycle.

There are various types of absorption chillers. One absorption chiller type uses ammonia and water as the refrigerant and the absorbent, respectively, and can be used to provide chilled water or other media well below 0°C for low-temperature applications such as ice making and district cooling [90,170–180,202]. In another common absorption chiller, water is used as the refrigerant with lithium bromide (LiBr) as the absorbent.

Similar to vapor-compression chillers, absorption chillers are also designed in single- or multiple stages, corresponding to single or multiple temperature levels, in order to provide cooling.

5.4.3.1 Description and operation

A single-stage absorption chiller is shown in Figure 5.3b. As can be seen, the compressor in the vapor-compression cycle is substituted with an absorber, a pump, a generator and a rectifier.

During operation of a single-stage absorption chiller, the following main steps occur:

- The refrigerant is vaporized in the evaporator, absorbing thermal power $\dot{Q}_{load}$ from the region being cooled.
- The refrigerant vapor then flows into the absorber (e.g., water), where it is absorbed by the liquid absorbent, to form a "weak solution." Cooling water is normally used to remove the heat which is released during absorption.
- The weak solution is then pumped to high pressure and passes through a heat exchanger to the generator.
- In the generator, thermal power $\dot{Q}_i$ at a relatively high temperature is added to drive refrigerant vapor out of the solution.

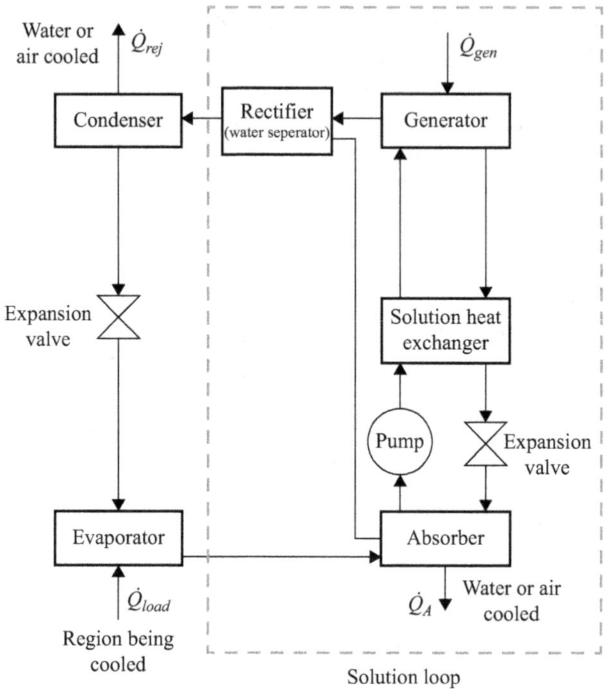

Figure 5.3b Simplified schematic diagram of a single-stage absorption chiller system. The cycle is similar to that for a conventional vapor-compression chiller, but with the devices enclosed by the dashed rectangle in this diagram replacing the compressor in the conventional system

- The refrigerant vapor flows into the condenser (after passing through a rectifier, which traps the water vapor and returns it as a liquid to the absorber) where heat is rejected.
- The condensed refrigerant then passes through an expansion valve to the evaporator.
- The strong solution remaining in the generator passes through the solution heat exchanger and an expansion valve, and returns to the absorber.

The heat required for the generator is often more than the electricity required for the compressor in the vapor-compression cycle and, therefore, the COP of vapor absorption cycles are often lower than that for vapor-compression cycles. However, this cycle becomes advantageous relative to the vapor-compression cycle when renewable or waste heat at low cost is available for use in the generator. For this reason, absorption cooling systems have been receiving much interest in recent years.

The cycle of a two-stage absorption chiller (Figure 5.3c) is similar, except relatively higher temperature heat input is required, at two temperatures.

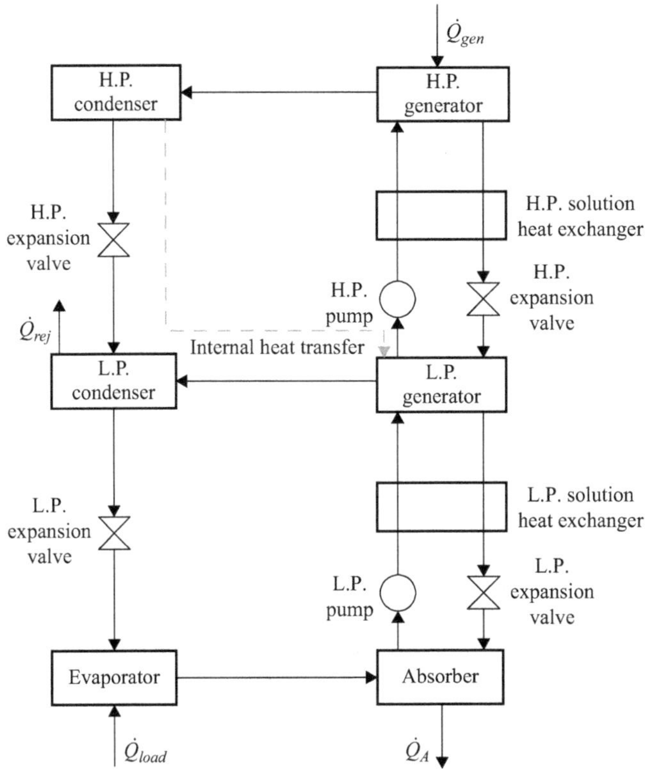

Figure 5.3c Simplified schematic diagram of a double-stage absorption chiller system. Notes: H.P. = high-pressure; L.P. = low-pressure

5.4.3.2 Performance and efficiency

For single-stage indirect-fired absorption chillers, the heat source can be low-pressure steam (62–83 kPa), or hot water (at approximately 116°C) [170]. The heat is used to separate the ammonia from the liquid mixture. Typical values of COP_{ch} of such a system range from approximately 0.65 to 0.80 [170,171,180]. Two-stage absorption chillers have higher COP_{ch} values (approximately 0.9–1.2) than the single-stage absorption chillers [170,180], but require higher grade heat, i.e., steam at higher pressures (296–896 kPa), or hot water at higher temperatures (at approximately 205°C) [170].

Typical COP_{ch} values for absorption chillers having varying stages can be plotted as a function of heat-input conditions (i.e., saturated steam pressure for steam-driven systems and water temperature for hot water-driven systems). One such plot (see Figure 5.4) clearly demonstrates that the efficiency of absorption chillers, as measured by their coefficients of performance, increases as the quality of the driving heat input, as measured by its temperature, increases.

Several modifications to the absorption system exist such as integrating the generator with a compressor to assist with vapor compression. Compared to the

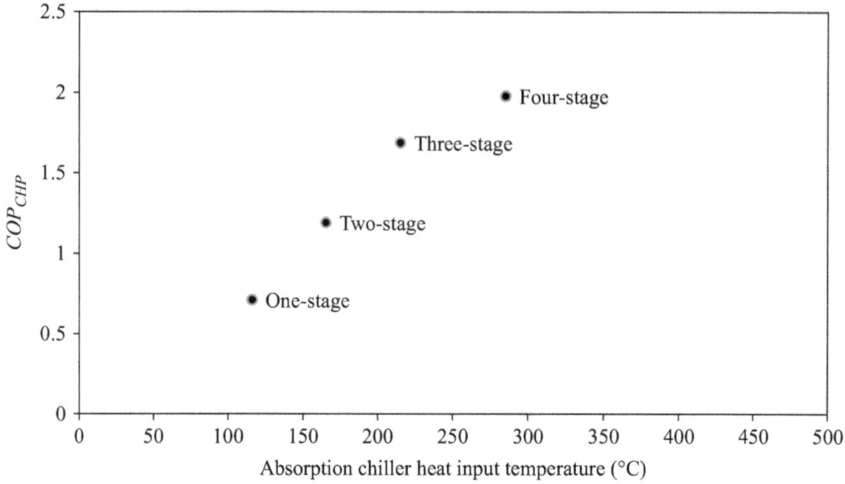

Figure 5.4 Variation of chiller COP with heat-input temperature, for absorption chillers having various stages. The data are based both on actual data for single- and double-stage absorption chillers, and on data obtained by numerical simulation [191, 192] for several multi-stage absorption chillers. The data shown represent approximate average conditions. (Heat-input temperatures actually vary for 1-, 2-, 3- and 4-stage absorption chillers between approximately 110–125°C, 150–180°C, 200–230°C and 270–300°C, respectively, while COP_{ch} values vary by about ±0.1 around the values shown here.)

conventional absorption cycle, operation of this hybrid system requires electricity and heat (in smaller amounts) for vapor compression.

Two-stage absorption chiller systems are not necessarily thermodynamically superior to single-stage systems when driven by cogenerated heat, because the heat is required at higher temperatures. Thus, the efficiency of electricity production is reduced by a relatively greater amount. Moreover, two-stage absorption chillers are more expensive than single-stage chillers [170,180].

Absorption chiller systems require much less electricity than conventional chillers. Although electricity is required to drive the solution and the refrigeration pumps in absorption chiller systems [170], this amount of electricity is very small compared to the electricity required by the compressor of conventional chiller systems.

5.4.4 Free cooling

In free cooling, a "freely" available low-temperature medium, often a natural resource such as lake water, is used as a heat sink to remove heat from a variety of higher temperature mediums where cooling is needed. The high-temperature medium includes building spaces, the district operating fluid used for cooling purposes or a storage medium, when the cooling is not immediately needed.

The low-temperature medium could be cool night air, an aquifer, and groundwater and water from lakes, rivers and oceans, as they remain cooler than the air.

Sometimes the temperature of such cool mediums is only adequate to provide precooling, which can nonetheless be beneficial. In other cases, the freely available low-temperature medium is used to replace air that is used in cooling towers as the high-temperature medium of a vapor-compression cycle. Such use, although not as economical as free cooling, can enhance the performance of the refrigeration cycle and thus lower refrigeration system operating costs. For example, cooling systems in the Paris district cooling network use river water from the Seine to replace cooling towers in their central cooling stations that operate based on the vapor-compression cycle.

Cooling to lower temperature can also be achieved using heat pumps operating in a cooling mode. Such cooling systems are associated with lower costs and environmental benefits compared to the refrigeration cycles described in the previous sections due to the significant energy savings they accrue, but are limited to the availability of a low-temperature medium. Furthermore, since the cooling provided by free cooling is limited to the temperature of the cool medium available, most free cooling systems are used with a heat pump or vapor-compression chiller to provide supplementary cooling.

5.5 Systems for district cooling

District cooling is a method for supplying cooling capacity typically for space cooling, but sometimes for industrial chilling, and is used in residential, institutional and commercial buildings as well as in industry.

5.5.1 Description and categorization

District cooling systems deliver chilled water from a central source to a network of residential, commercial and institutional buildings and thus collect heat from buildings. They have become increasingly popular in densely populated areas due to their high efficiencies.

District cooling systems can be categorized according to their refrigeration principle: vapor-compression, absorption and free cooling. A district cooling system using heat-driven absorption chillers requires two sets of distribution systems: hot pipes in which steam or hot water is transported from the heat source (e.g., a cogeneration system) to the generator portion of the absorption chiller, and cold pipes in which chilled water (at as low as 2°C [170–180]) is carried out along the district cooling network.

The use of district cooling for the provision of cooling to a community is illustrated in Figure 5.5. In the thermal transport loop, the cooled fluid is transported to the community after exiting the cooling plant and then returned after use in the community for recooling. Once-through systems exist in which the fluid is water and it is discarded after use, but these are increasingly uncommon due to the cost of losing of the transport fluid and the associated environmental impact of

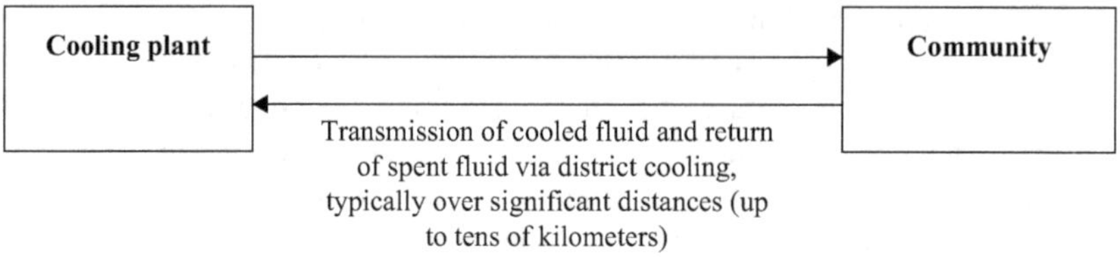

Figure 5.5 The use of district cooling for the provision of cooling to a community. The fluid is transported to the community after it is cooled in the cooling plant and the spent fluid is returned after use in the community for recooling. Less commonly, once-through systems exist in which the fluid is water and it is discarded after use

discarding it and extracting new water. The transmission of the cool transport fluid in district cooling is typically over significant distances, from tens of meters to up to tens of kilometers.

5.5.2 Distribution and capacity

Chilled water from central cooling plant is distributed via pumps through supply and return pipes. The pump is designed to overcome the pressure loss through distribution pipes as well as to provide a pressure differential between the supply and return pipes to ensure adequate water flow through consumer's heat exchanger.

The central cooling system provides cooled water for distribution at 4–7°C. Ice water and ice slurry systems provide cooling temperatures as low as 1°C and −1°C, respectively. The supply water temperature of a district cooling system used for community space cooling is typically varied according to the outside temperature and space cooling needs. For moderate outdoor temperatures associated with smaller space cooling loads, cooling water at higher temperatures is used to increase the central cooling plant efficiency as well as to reduce distribution heat gains. The distribution pipes are sometimes insulated to prevent heat gains or losses through the network. In northern climates, ground temperatures at the installation depth (larger than 0.6 m) are close to the chilled water temperatures of the district systems, so the heat gains of the district cooling systems are often small.

The thermal capacity of a district cooling system is affected by the rate of water flow as well as the temperature difference between the supply and return pipes, called the temperature differential. Distribution networks are often designed to ensure a large temperature differential between the supply and return water as it is associated with smaller pipes and pumping energy. However, the temperature differential is limited and affected by the energy transfer station at the consumers' side. The maximum temperature differential between the supply and return pipes for cooling systems, achieved at peak-load conditions, is about 11°C for ice-based systems with a 1°C supply temperature, and approximately 8°C for conventional chiller systems with a 4°C supply temperature [180]. To maintain such high-temperature differentials at partial load conditions, the distribution flow rate can be varied if variable flow rate facilities at the consumers' side are available.

5.5.3 Applications and examples

Similar to electric utilities, integration of energy storage systems with district energy systems assists in load leveling and peak shaving. Cold storage technologies integrated with district cooling systems allow the cooling technologies in the central cooling station to be sized closer to the average load than peak load, thus reducing the capital costs of the central station. In addition, the electricity use for pumping water through the district network as well as operating electricity-based chillers can be shifted to off-peak hours, thus reducing operating costs.

Many district cooling configurations are possible. Sometimes steam is delivered to buildings that each has heat-driven absorption chillers (e.g., the New York City district energy system operated by Consolidated Edison) [99]. District cooling networks are integrated with district heating systems in some designs and separate in others, e.g., the chilled water network in Indianapolis, Indiana, is independent of the district heating network [99]. Sometimes district cooling is integrated with a district heating system.

Several selected applications are discussed below:

- A district heating and cooling system operated by Energy Networks Incorporated since 1962 serves over 70 per cent of the buildings in downtown Hartford, Connecticut [99,108]. A natural gas-fired cogeneration plant, completed in 1990, produces the hot and cold water required for the system, as well as electricity [99,108].
- The feasibility of district heating and cooling was assessed for downtown Des Moines, Iowa [98], considering a mothballed 210 MW (electrical) coal-fired power plant as the source of heating and cooling fluids. The system was predicted to break even economically in 20 years, and have a lifetime of 40 years. District cooling using electrically driven centrifugal chillers was estimated to provide cooling at competitive prices; the use of absorption chillers driven by hot water from the district heating system was estimated to be slightly more expensive [98].
- Waste heat can also drive absorption chillers. At a laboratory in Ft. Worth, Texas, the exhaust gases from a 3.7 MW gas turbine drive a 1,500 ton (5,280 kW) chiller [109]. A system in Phoenix, Arizona, contains two 300 kW natural gas engine generators, the exhaust and jacket cooling water from which feed a 250 ton (880 kW) chiller [109].
- Edmonton Power proposed for downtown Edmonton, Alberta, a major cogeneration-based district heating and cooling project having (i) an initial supply capacity of 230 MW (thermal) for heating and 100 MW (thermal) for cooling, with the potential to expand to about 400 MW (thermal) for heating over the next ten years; (ii) the capacity to displace about 15 MW of electrical power used for electric chillers through district cooling and (iii) the potential to increase the efficiency of the Rossdale power plant that would cogenerate to provide the steam for district heating and cooling from about 30 per cent to 70 per cent [99]. Screw vapor-compression chillers were to be used originally, and absorption chillers in the future.

5.5.4 Types of district cooling systems

While there are various methods to provide cooling for individual buildings, additional sources are available for district cooling. The main types of district cooling systems are described in the subsequent subsections, pointing out their similarities and differences. These systems are also illustrated in Figure 5.6, e.g., steam turbine cycles, gas turbine cycles and reciprocating engines. The principal types of cogeneration systems are illustrated in Figure 5.6.

5.5.4.1 Hydrocarbon fuels

Hydrocarbon fuels can be used in a central station to provide a chilled fluid for district cooling. A hydrocarbon fuel, such as natural gas, is used to provide hot steam that can be used in absorption chillers. Such a technique typically is not used for refrigeration or small-scale cooling systems.

5.5.4.2 Electricity

There are several cooling technologies that operate using electricity. These include, electric boilers integrated with vapor-compression chillers and electric chillers.

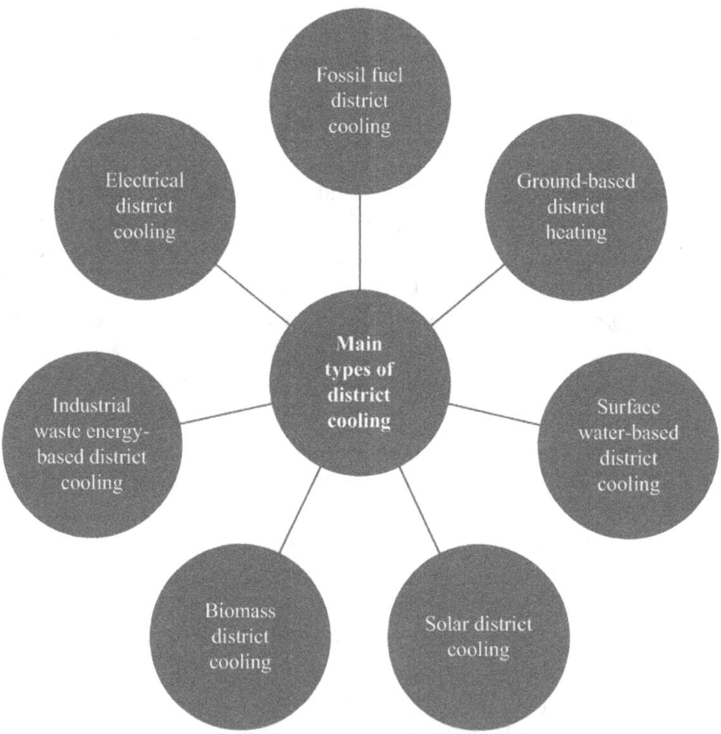

Figure 5.6 The main types of district cooling, including traditional methods using fossil fuels, as well as new methods based on renewable energy and other energy sources

Various types of electric boilers can be used in district heating and industries. However, their uses have been limited due to availability of technologies, such as vapor-compression chillers and heat pumps that use electrical power more efficiently to operate.

Central cooling plants that operate based on the principles of the vapor-compression cycle (see section 5.4.2) can be integrated with a district cooling plant to provide large-scale cooling. For instance, a district cooling plant in Pearl-Qatar, an artificial island in Doha, Qatar, uses 52 chillers to supply chilled water to 41,000 residents in 15,000 buildings. The system contains 26 cooling towers, 26 chilled water pumps (each with a capacity of 7,500 GPM) and 26 condenser water pumps (each with a capacity of 11,170 GPM) [181]. The system provides more than 130,000 tons of refrigeration for 50,000 residents in more than 80 apartment towers, beachfront villas, townhouses, shopping complexes, offices, schools and hotels throughout the island. The plant represents a successful exploitation of the concept of economy of scale in district cooling systems.

5.5.4.3 Surface waters

Surface water is frequently used in district cooling systems due to lower costs and environmental benefits. Surface waters used for district cooling systems include the following:

- sea water (e.g., in Hong Kong and Stockholm),
- river water (e.g., in Paris) and
- lake water (e.g., in Toronto).

Depending on the required temperature of the surface water for district cooling, water from deep levels need to be used, increasing the capital costs of the system.

In Stockholm, one of the largest district cooling distribution networks worldwide uses sea water directly to provide district cooling. This is a relatively shallow system (<20 m) due to the low temperature of the water in this region. Throughout the year, seawater enters the cooling plant's heat exchangers and collect heat from the city network, providing the cooling effect at 7–9°C. The warmed water is then returned to the sea. If further cooling is required, heat pumps are used to deliver more heat to the sea. With 220 MW of installed capacity and 350 GWh per annum output (based on data for the year 2011), the cooling system provides a significant saving in electrical energy and reduces carbon dioxide emissions in Stockholm by about 50,000 tonnes per year.

Another example of free cooling is a deep lake water cooling system installed in Lake Ontario to utilize the steady low temperature of this lake at a depth of 83 m. Heat is exchanged using polyethylene pipes held on the bottom of the lake. The lake water used for cooling enters the distribution network heat exchangers at 4°C and leaves at 13°C. Using the lake cooling capacity, the plant provides approximately 260 MW (75,000 tonnes) of refrigeration to the residential and commercial sector through a district cooling network [183].

A case study on the free cooling plant in the Paris district cooling network is included in section 5.6.2.

5.5.4.4 Ground

The heat extracted from a central cooling station can be delivered to the ground using a heat pump. Being cooler than the outside air in warm seasons, the ground is a more efficient alternative to the outside air for delivering heat that is extracted from the cooled spaces via a district network. For example, the aquifer at Gardermoen, Norway, which is one of the largest groundwater reservoirs in that country, is used to heat and cool the Gardermoen Airport. The system operation changes with season as follows:

- Ground water from the hot aquifer wells is used as heat source for the heat pump in the winter. The ground water is then returned to the cold wells
- Ground water is pumped from the cold aquifer wells and used for cooling in the summer, essentially reversing the process. The ground water is then returned to the hot wells [182].

Ground can also act as a source of heat to provide the input energy (i.e., heat) for absorption chillers in a central station. More information on utilizing the ground as a source of heat is given in section 4.5.6.

The main advantage of the use of geothermal sources for cooling (and heating) purposes is their availability and low operating costs. However, they are associated with large capital investments and often have longer payback periods. This challenge can be partially overcome when geothermal energy sources are used in both heating and cooling. Furthermore, geothermal energy sources and their temperature vary by location which can greatly influence the cost of the technologies needed to extract the heat at various locations.

5.5.4.5 Solar energy

The solar energy collected by thermal collectors in the form of heat can be used in absorption chillers to provide cooling. For example, the solar thermal cooling plant at the United World College in Singapore uses large-scale flat plate collectors, operating at a temperature of about 100°C [184]. The system's collector surface area is 3,900 m^2. The collectors power a 1,500 kW cooling unit, which is based on a single-effect lithium bromide absorption chiller operating at a nominal working temperature of 88°C.

5.5.4.6 Biomass

The use of biomass, including municipal waste, provides heat that can be used in absorption chillers to provide cooling. Various options regarding production of heat from biomass are given in section 4.5.6.4.

5.5.4.7 Industrial waste energy

Industrial processes, such as gasification of liquefied natural gas, that absorb a large amount of energy (i.e., provide a large amount of cold) can be integrated with district cooling systems to provide cooling at low costs. Such cold energy would otherwise be wasted or not used.

5.6 Case studies

5.6.1 *Case study: district cooling network in Paris, France*

The Climespace district cooling network in Paris, France, was opened in 1991 and is the largest district cooling network in Europe, distributing chilled water over 71 km to customers equipped with energy transfer stations. The Climespace district cooling network is connected to eight cooling plants, three energy storage systems and a geothermal energy plant. The entire system has a cooling capacity of 285 MW and produced a total of 420 GWh of chilled water for district cooling in 2015 [185,186].

5.6.1.1 Cooling plants

The Climespace system has eight cooling stations, including three that use free cooling from the Seine River to improve their cooling capacity. Details on the cooling stations are listed in Table 5.1.

Originally, the Climespace district cooling network provided cooling using outdoor air as the cooling medium. Cooling towers and fans were installed throughout the network. However, due to the high-energy consumption of the fans and the relatively higher temperatures of the air compared to river water, no significant savings in energy consumption resulted and investments were made in river water cooling to improve system efficiency. Details on one of the free cooling plants in the Paris district cooling network are included in section 5.6.2.

5.6.1.2 Energy storage systems

The Climespace network is equipped with three energy storage systems, including one chilled water storage system and two ice storage plants. Due to the low nighttime cooling demand, the cooling provided by the chillers is stored in the tanks during nights to cover peak cooling demands during days. The tanks also provide a continuous supply of cooling when the cooling plants are offline due to unexpected production disruptions. In emergency cases, the storage systems can respond quicker than conventional chillers (in 10–20 minutes depending on the storage type).

The chilled water storage system is installed at plant Halles and has a 90-MWh energy storage, with a 17 MW capacity at 5°C. The storage comprises thirteen 1,000-m^3 tanks storing a total of 12,000 m^3 of chilled water at approximately 3°C (usually between 2°C and 5°C). One tank always remains empty and is only used in charging and discharging modes. For instance, when charging, the empty tank is filled with cold water while warm water from another tank is drained to the network return.

The two ice storage systems store 20 and 30 MWh of energy. The 20-MWh ice storage consists of two internal and external melting units which are installed at plant Opera to improve the plant cooling capacity during peak times by approximately 5 MW (at 1°C). The 30-MWh ice storage consists of four internal and external melting units which are installed at plant Halles. They supply an additional

Table 5.1 Cooling plants in Paris district cooling network[1]

Plant name	Cooling capacity (MW)	Temperature delivered to network (°C)	Plant components
Halles	42	2–5	10 Refrigeration units – air cooled 90 MWh chilled water storage 30 MWh ice storage
Canada	52	2[2]	8 Refrigeration units – free cooled A free cooling station[3]
Tokyo	52	0.7	8 Refrigeration units – free cooled
Auber	24	–	
Opera	33	1	9 Refrigeration units – air cooled 20 MWh ice storage
Etiole	8	–	2 Refrigeration units – air cooled
Bercy	44	3	7 Refrigeration units – free cooled
PNE-Philharmonie	0.8	–	1 Refrigeration unit – air cooled A geothermal plant

[1]Data sources: References 185–188. Dashed entries indicate that data were not readily available.
[2]Four lines of two groups are installed on two levels. The two groups on the same line operate in series: the water temperature is lowered from 10°C to 6°C in the first group and then from 6°C to 2°C, the temperature delivered to the network, in the second group.
[3]A free cooling station with an 11,200 m^3/h capacity using five plate heat exchangers. Two cooling circuits are used, one of which is integrated with the refrigeration condenser.

5–13 MW cooling capacity to the plant Halles capacity. They allow an increase in the density of the transported energy by the network by lowering the network water temperature from 4°C to 2.2°C [187].

5.6.1.3 Distribution network

The Climespace energy network supplies chilled water to its cooling consumers at 1–4°C and operates at network return temperatures of approximately 12–14°C [185]. This is achieved through two separate supply and return pipelines. The network is connected to approximately 600 energy transfer stations where heat is exchanged between the main distribution network and a secondary network. The district cooling network distributes cooling mainly among offices, representing 65.4 per cent of the cooling while residential buildings constitute only 0.5 per cent of the cooling. A breakdown of the cooling consumers of the district energy network is shown in Table 5.2.

5.6.1.4 Geothermal energy plant

The geothermal energy plant was installed in 2013 and uses boreholes. Water at 60°C is extracted from a geothermal well at depths of 1,600–1,700 m and returned to another well. This hot water is used to meet some of the heat demand for a 1,000,000-m^2 development and to produce cooling for the development in summer via absorption chillers [188].

Table 5.2 Breakdown of cooling consumers of the Climespace district energy network in Paris [186]

Consumer	Fraction of district cooling use (%)
Offices	65.4
Businesses	10.4
Financial organizations	9.4
Hotels and restaurants	7.6
Theaters and museums	6.7
Housing	0.5

5.6.1.5 System benefits

The Climespace district cooling network in Paris provides various benefits, such as improved efficiency and lower electricity consumption and emissions, as listed below:

- **Efficiency**. Compared to standalone installations providing equivalent cooling, the Climespace system achieves an efficiency that is 50 per cent higher and a 35 per cent reduction in electricity consumption [186]. The system COP increased by 30 per cent to 4.03 between 2002 and 2012, as a result of operation of three free cooling stations. Furthermore, for plants using cooling towers, cool storage during the night allows the lower outside temperatures (i.e., lower condensing temperatures) to be exploited to improve chiller performance.
- **Emissions**. The Climespace system has reduced CO_2 emissions by 50 per cent, compared to standalone installations providing equivalent cooling, and has reduced refrigerant emissions and chemical use by 90 and 80 per cent, respectively [186]. Furthermore, cool storage during electricity off-peak hours reduces CO_2 emissions associated with electricity use for cooling, e.g., the reduction in CO_2 emissions has been evaluated as 200 kg CO_2 for one day using storage, based on 140 MWh of cooling produced by the storage units [187].
- **Economics**. Storing energy during electricity off-peak hours reduces the electricity costs for cooling generation during off-peak periods.
- **Environmental resources**. Free cooling used in three of the network plants reduces the treatment and use of fresh water from the city's fresh water supply. In plant Canada for instance, this water saving amounts to up to 500,000 m^3 per year.

5.6.2 Case study: free cooling plant in Paris, France

The Bercy cooling plant is a district cooling facility in Paris, France, that uses river water from the Seine River as a free cooling source. It was built in six stages between 1995 and 2009, when the free cooling was added. The plant is an independent extension of the Climespace district cooling network (see section 5.6.1) in Paris. It has a cooling capacity of 44 MW and supplies chilled water to customers,

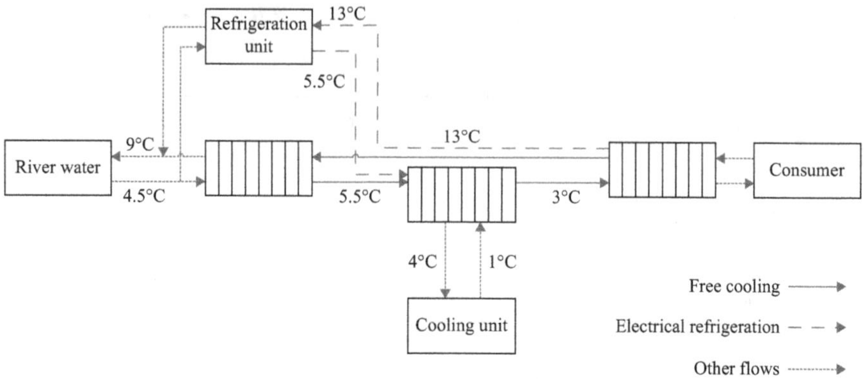

Figure 5.7 Process diagram representing the Bercy cooling plant in Paris (temperature data used in the figure are derived from Reference 165)

including offices and institutional and commercial buildings, along 10 km of network pipes. The plant consists of seven electrical centrifugal chillers, four river water pumps, condensing and evaporating circuits, and onsite generator equipment for emergency and ancillary electricity demands.

Since the availability of free cooling depends on local conditions (i.e., the river water temperatures) and regulations govern the temperature of the used river water that can be returned to the river, free cooling production varies. For example, free cooling reduced annual energy consumption by 11 per cent in 2010, but by only 3 per cent in 2011 when the river water temperatures were higher. Chillers are installed to provide cooling when free cooling is not available or is only partially available. A process diagram for the plant is shown in Figure 5.7. It is seen that there are two flow circuits representing the electrical refrigeration mode and free cooling mode. Note that when river water temperatures are not sufficiently low to provide free cooling, the river water is used as the cooling medium for electrical refrigeration.

5.6.2.1 Chillers

The plant consists of seven electrical centrifugal chillers, each with a 3.75–9.5 MW cooling capacity corresponding to a 0.8–1.9-MW electric power consumption. The chillers are used to provide all of the plant cooling when the river water temperature is above 8°C. At river water temperatures between 5°C and 8°C, the chillers are used with partial free cooling from the river. Depending on the river water temperatures, annual electricity consumption of the plant is approximately 13 GWh. For a total annual COP of 4.2, approximately 55 GWh in annual sales are achieved.

5.6.2.2 Free cooling circuits

The electrical chillers are not used during winter when the river temperature drops below 5°C and 100 per cent of the cooling can be achieved using river water pumps and heat exchangers. According to data from 2003 to 2008, the river water temperatures were below 5°C for 6 per cent of the year and between 5°C and 8°C for 17 per cent of the year. Compared to options such as using cooling tower capacity, free cooling provides greater reductions in electricity consumption by the chillers. For environmental reasons, the highest rejected water temperatures cannot exceed 30°C, and the temperature difference between river water and the rejected water cannot exceed 5°C.

5.6.2.3 District network

Polyurethane- and glass-insulated pipes of 0.6 m diameter with polymer casings are used for both the supply and return lines in the district network. Further information on the district network connected to the Bercy plant is provided in section 5.6.1.3.

5.6.2.4 System benefits

The free cooling circuit has provided several benefits, including improved efficiency and reduced electricity consumption and emissions, as detailed below:

- **Efficiency**. Since the installation of the free cooling component in the Bercy plant, the average COP of the plant's chillers increased by 34 per cent, with maximum COP's of 20 having been achieved.
- **Emissions**. The use of free cooling in the plant has reduced CO_2 emissions by approximately 7,400 tonnes annually.
- **Economics**. Capital costs for the Bercy cooling plant were EUR 34 million, EUR 350,000 of which was spent on the free cooling part of the plant. A 15-year payback period is expected. In the Bercy plant, free cooling contributes to electrical savings as high as 400 MWh per month, typically in winter when the river temperature decreases to levels where free cooling becomes available. This is equivalent to almost 60 per cent of the average monthly electricity consumption during winter without free cooling. When the river temperature is higher and the plant is used for partial cooling, the savings decline to approximately 8 per cent of total energy consumption, according to plant operation data from Jan 2010 to March 2013.

5.7 Closure

Systems for cooling and district cooling are described, modeled and assessed, accounting for applications, types, classifications, operation, performance and efficiency. The relation between chillers and heat pumps is outlined. District cooling is shown sometimes to be able to meet the space cooling requirements of buildings in the residential, commercial and institutional sectors with advantages over conventional cooling systems.

Chapter 6

Integrated systems for cogeneration and district energy

Overview

Systems that integrate cogeneration and district energy are described. Included are descriptions of district energy (i.e., district heating and cooling), showing how it builds on and combines technologies for district heating and district cooling. Cogeneration-based district energy is defined and many facets of it are discussed, including possible variations of cogeneration-based district energy. Cogeneration-based district energy is compared with alternative and conventional systems for electrical and thermal energy, which constitute the main competing technologies. A general model for cogeneration-based district energy suitable for engineering and thermodynamic assessments is presented and illustrated and various operation modes for it identified. A detailed thermodynamic description of a combined cogeneration-based district energy system is given, along with corresponding energy and exergy balances and efficiencies. Alternative measures of system efficiency are also provided. The need for proper measures of merit for systems for cogeneration-based district energy is emphasized, especially when they include chiller systems, since useful, meaningful and logical measures are lacking. Finally, systems for integrated cogeneration-based district energy are described, by drawing on the material presented for both cogeneration and district energy. It is explained how district cooling systems using absorption chillers often complement district heating systems when both use heat supplied from a cogeneration plant because the demand for heat in a district heating system is lower in summer than in winter and heat-driven district cooling, which requires heat mainly in the summer, can help to balance the seasonal demands for cogeneration-derived heat.

6.1 Introduction

Some information is available on the integration of cogeneration and chiller systems, particularly for district cooling, but much more is needed to ensure that the systems achieve their full potential. The material presented in this section, especially the development of measures of efficiency, was developed by the present authors to provide useful, meaningful and logical measures.

Increased utilization of integrated cogeneration and district energy systems requires good experience with and knowledge of their characteristics and behaviors. In particular, facets related to the integration itself are needed, given the potential benefits it offers are not reaped if the integration is poorly done. Knowledge related to integration is different from information on each involved technology (heating, chilling, cogeneration, district heating and district cooling) considered independently, since the behaviors and characteristics of these separate technologies when combined to form integrated energy systems are often complex and sometimes confusing.

Significant implications of the integration of cogeneration with district energy include the fact that the efficiency of an integrated system is not necessarily maximized if the efficiencies of the components from which it comprised are each maximized. The counter-intuitive nature of this observation often causes it to be overlooked, but emphasizes the need to focus on it and explain it fully.

This implication alone highlights the need for appropriate integration of cogeneration and district energy in general and a sound understanding of the systems and factors affecting their integration. This need is especially valued when cooling and district cooling are involved, as their inclusion in integrated systems raises the complexity notably and increases the degree of counter-intuitiveness of some results.

By investigating the implications of various integration mechanisms for cogeneration and district energy technologies, understanding of cogeneration and district energy systems and their integration can be improved and their utilization expanded beneficially.

6.2 District energy

The use of district energy (i.e., district heating and cooling) for the provision of heating and cooling to a community is illustrated in Figure 6.1. There, independent processes are used to provide the heat and cool. The thermal products are used to satisfy all or some of the thermal energy needs of a community. In each thermal transport loop, the processed fluid is transported to the community after exiting its central plant and then returned after use in the community for reprocessing (i.e., reheating or recooling). Once-through systems exist in which the fluid is water and it is discarded after use, but these are increasingly uncommon due to the cost of losing of the transport fluid and the associated environmental impact of discarding it and extracting new water. The transmission of the heat transport fluids in district energy is typically over significant distances, from tens of meters to up to tens of kilometers.

6.3 Trigeneration and multigeneration

Cogeneration can be extended from processes that simultaneously provide two products, more often than not heat and electricity in which case it is also referred to as CHP, by adding the production of thermal energy in the form of “cool” as a third

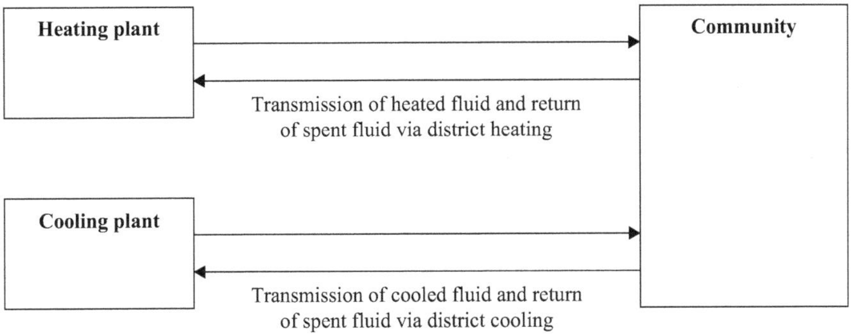

Figure 6.1 District energy (i.e., district heating and cooling) system based on independent processes for providing heat and cool. The thermal products are used to satisfy all or some of the energy needs of a community. The transport of the heated fluid and its return after use constitutes district heating, while the transport of the cooled fluid and its return after use constitutes district cooling. The heated and cooled fluids are usually returned after use for reheating and recooling, respectively, although once-through systems exist in which the fluid is water and it is discarded after use. The transmission of the thermal transport fluids in district heating and district cooling is typically over significant distances (up to tens of kilometers)

product. This type of extended cogeneration is often referred to as trigeneration CCHP.

Taking the concept of extended cogeneration one step further leads to the general concept of multigeneration or polygeneration, which involves the simultaneous production of multiple products in a single process or system. Multigeneration in principle includes both cogeneration and trigeneration, although it is also sometimes viewed as processes that produce four or more processes.

Trigeneration and multigeneration are covered throughout this book, often as an extension of the cogeneration concept.

6.4 Cogeneration-based district energy

The use of cogeneration-based district energy for the provision of electrical, heating and cooling services to a community is illustrated in Figure 6.2. This system can also be referred to as a trigeneration-based district energy. The electrical and thermal products are used to satisfy all or some of the energy needs of a community. In each thermal transport loop, the processed fluid is transported to the community after exiting the central cogeneration plant and then returned after use in the community for reprocessing (i.e., reheating or recooling). Again, it is noted that once-through systems exist in which the fluid is water and it is discarded after use, but these are increasingly uncommon due to the cost of losing of the transport

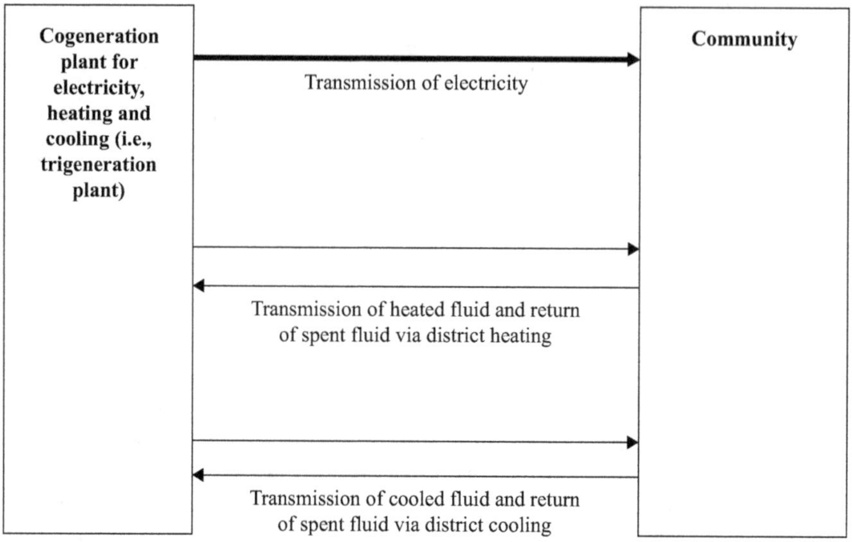

Figure 6.2 Cogeneration-based district energy where thermal products include both heat and cool, i.e., trigeneration-based district energy. The electrical and thermal products are used to satisfy all or some of the energy needs of a community. The transport of the heated fluid and its return after use constitutes district heating, whereas the transport of the cooled fluid and its return after use constitutes district cooling. Together, the two thermal transmission loops constitute district energy. The heated and cooled fluids are usually returned after use for reheating and recooling, respectively, although once-through systems exist in which the fluid is water and it is discarded after use. The transmission of the thermal transport fluids in district heating and district cooling is typically over significant distances (up to tens of kilometers). Heavy lines denote electricity flows, and light lines material flows for the transmission of thermal energy

fluid and the associated environmental impact of discarding it and extracting new water. The transmission of the heat transport fluids in district energy is typically over significant distances, from tens of meters to up to tens of kilometers.

6.4.1 Variations of cogeneration-based district energy

The general cogeneration-based district energy in Figure 6.2 can be simplified to provide only one thermal product, leading to cogeneration-based district heating (Figure 6.3) and cogeneration-based district cooling (Figure 6.4). In both cases, the transmission of the thermal product in district heating or cooling is typically over significant distances, from tens of meters to up to tens of kilometers.

The cogeneration-based district heating system in Figure 6.3 simply omits the cooling plant and district cooling infrastructure found in a cogeneration-based

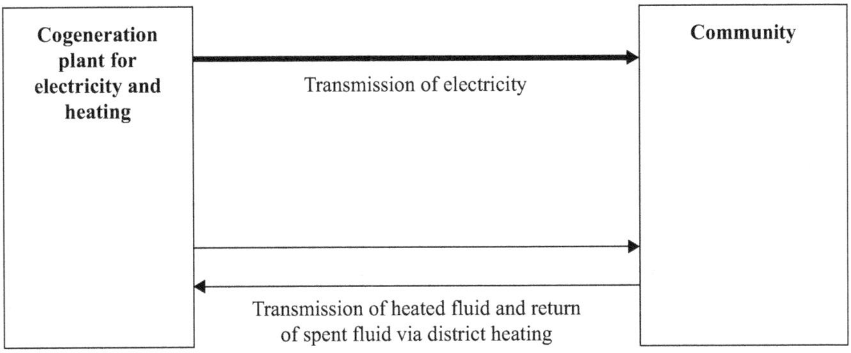

Figure 6.3 *Cogeneration-based district heating. The electrical and thermal products are used to satisfy all or some of the energy needs of a community. District heating usually encompasses the transport of the heated fluid from the cogeneration plant and its return for reheating after use in the community, although once-through systems exist in which the fluid is water and it is discarded after use. The transmission of the thermal transport fluids in district heating is typically over significant distances (up to tens of kilometers). Heavy lines denote electricity flows, and light lines material flows for the transmission of thermal energy*

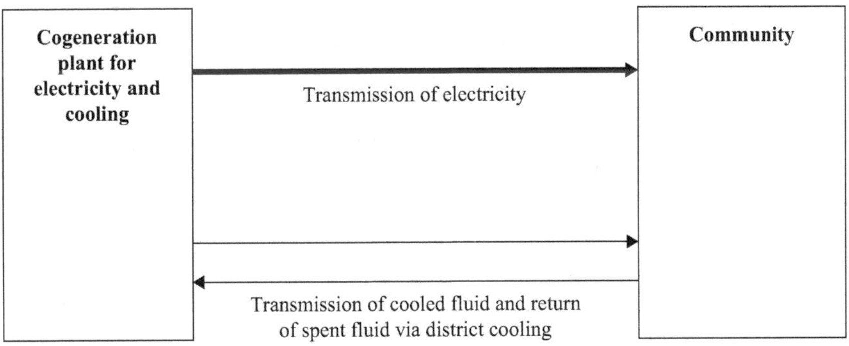

Figure 6.4 *Cogeneration-based district cooling. The cooled fluid is obtained using the heat from cogeneration to drive an absorption chiller. The electrical and thermal products are used to satisfy all or some of the energy needs of a community. District cooling usually encompasses the transport of the cooled fluid from the cogeneration plant and its return for recooling after use in the community, although once-through systems exist in which the fluid is water and it is discarded after use. The transmission of the thermal transport fluids in district cooling is typically over significant distances (up to tens of kilometers). Heavy lines denote electricity flows, and light lines material flows for the transmission of thermal energy*

district energy system, and correspondingly provides some or all of the electrical and heating needs of a community. In the heat transport loop, the processed fluid is transported to the community after exiting the central cogeneration plant and then returned after use for reheating, or it is discarded after use in a once-through mode of operation.

The cogeneration-based district cooling system in Figure 6.4 provides some or all of the electrical and cooling needs of a community. It omits district heating infrastructure found in a cogeneration-based district energy system and instead uses the heat from cogeneration to drive an absorption chiller the produces the cooled fluid. In the cool transport loop, the processed fluid is transported to the community after exiting the central plant and then returned after use for recooling, or it is discarded after use in a once-through mode of operation.

6.4.2 Comparison of cogeneration-based district energy with alternative and conventional systems

It is helpful to compare and contrast the general cogeneration-based district energy in Figure 6.2 with alternative and conventional systems for providing the same services (electrical, heating and cooling).

A system with independent processes for providing electricity, heat and cool to a community, but still using district energy (i.e., district heating and district cooling) to transport the two thermal products, is shown in Figure 6.5. The electrical and thermal products are used to satisfy all or some of the energy needs of a community. The heated and cooled fluids are usually returned after use for reheating and recooling, respectively, although once-through systems exist in which the fluid is water and it is discarded after use. The transmission of the thermal transport fluids in district heating and district cooling is typically over significant distances (up to tens of kilometers).

A more conventional approach for providing electrical, heating and cooling services to a community is shown in Figure 6.6, where multiple independent electrical, heating and cooling plants are located within the community. Commonly, each building has its own heating and cooling plant. Local distribution of the electrical product, but not lengthy transmission, is necessary in this system. But it is noted that some buildings can have electrical generating capacity, for instance by using solar photovoltaics, wind generation or a micropower system.

An alternative of this more conventional approach for providing electrical, heating and cooling services to a community is shown in Figure 6.7, where single independent electrical, heating and cooling plants are located within the community. Local distribution of the electrical and thermal products, but not lengthy transmission, is likely necessary in this system.

In all cases considered in this subsection, the same electrical, heating and cooling services are provided to the community, but the efficiencies, costs, environmental impacts and other performance parameters vary. Typically, the use of a single production plant is more efficient than the use of separate plants for electricity, heat and cool.

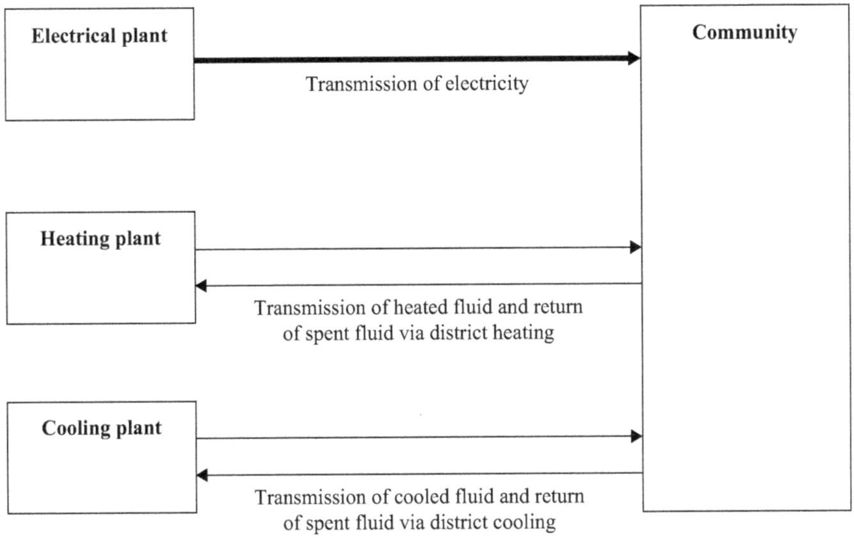

Figure 6.5 *Independent processes for providing electricity, heat and cool to a community, using district heating and district cooling to transport the latter two products. The electrical and thermal products are used to satisfy all or some of the energy needs of a community. The transport of the heated fluid and its return after use constitutes district heating, whereas the transport of the cooled fluid and its return after use constitutes district cooling. Together, the two heat transmission loops constitute district energy. The heated and cooled fluids are usually returned after use for reheating and recooling, respectively, although once-through systems exist in which the fluid is water and it is discarded after use. The transmission of the thermal transport fluids in district heating and district cooling is typically over significant distances (up to tens of kilometers). Heavy lines denote electricity flows, and light lines material flows for the transmission of thermal energy*

6.5 General system model and operation modes

Figure 6.8 illustrates an integrated system for cogeneration and district energy. The system combines the general cogeneration unit from Figure 3.1 and the general chiller from Figure 5.1.

During operation, fuel drives the cogeneration unit. Its electrical and/or thermal energy products drive the chiller. Additional energy may be externally imported to the chiller in the form of electrical power ($\dot{W}_{im}$) and in the form of thermal power ($\dot{Q}_{im}$) at a temperature T_{im}. Also, the cogeneration unit may produce for export from the system additional electrical power ($\dot{W}_{ex}$) and thermal power ($\dot{Q}_{ex}$) at a temperature T_{ex}. Wastes can come from both the cogeneration and chiller systems.

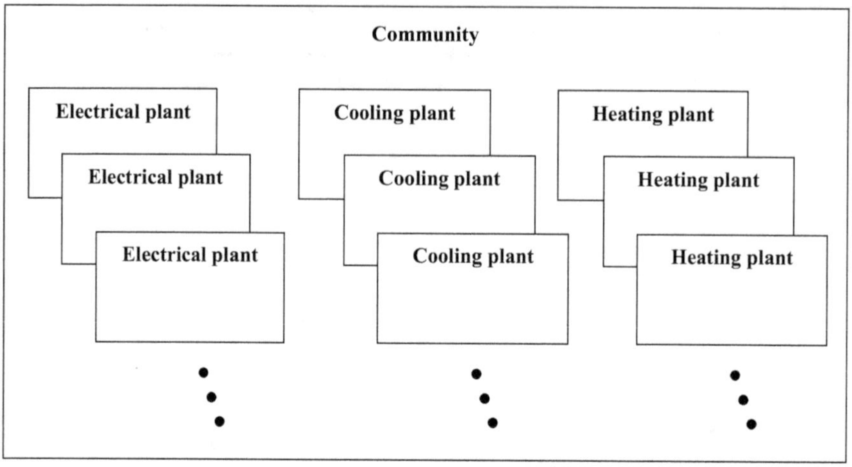

Figure 6.6 Provision of electricity, heat and cool to a community using multiple independent plants within the community to satisfy its energy needs. Commonly each building has its own heating and cooling plant, and some buildings also have electrical generating capacity

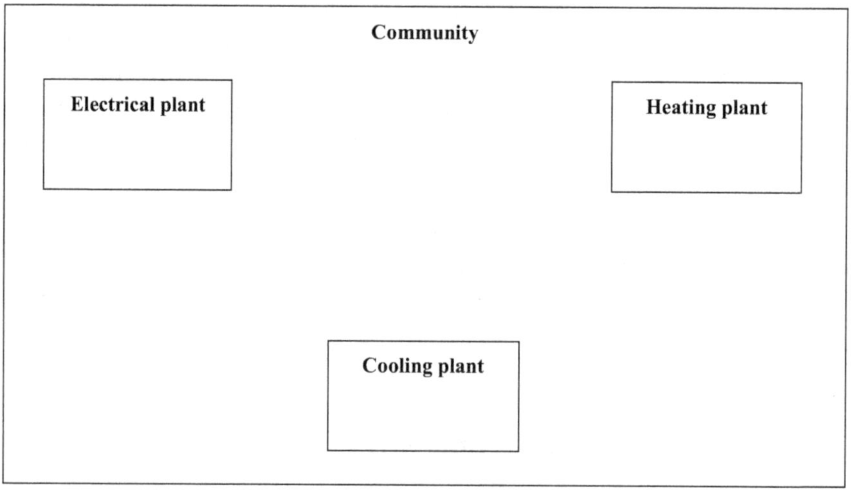

Figure 6.7 Provision of electricity, heat and cool to a community using independent plants within the community to satisfy its energy needs

Several interesting cases for system operation are noted:

- When all import and export terms are zero, the combined system represents an engine-driven chiller. In this case, the thermal and/or electrical outputs of the engine exactly match the requirements of the chiller. Such a system is illustrated in Figure 6.9. In this subcase, it can be seen that the system is sized to

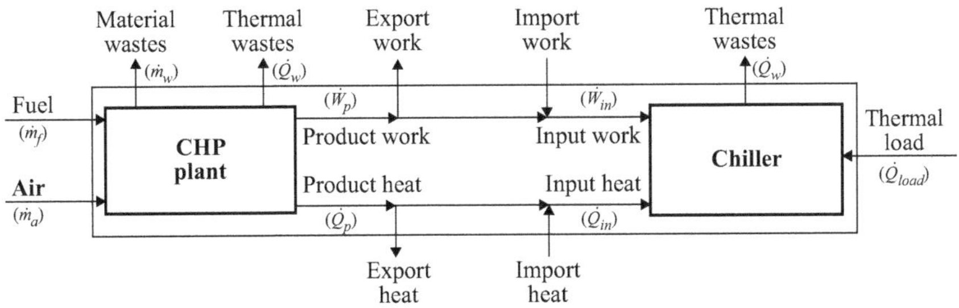

Figure 6.8 General model for a combined cogeneration-based district energy system

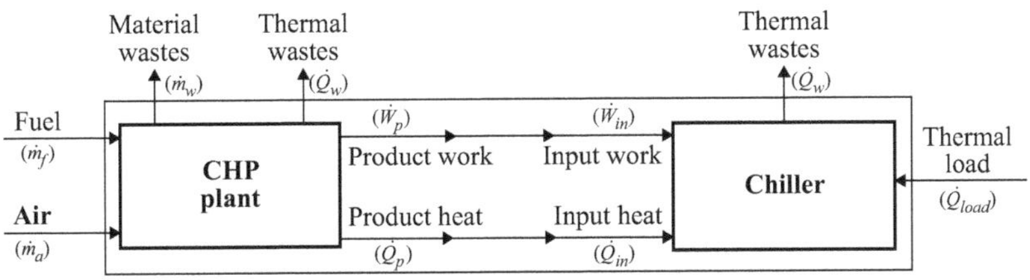

Figure 6.9 Model for a combined cogeneration-based chiller system when the cogeneration unit is sized to meet work and thermal needs of the chiller

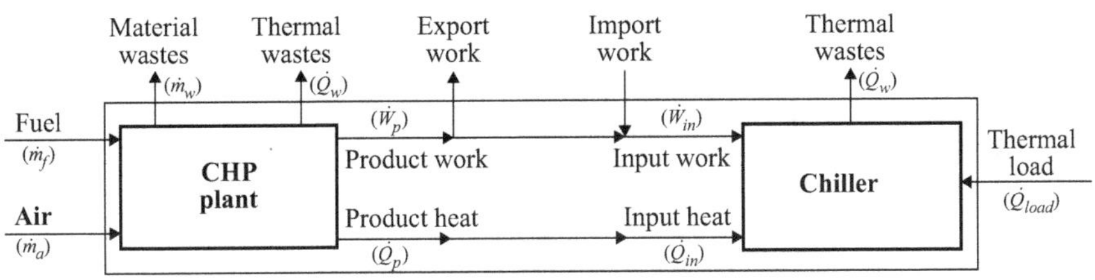

Figure 6.10 Model for a combined cogeneration-based chiller system when the cogeneration unit is sized to meet only thermal needs of the chiller

meet work and heat needs precisely, so the cogeneration or CHP plant is selected or designed with the exact ratio of work to product heat output required. Thus, there can be neither a surplus nor a deficiency of work or of heat in this system.

- When the cogeneration unit is designed to match the thermal needs of the chiller, all heat import and export terms are zero. Such a system is illustrated in Figure 6.10. There, it can be seen that with the system sized to meet thermal needs, since the cogeneration plant has a fixed ratio of work to product heat output, there can only be a surplus or deficiency of work.

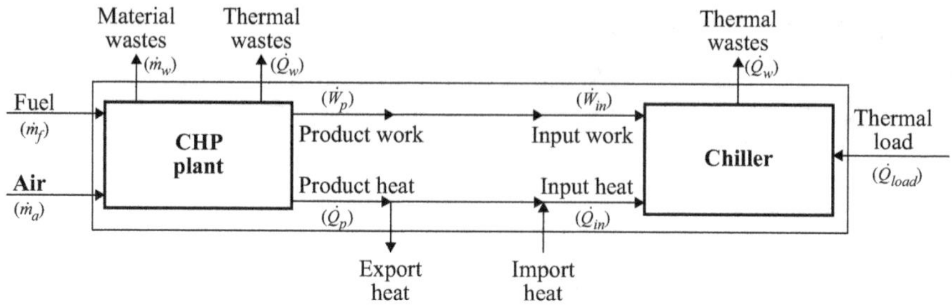

Figure 6.11 Model for a combined cogeneration-based chiller system when the cogeneration unit is sized to meet only work needs of the chiller

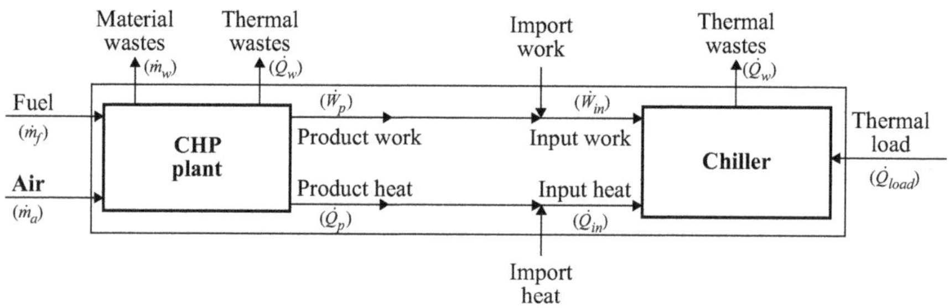

Figure 6.12 Model for a combined cogeneration-based chiller system when the cogeneration unit does not meet thermal and work needs of the chiller

- When the cogeneration unit is designed to match the work or electrical needs of the chiller, all work import and export terms are zero. Such a system is illustrated in Figure 6.11. There, it can be seen that with the system sized to meet work needs, since the cogeneration plant has a fixed ratio of work to product heat output, there can only be a surplus or deficiency of heat.
- When the cogeneration plant is so small relative to the chiller that it produces less product heat and electricity than is required by the chiller, all work and heat export terms are zero. Such a system is illustrated in Figure 6.12.
- When the cogeneration plant is so large relative to the chiller that it produces more product heat and electricity than required by the chiller, all work and heat import terms are zero. Such a system is illustrated in Figure 6.13.

In the last case, the system is referred to as a trigeneration or CCHP system irrespective of the cooling equipment or the way the cooling equipment is linked to the cogeneration system [193].

The first three cases in the immediately above set of bullets address the different ways a cogeneration plant can be linked to or follow the load of a chiller.

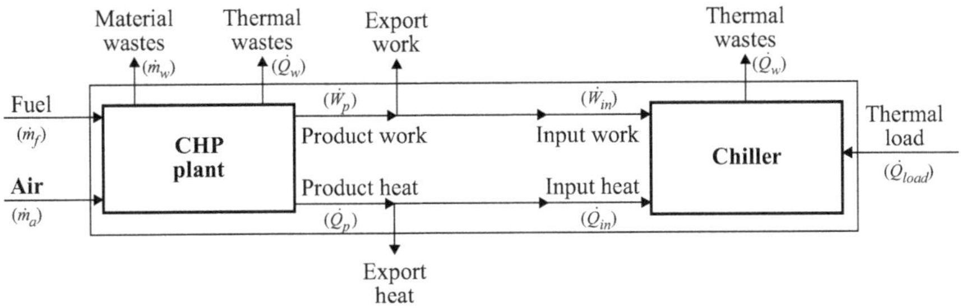

Figure 6.13 Model for a combined cogeneration-based chiller system when the cogeneration unit produces more thermal and work products than needed by the chiller

In the first case, the cogeneration plant exactly follows the chiller loads, both work and heat. The cogeneration plant follows the thermal load of the chiller in the second case, and the thermal load of the chiller in the third case.

The last two cases address what happens when the scales or sizes of the cogeneration and chiller plants are mismatched. In the fourth case, the chiller plant is relatively very small, so the cogeneration plant can easily meet all of its work and heat needs and still have an excess of these commodities to export from the system. In the fifth case, the chiller plant is relatively very large, so the cogeneration plant cannot satisfy totally any of its work or heat needs, so these commodities must be imported to the system.

The cases described above are applicable when cooling is produced using the thermal or mechanical power from cogeneration plant. When heat is employed, the integration usually involves the use of an absorption chiller with a cogeneration plant. This approach can lead to an enhanced efficiency [194,195].

6.6 Description of a combined cogeneration/district energy system

A thermodynamic analysis is performed for a general combined cogeneration and district energy system. The thermodynamic analysis is similar to that in section 3.5 for a general cogeneration system, in section 4.3 for a general heating system, and in section 5.3 for a general chiller. Similar assumptions are applied.

6.6.1 Energy and exergy balances

An energy balance for the system in Figure 6.8 can be written as:

$$\dot{m}_f h_f + \dot{m}_a h_a + \dot{W}_{im} + \dot{Q}_{im} + \dot{Q}_{load} = \dot{W}_{ex} + \dot{Q}_{ex} + \sum_w \dot{Q}_w + \sum_w \dot{m}_w h_w \quad (6.1)$$

The summations are over all material and thermal waste streams for the combined system. The terms on the left-hand side of (6.1) represent the total energy input rate, and are associated in order with (i) the fuel and air supplied to the cogeneration unit, (ii) the supplementary work and heat supplied to the chiller and (iii) the cooling load. The terms on the right-hand side represent the energy output rate associated with excess products (electrical and thermal) from the cogeneration unit, and the energy output rate associated with wastes (thermal and material).

A corresponding exergy balance can be written as:

$$\dot{m}_f\varepsilon_f + \dot{m}_a\varepsilon_a + \dot{W}_{im} + \dot{Q}_{im}\tau_{im} + \dot{Q}_{load}\tau_{load} = \dot{W}_{ex} + \dot{Q}_{ex}\tau_{ex} + \sum_w \dot{Q}_w\tau_w + \sum_w \dot{m}_w\varepsilon_w + \dot{I} \tag{6.2}$$

The terms on the left-hand side of (6.2) represent the exergy input rates with fuel, air, import work and heat, and the heat load, while, on the right-hand side, the first two terms represent the exergy output rates associated with export electrical and thermal energy products, the second two terms represent the exergy output rates associated with thermal and material energy wastes, and the final term represents the exergy consumption rate.

6.6.2 *Efficiencies*

Following (2.19) and (2.20), an energy-based figure of merit, *f*, for the integrated cogeneration/chiller system in Figure 6.8 can be written as:

$$f_{sys} = \frac{(\dot{Q}_{load}) + (\dot{Q}_{ex} + \dot{W}_{ex,net})}{(\dot{m}_f h_f) + (\dot{Q}_{in} + \dot{W}_{im,net})} \tag{6.3}$$

where the network term that is negative is set to zero in this expression. The network terms are defined as follows:

$$\dot{W}_{ex,net} = \dot{W}_{ex} - \dot{W}_{im} \tag{6.4}$$

and

$$\dot{W}_{im,net} = \dot{W}_{im} - \dot{W}_{ex} = -\dot{W}_{ex,net} \tag{6.5}$$

It has been presumed here that heat will not be simultaneously imported and exported at the same temperature, although imports and exports of heat at different temperatures have been accounted for in (6.3). The numerator in (6.3) includes the products for the chiller (chilling effect) and the cogeneration unit (export heat and network). The denominator includes the external driving inputs for the cogeneration unit (fuel) and the chiller (import heat and network). An energy-based "figure of merit" rather than energy efficiency is used because the latter is too restrictive. The figure of merit must be sufficiently general that it applies to a system which, in the extremes, can be a cogeneration unit or a chiller. The figure f_{sys} can take on numerical values between 0 and infinity.

A corresponding exergy-based figure of merit, which is simply an exergy efficiency (and thus varies between zero and unity), can be written as:

$$\psi_{sys} = \frac{(-\dot{Q}_{load}\tau_{load}) + (\dot{Q}_{ex}\tau_{ex} + \dot{W}_{ex,net})}{(\dot{m}_f\varepsilon_f) + (\dot{Q}_{im}\tau_{im} + \dot{W}_{im,net})} \tag{6.6}$$

where again the network term that is negative is set to zero in this expression.

It is noted that (i) when the combined system contains no chiller (i.e., it is just a cogeneration unit), the measures of efficiency in (6.3) and (6.6) reduce to those in (3.5) and (3.6) for a cogeneration plant and (ii) when the combined system contains no cogeneration unit (i.e., it is just a chiller), the measures in (6.3) and (6.6) reduce to those in (5.3) and (5.4) for a chiller. Whether the measures in (6.3) and (6.6) behave more like those for a chiller or a cogeneration unit depend on which device in Figure 6.8 is dominant.

It is also noted that the condenser temperature in the chiller is an important quantity to consider in optimizing the design of integrated cogeneration/chiller systems. A lower condenser temperature is desirable and can be achieved if a water body is available for cooling in the condenser, so that air cooling is not required. For integrated systems which form part of district cooling networks, it can be very beneficial to use central chillers and district cooling grids, rather than producing the cooling at the user site (using electricity and/or heat) because

- the seasonal efficiency of the chiller can increase due to the ability to operate at peak efficiency more often in a central large plant and
- lower chiller condenser temperatures (e.g., 20°C) can be used if cooling water is available to the central plant, relative to the condenser temperatures of approximately 35°C needed for air-cooled building chillers.

These two effects can lead to central large chillers having double the efficiencies of distributed small chillers.

6.6.3 Alternative measure of system efficiency

An alternative method of assessing the efficiency of an integrated system is now considered, which is based on and extends the concepts introduced in section 3.5.3 regarding COP_{CHP} values for cogeneration plants. Consider the case of an integrated system in Figure 6.8 in which there are no import or export heat terms, and no import work terms. The product heat from the cogeneration plant matches that needed by the chiller. For this integrated system, a coefficient of performance, COP_{sys} can be defined as:

$$COP_{sys} = \frac{\text{Cooling effect}}{\text{Export electricity curtailed}} = \frac{\dot{Q}_{load}}{(\dot{W}_{ex})_c} \tag{6.7}$$

where the denominator is the difference between $(\dot{W}_p)_{elec}$, the export work from the cogeneration plant (in the all-electric mode) before a chiller is attached, and $\dot{W}_{ex}$, the export work after a chiller is attached and the cogeneration plant is modified so that it satisfies the heat requirement of the chiller, if any. That is:

$$(\dot{W}_{ex})_c = (\dot{W}_p)_{elec} - \dot{W}_{ex} \tag{6.8}$$

The term COP_{sys} is analogous to a refrigeration COP.

When the chiller is electrically driven, the cogeneration plant is simply an electrical generating station. When the chiller is a heat-driven absorption machine, the cogeneration plant cogenerates the heat required by the chiller (in quantity and quality) as well as electricity.

For electrically driven chillers, the export electricity curtailed is simply the amount of electricity used by the chiller, i.e., $(\dot{W}_{ex})_c = \dot{W}_i$. Thus:

$$COP_{sys} = COP_{ch} \text{ (for electrically driven chillers)} \tag{6.9}$$

For heat-driven absorption chillers, the export electricity curtailed is the electrical output of the cogeneration plant in the all-electric mode that is curtailed in producing the heat needed by the absorption chiller, i.e., $(\dot{W}_{ex})_c = \dot{W}_c$. The small amount of electricity required to operate pumps in the absorption chiller is neglected here, but can straightforwardly be accounted for. Then:

$$COP_{sys} = \frac{\dot{Q}_{load}}{\dot{W}_c} \text{ (for heat-driven absorption chillers)} \tag{6.10}$$

By noting that the product heat from the cogeneration plant is the input heat to the absorption chiller (i.e., $\dot{Q}_i = \dot{Q}_p$) and recalling the COP definitions in (3.10) and (5.3) (with input electricity neglected with respect to input heat, as is approximately true for absorption chillers), the COP_{sys} expression above can be written as:

$$COP_{sys} = \frac{\dot{Q}_p}{\dot{W}_c}\frac{\dot{Q}_{load}}{\dot{Q}_i}$$

$$= (COP_{CHP})(COP_{ch}) \text{ (for heat-driven absorption chillers)} \tag{6.11}$$

It is interesting to consider the case where the cogeneration unit is a cogenerating steam power plant. In that case, the plot of COP_{CHP} vs. temperature in Figure 3.6 can be considered, as can the plot of COP_{ch} (for absorption chillers) vs. temperature in Figure 5.4. Using the information in these plots and (6.11), COP_{sys} is plotted in Figure 6.14. The trade-off between efficiencies of the cogeneration and chiller plants is illustrated in Figure 6.14. As the temperature at which heat is required by the chiller (and produced by the cogeneration plant) increases, (i) the efficiency of the absorption chiller, as measured by its COP_{ch}, increases and (ii) the efficiency of the cogeneration plant, as measured by COP_{CHP}, decreases. For the data used to develop Figure 6.14, COP_{sys} increases slightly as the temperature at which steam is extracted from the cogeneration plant increases. In general, however, depending on which effect is predominant, the efficiency of the integrated

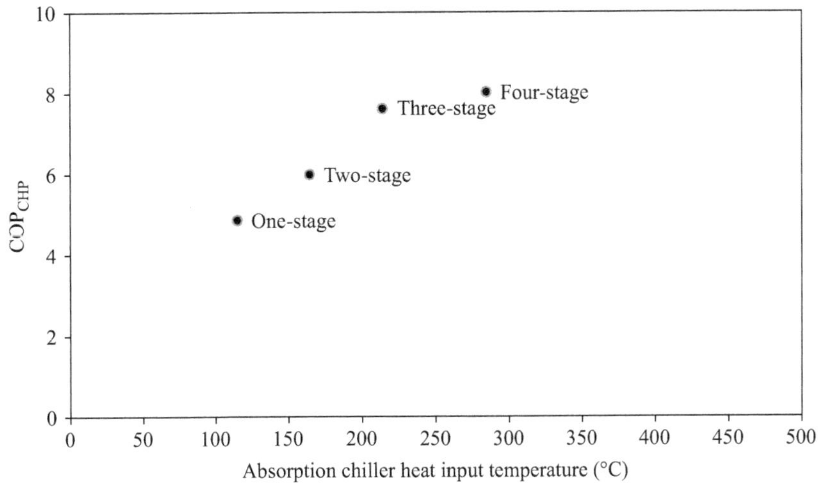

Note: For electric chillers, $COP_{sys} = COP_{ch} \approx 4.5$

Figure 6.14 Variation of system COP with absorption-chiller heat-input temperature, for some systems incorporating steam turbine-based cogeneration and absorption chillers having various stages, and comparison with COP_{sys} for similar systems incorporating electric chillers. Based on data in Figures 3.6b and 5.4

system, as measured by COP_{sys} can increase, decrease or remain constant as the temperature (and pressure) at which steam is extracted from the cogeneration plant increases.

This trade-off is less visible for systems where little electrical output is curtailed as useful product heat is produced (e.g., for gas turbine and diesel engines). For these, COP_{sys} is generally high (approaching infinity where the electrical output curtailed approaches zero as useful heat is produced from the cogeneration plant).

It is useful to compare the values of COP_{sys} in Figure 6.14 for absorption chillers driven by cogeneration-derived heat, with the values of COP_{sys} for electrically driven chillers. In general, the values of COP_{sys} for electrical chillers (for which COP_{ch} ranges from approximately 3 to 6) are similar to those for one- and two-stage absorption chillers, for which COP_{ch} ranges from 0.6 to 1.8 (see Figure 5.4). Although at first one might expect that the electrically driven chiller system would be more efficient, since COP_{ch} values for such chillers are several times greater than those for absorption chillers, the efficiency of the overall system when absorption chillers are used is increased by the high values of COP_{CHP} for cogeneration plants.

In general, the value of COP_{sys} exhibits a tendency to increase slightly with heat-supply temperature or to remain constant, for systems using absorption chillers (of any number of effects) or electrical chillers.

6.7 Systems for integrated cogeneration/district energy

Most of the cogeneration and district energy systems described in sections 3.6 and 5.5 utilize cogeneration and heat-driven absorption chillers and/or electrically driven chillers. Those descriptions of these systems are adequate so they are not repeated here.

But it is pointed out that the cogeneration systems include both topping and bottoming cycles, and that the most common cogeneration systems include the following: steam-turbine cogeneration systems, gas-turbine cogeneration systems, combined-cycle cogeneration systems, reciprocating engine-based cogeneration systems and others. Systems used for cooling mainly include vapor-compression chillers, which are driven by work or electricity, and absorption chiller systems, which are mainly driven by heat.

6.8 Case study: cogeneration for district heating in Houston, Texas, USA

A large-scale cogeneration system is operated by Thermal Energy Corp. (TECO) at the Texas Medical Center in Houston, Texas, to provide thermal services. The medical center consists of 49 member institutions including hospitals, academic institutions, medical and nursing schools. The system includes a cogeneration system, a chilled water system, an energy storage system and a district energy network. The main system components are described in the following subsections.

6.8.1 Cogeneration system

A high-efficiency gas turbine uses natural gas to produce 48 MW of electrical power. In summer, heat from the turbine exhaust gases is captured via eight heat recovery steam generators to produce 56,700 kg/hr of steam. The heat recovery steam generators are sized for the low heating loads in the summer. In the winter, natural gas duct burners are used to produce 150,000 kg/hr of steam [196] increasing the overall cogeneration efficiency by 15 per cent. Future expansions of the plant are expected to include an increased power generation capacity of 100 MW and steam production capacity of 426,000 kg/hr [196].

In hot and humid periods, chilled water from supply or return lines is used to cool turbine inlet air to enhance power production. The cogeneration plant can operate to serve 100 per cent of its own peak electrical load used to power equipment and 100 per cent of the customers' peak process and space heating loads. The customers use electricity from the grid and operate their own independent emergency generators, although the electricity from the cogeneration plant can be used for emergency backup. Since they have no on-site chillers (see the next section), their need for emergency generator capacity is reduced by 50 per cent, thus reducing the electric distribution load.

Table 6.1 Details of chilled water and steam distribution system for the district energy system in Houston, Texas[1]

	Chilled water	Steam
Piping diameter range (cm)	15–152	5–41
System pressure (kPa)	450	1,720
Supply temperature (°C)	4.4	232.2
Return temperature (°C)	11.7	65.6

[1]Data source: Reference 196.

6.8.2 Chilled water system

A total of 27 centrifugal chillers provide approximately 120,000 tons of chilled water capacity. They are used at night during periods of low electricity rates to cool the return water from the network for cool water supply during the next day. Additional chillers are planned to be installed to increase the cooling capacity to 168,000 tons [196].

6.8.3 Energy storage

A 75,000 ton-hour (8.8 million gallon) stratified TES tank is installed in the plant. The TES allows for increased peak chilled water capacity in a cost-effective manner by producing chilled water via centrifugal chillers at low cost, generally at night, from the grid, storing it and using it for cooling purposes when it is needed. When cooling is needed at peak hours, the power for the chillers is supplied via the cogeneration system. Future expansions of the plant are expected to include an increased chilled water storage of up to 152,000 ton-hours [196].

6.8.4 District energy network

Chilled water and steam are distributed through approximately 56 km of underground piping [197] among 162 buildings in almost 1,000 acres of land. Details of the chilled water and steam distribution system are included in Table 6.1. Chilled water annual energy sales are approximately 267,750,000 ton-hour, making it the largest district cooling system in the U.S. [196].

Before investing in cogeneration, TECO previously operated a district energy network that was coupled to electric chillers, gas-driven steam generators and emergency generators for backup. Only about 30 per cent of the input energy was converted to electricity and the remainder was wasted. Installation of the cogeneration district energy plant to improve efficiency started in 2007 and continued in stages until the first phase of the system started operating in 2010.

6.8.5 System benefits

The cogeneration system has yielded numerous benefits:

- **Efficiency**. The plant has increased overall fuel conversion efficiency from 42 per cent to a minimum of 68 per cent often as high as 80 per cent [198], mainly due to the use of cogeneration and TES.

Table 6.2 Emissions reductions from use of cogeneration and energy storage in the TECO plant

Constituent	Annual emission reduction (tons)
NO_x	302,000
CO_2	305,000
Carbon	83,000

Data source: Reference 198.

- **Emissions**. As a result of the increased fuel conversion efficiency and use of TES, approximately 30–61 per cent less fossil fuel is used [197,199], compared to that used for electricity supplied by the grid and conventional steam production, resulting in reduced carbon dioxide emissions (see Table 6.2) [198]. The emission reductions are the equivalent of avoiding the use of 52,000 fossil fuel-driven cars or planting 83,000 acres of new forest annually.
- **Economics**. The increased system efficiency and system additions such as TES have resulted in an annual average financial savings of $4,345,000 from 2011 to 2015 [197]. Savings are expected to increase to more than $200 million for the next 15 year period. As a result, customer rates declined by 2.0 and 1.4 per cent in 2010 and 2011, respectively. TECO operates to generate heating and cooling for the medical center independent of varied and often high-grid electricity costs. The expansion of the plant in stages enables system additions as demand increases, contributing to long-term savings of the plant.
- **Reliability**. TECO aims to generate 100 per cent of the needed power in the medical center and does not depend on electricity from the grid. In the event of a prolonged grid outage, the cogeneration system is able to provide uninterrupted energy services to the customers. This is especially important in hospital operations during emergency incidents to ensure that essential emergency response can be provided. Also, when there is a high demand on the electrical grid, such as the record high demand of 66,000 MW in August 2011, the cogeneration system installed at the Texas medical center can operate independent of the grid.
- **Flexibility**. The plant construction was planned for completion in stages allowing for operational flexibility. The use of TES within the plant operation allows more flexibility in use of electricity to run chillers for cooling.

6.8.6 Operational performance

The operational performance of TECO cogeneration-based district energy system was monitored from July 2011 to September 2011, after Phase 1 completion. While the plant had been operational only for a few months, early assessments demonstrated that the new facilities were operating according to the design. During the 967 hrs of operation in this three-month period, 111 GWh of natural gas were used to produce 40,442 MWh of electricity and 39 GWh of heat in the form of steam.

To produce the same amount of electricity and heat conventionally (i.e., with a utility plant and a natural gas central package boiler) with a plant efficiency of 38.5 per cent, 393 GWh of natural gas would have been used [199]. During this period, CO_2 emission declined 52 per cent to 25,007 tons.

6.9 Closure

Systems that integrate cogeneration and district energy are described, and contrasted with alternative and conventional systems for electrical and thermal energy. A general model for cogeneration-based district energy is presented and various operation modes are illustrated. The need for proper measures of merit for systems for cogeneration-based district energy is emphasized, especially when they include chiller systems. It is explained how district cooling systems using absorption chillers often complement district heating systems when both use heat supplied from a cogeneration plant, by balancing the seasonal demands.

Chapter 7

Comparison of systems for integrated cogeneration and district energy

Overview

Systems for integrated cogeneration and district energy are compared, pointing out similarities and differences. The cases considered in the comparison exercise are carefully described, including the specification of efficiency measures. Performance and efficiency measures are defined and discussed for the chillers, for cogeneration and heating, and for integrated cogeneration-based district energy. The comparison of several systems for integrated cogeneration and district energy demonstrates the merits of different integrated systems for cogeneration and district energy, and also clarifies the separate technologies involved and how they operate in tandem in mutually beneficial manners. The comparison highlights the need for proper measures of merit for systems for cogeneration-based district energy, particularly when they include cooling, and the way in which district cooling systems using absorption chillers can complement district heating systems when both use heat supplied from a cogeneration plant because the demand for heat become more balanced seasonally. Comparative assessments are demonstrated to be both useful and necessary, given the breadth of technologies available for cogeneration, heating, cooling, district heating and district cooling, particularly as a way of assisting designers in selecting parameter values and system configurations, and in helping decision-makers choose among competing options. The comparisons in this chapter also serve to illustrate the technologies that are involved in cogeneration-based district energy, and to help demonstrate the merits of different integrated cogeneration and district energy systems by providing fair comparative assessments.

7.1 Introduction

This chapter is intended to help demonstrate the merits of different integrated cogeneration and district energy systems, by providing a comparison of several such systems. Such a comparison is useful and necessary, given the breadth of technologies available for cogeneration, heating, cooling, district heating and district cooling. Comparisons are particularly important for helping designers select parameter values and system configurations, and for helping decision-makers choose among competing options.

The chapter also serves to illustrate the material discussed previously in this book, as it encompasses many of the technologies and systems already discussed as well as ways to integrate them.

7.2 Descriptions of cases considered

In this comparison, integrated cogeneration/chiller systems are considered which are made up of all possible combinations of the following four cogeneration options:

- diesel,
- gas turbine,
- steam turbine and
- combined cycle (i.e., a cogeneration plant integrating gas turbine and steam turbine power cycles);

and the following three chiller options:

- electric,
- single-stage absorption and
- two-stage absorption.

For each of the 12 resulting combinations, 3 modes of operation are considered, in which the integrated system is designed for:

(a) cooling,
(b) cooling and heating (with the characteristics of the export heat identical to those of the heat used to drive the absorption chillers) and
(c) cooling and heating (with the export heat at a relatively low temperature, such as that applicable for low-temperature district heating).

The three modes of operation are intended to allow a wide range of possible operating scenarios to be considered, while not including so many that the comparisons become too confusing. In each mode of operation, excess electricity, when it is generated, is exported from the system as a byproduct. For systems involving an electrical chiller:

- the cogeneration system operates in an all-electric mode (i.e., as an electrical generating station),
- there is, consequently, no export electricity, as the cogeneration system is sized to meet the electrical needs of the chiller and
- the last two of the three modes of operation above (i.e., the ones involving heating) are not applicable as the cogeneration plant produces no thermal product.

Consequently, 12 cases are considered for operation mode (a) identified above, and 8 cases for each of modes of operation (b) and (c) (corresponding to the eight combinations involving absorption chillers), for a total of 28 cases.

In the comparison, the cooling load is set for all cases at 100 units and the export heat, where applicable, at 100 units. Efficiencies and characteristics of the cogeneration devices when operating at full load, as reported in a report [190] and summarized in Table 7.1, are used. The efficiencies in Table 7.1 are based on full-load conditions and lower heating values of fuels. Table 7.1 includes, for each cogeneration option, efficiencies for all-electric operation, and electrical and total (accounting for electrical and thermal products) efficiencies for each of the three product heat levels of interest (i.e., for one- and two-stage absorption chillers, and for district heating). The chillers are assumed to produce chilled fluid at 6.5°C. The electrical chiller is assumed to have a COP_{ch} of 4.5, while the one- and two-stage absorption units, respectively, are assumed to have COP_{ch} values of 0.67 and 1.2, and to require steam at 120°C/2 bar and 170°C/8 bar. Note that the efficiency in the electric mode of operation is the technology's electrical generation efficiency and that the cogeneration total efficiency is the sum of the cogeneration electrical and heat efficiencies.

7.3 Efficiency measures for cases considered

For each case considered, Table 7.2 presents the fuel energy used, the export electricity generated, and, for the chiller, the cogeneration plant and the integrated system, respectively, energy-based measures of efficiency (COP_{ch}, η_{CHP}, f_{sys}) and exergy-based measures of efficiency (ψ_{ch}, ψ_{CHP}, ψ_{sys}). For those cases producing only cooling and, where appropriate, byproduct electricity, Table 7.3 presents for the chiller, the cogeneration plant and the integrated system, respectively, alternative measures of efficiency using coefficients of performance (COP_{ch}, COP_{CHP}, COP_{sys}) and exergy equivalents of these energy-based measures. For all cases considered, Table 7.4 presents coefficients of performance for the cogeneration portions of the integrated systems, and exergy equivalents of these energy-based efficiency measures.

Discussed below in three parts are those aspects of the analyses relevant to (i) the chiller portions of the integrated systems, (ii) the cogeneration portions of the integrated systems and (iii) the integrated cogeneration/district energy systems.

7.4 Performance and efficiency of chillers

The chiller COP values in Tables 7.2 and 7.3 range from 0.67 to 4.5, and are specified as input data. These energy-based efficiency measures indicate that electrical chillers are approximately four times more efficient than two-stage absorption chillers, which are approximately two times as efficient as one-stage absorption units. That is, one- and two-stage absorption chillers, respectively, require eight and four times as much driving input energy as electrical chillers. However, this energy information neglects the fact that the chillers are driven by different forms and qualities of energy, and thus can be misleading.

Table 7.1 Various energy efficiencies for several cogeneration systems[1]

Cogeneration plant	Efficiency in electric mode of operation, $(\eta_{CHP})_{elec}$	Efficiencies in cogeneration mode of operation, by heat application					
		District heating (using hot water supplied at 100°C and returned at 75°C)		Single-stage absorption chilling (driven by steam at 120°C/2 bar)		Double-stage absorption chilling (driven by steam at 170°C/8 bar)	
		Cogeneration total efficiency, η_{CHP}	Cogeneration electrical efficiency	Cogeneration total efficiency, η_{CHP}	Cogeneration electrical efficiency	Cogeneration total efficiency, η_{CHP}	Cogeneration electrical efficiency
Diesel	0.42	0.79	0.40	0.68	0.40	0.68	0.40
Gas turbine	0.38	0.88	0.35	0.85	0.35	0.82	0.35
Steam turbine	0.38	0.90	0.35	0.90	0.29	0.90	0.23
Combined cycle	0.50	0.87	0.45	0.87	0.43	0.87	0.39

[1]All efficiencies are based on full-load conditions and lower heating values of fuels. The efficiency in the electric mode of operation is simply the electrical generation efficiency for the technology. The cogeneration total efficiency is the sum of the cogeneration electrical efficiency and the cogeneration heat efficiency.

Table 7.2 Energy quantities and system and device efficiency measures for several integrated cogeneration/district energy systems

Chiller	Cogeneration plant	Energy quantities[1]		Efficiency measures, by device (dimensionless)					
				Energy-based			Exergy-based		
		Fuel energy used	Export electricity, W_{ex}	Chiller: COP_{ch}	Cogeneration: η_{CHP}	System: f_{sys}	Chiller: ψ_{ch}	Cogeneration: ψ_{CHP}	System: ψ_{sys}
(a) Systems for cooling and byproduct electricity									
Electric[2]	Diesel	52.9	0	4.5	0.42	1.892	0.218	0.42	0.091
	Gas turbine	58.4	0	4.5	0.38	1.712	0.218	0.38	0.083
	Steam turbine	58.4	0	4.5	0.38	1.712	0.218	0.38	0.083
	Combined cycle	44.0	0	4.5	0.50	2.252	0.218	0.50	0.109
Absorption (1-stage)	Diesel	533	213	0.67	0.68	0.588	0.127	0.47	0.409
	Gas turbine	299	105	0.67	0.85	0.685	0.127	0.48	0.366
	Steam turbine	245	71.0	0.67	0.90	0.699	0.127	0.45	0.310
	Combined cycle	339	146	0.67	0.87	0.725	0.127	0.54	0.444
Absorption (2-stage)	Diesel	298	119	1.2	0.68	0.736	0.171	0.50	0.416
	Gas turbine	177	62.1	1.2	0.82	0.914	0.171	0.51	0.377
	Steam turbine	124	28.6	1.2	0.90	1.034	0.171	0.46	0.269
	Combined cycle	174	67.7	1.2	0.87	0.966	0.171	0.55	0.418
(b) Systems for cooling, heating and byproduct electricity									
Absorption (1-stage)	Diesel	890	356	0.67	0.68	0.625	0.127	0.47	0.434
	Gas turbine	499	175	0.67	0.85	0.751	0.127	0.48	0.411
	Steam turbine	409	119	0.67	0.90	0.779	0.127	0.45	0.364
	Combined cycle	567	244	0.67	0.87	0.783	0.127	0.54	0.483
Absorption (2-stage)	Diesel	655	262	1.2	0.68	0.705	0.171	0.50	0.459
	Gas turbine	390	137	1.2	0.82	0.863	0.171	0.51	0.449
	Steam turbine	274	62.9	1.2	0.90	0.961	0.171	0.46	0.371
	Combined cycle	382	149	1.2	0.87	0.914	0.171	0.55	0.491

(Continues)

Table 7.2 (Continued)

Chiller	Cogeneration plant	Energy quantities[1]		Efficiency measures, by device (dimensionless)					
		Fuel energy used	Export electricity, W_{ex}	Energy-based			Exergy-based		
				Chiller: COP_{ch}	Cogeneration: η_{CHP}	System: f_{sys}	Chiller: ψ_{ch}	Cogeneration: ψ_{CHP}	System: ψ_{sys}
(c) Systems for cooling, low-temperature heating and byproduct electricity									
Absorption	Diesel	790	316	0.67	0.72	0.653	0.127	0.48	0.433
(1-stage)	Gas turbine	487	171	0.67	0.86	0.760	0.127	0.47	0.404
	Steam turbine	427	135	0.67	0.90	0.784	0.127	0.46	0.377
	Combined cycle	577	253	0.67	0.87	0.785	0.127	0.54	0.484
Absorption	Diesel	554	222	1.2	0.73	0.761	0.171	0.49	0.447
(2-stage)	Gas turbine	366	128	1.2	0.85	0.896	0.171	0.49	0.422
	Steam turbine	316	92.2	1.2	0.87	0.924	0.171	0.45	0.375
	Combined cycle	404	175	1.2	0.89	0.929	0.171	0.56	0.498

[1]The dimensions of these quantities are relative energy-rate units, where the cooling load has been fixed at 100 for all cases, and the heating load at 100 for all cases in which the products include heat.

[2]The cogeneration plant operates in an electric-only mode in this case.

Notes: For the heating in Part (b), the export heat conditions match those of heat supplied to the absorption chiller, while for the low-temperature heating in Part (c), the export heat conditions are fixed at 100°C, a temperature suitable for district heating.

Table 7.3 System and device COP-based efficiency measures for several integrated cogeneration/district energy systems producing only cooling and byproduct electricity

		Efficiency measures, by device (dimensionless)					
		Energy-based measures			Exergy-equivalent of energy-based measures		
Chiller	**Cogeneration plant**	**Chiller: COP_{ch}**	**Cogeneration:[2] COP_{CHP}**	**System:[3] COP_{sys}**	**Chiller**	**Cogeneration[2]**	**System[3]**
Electric	Diesel	4.50		4.50	0.218		0.218
	Gas turbine	4.50		4.50	0.218		0.218
	Steam turbine	4.50		4.50	0.218		0.218
	Combined cycle	4.50		4.50	0.218		0.218
Absorption (single-stage)	Diesel	0.67	14.00	9.38	0.127	3.56	0.453
	Gas turbine	0.67	16.67	11.16	0.127	4.24	0.539
	Steam turbine	0.67	6.78	4.54	0.127	1.73	0.219
	Combined cycle	0.67	6.29	4.21	0.127	1.60	0.203
Absorption (double-stage)	Diesel	1.20	14.00	16.80	0.171	4.74	0.811
	Gas turbine	1.20	15.67	18.80	0.171	5.31	0.908
	Steam turbine	1.20	4.47	5.36	0.171	1.51	0.259
	Combined cycle	1.20	4.36	5.24	0.171	1.48	0.253

[1]For the electric-chiller cases, the cogeneration plants produce only electricity, in the exact quantities required by the chiller. For the absorption-chiller cases, the cogeneration plants produce exactly enough heat to meet the chiller requirements.
[2]COP_{CHP} and its exergy equivalent are undefined for the electric-chiller cases, as no work is curtailed to produce no heat.
[3]$COP_{sys} = (COP_{ch})(COP_{CHP})$ for the absorption-chiller cases, and $COP_{sys} = COP_{ch}$ for the electric-chiller cases. Analogous relations apply for the exergy equivalents of these measures.

Table 7.4 Coefficients of performance on energy bases for the cogeneration portions of several integrated cogeneration/district energy systems

	Cogeneration coefficient of performance, by cogeneration type (dimensionless)							
	COP_{CHP}				COP_{CHP} on an exergy basis			
	Diesel	**Gas turbine**	**Steam turbine**	**Combined cycle**	**Diesel**	**Gas turbine**	**Steam turbine**	**Combined cycle**
(a) Systems for cooling and byproduct electricity								
Absorption (1-stage)	14.00	16.67	6.78	6.29	3.56	4.24	1.73	1.60
Absorption (2-stage)	14.00	16.67	4.47	4.36	4.74	5.31	1.51	1.48
(b) Systems for cooling, heating and byproduct electricity								
Absorption (1-stage)	14.00	16.67	6.87	6.29	3.56	4.24	1.73	1.60
Absorption (2-stage)	14.00	15.67	4.47	4.36	4.74	5.31	1.51	1.48
(c) Systems for cooling, low-temperature heating and byproduct electricity								
Absorption (1-stage)	14.00	17.05	9.07	6.99	3.76	4.07	2.16	1.67
Absorption (2-stage)	14.00	16.70	7.60	5.99	4.48	4.52	2.06	1.62

The chiller exergy efficiencies in Tables 7.2 and 7.3 provide different insights. Although electric and one-stage absorption chillers are still shown to be the most and least efficient chillers, respectively, the ψ_{ch} values are all less than unity and range approximately from 0.1 to 0.2. Values of ψ_{ch} are less than those for COP_{ch} because the exergy associated with the heat removed during chilling is much less than the energy (since the heat is removed at near-environment temperatures). The range in ψ_{ch} values is much less than that for COP_{ch} values because the COP values is based on inputs of electricity and/or heat (two different energy forms), while the exergy efficiencies are evaluated on a more common basis, using electrical equivalents for all energy forms.

In other words, although absorption chillers require much more input energy than electrical chillers, the energy required in the former case is of lower quality (low-temperature heat) than that required in the latter (electricity).

7.5 Performance and efficiency for cogeneration and heating

The cogeneration energy efficiencies in Tables 7.2, which are based on the values specified in Table 7.1, range approximately from 0.7 to 0.9 when both electricity and heat are being generated (as required for the absorption chiller-based cases), and from 0.4 to 0.5 when they operate in an all-electric mode (as required for the electrical chiller-based cases). The cogeneration exergy efficiencies in Table 7.2 are the same as the energy efficiencies in the latter cases, and less than the energy efficiencies in the former cases (since the exergy associated with the product heat from a cogeneration plant is less than the energy). In fact, ψ_{CHP} values when heat is cogenerated tend to be only slightly greater than ψ_{CHP} values when only electricity is generated, for the heat-production temperatures of interest here.

The variations in energy efficiencies between the four cogeneration options considered (diesel, gas turbine, steam turbine, combined cycle) mirror the variations in Table 7.1 (which lists some of the specified data used in this section). The variations in exergy efficiencies between the cogeneration options also follow the trends exhibited by the energy efficiencies; however, the range in η_{CHP} values is in general much greater than the range in ψ_{CHP} values, again because the exergy efficiencies weight the thermal and electrical energy products on equivalent bases. That is, exergy efficiencies recognize that, although much product heat can be produced at the cost of slightly reduced electrical output, the quality of the product heat is much lower than the quality of the electricity. The higher quality (i.e., temperature) of the heat produced for two-stage absorption chillers compared to one-stage units is reflected in higher ψ_{CHP} values, but is not reflected in the energy efficiencies.

The COP_{CHP} values presented in Table 7.4 demonstrate different characteristics of the cogeneration options considered. For diesel- and gas turbine-based cogeneration systems, which lose little electrical output when cogenerating product heat, COP_{CHP} values range approximately from 14 to 17, and are much greater than COP_{CHP} values for steam turbine- and combined cycle-based systems, which range approximately from 4 to 9. Since COP_{CHP} values exceed unity for all cases, more than one unit of useful thermal energy is produced for every unit of electrical energy sacrificed. The COP_{CHP} values for the cases in which heat is produced at higher temperatures (i.e., for two- rather than one-stage absorption chillers) are generally lower than the COP_{CHP} values for cases in which lower temperature heat is produced, because more electrical output must be curtailed to achieve the greater required temperature. The trends exhibited by the COP_{CHP} values are also visible when these values are re-evaluated on an exergy basis, but the magnitude of the values are much lower, ranging approximately from 4 to 5 for the diesel and gas turbine cases, and from 1.5 to 2 for the steam turbine and combined cycle cases. The energy-based values exceed the exergy-based ones because the energy of the cogenerated heat is greater than the corresponding exergy.

It is interesting to note that, in all cases, cogeneration appears to be worthwhile thermodynamically, as more than one unit of thermal exergy is rendered useful via cogeneration for each unit of electrical exergy sacrificed.

The only change in a given cogeneration plant in switching from the case for the production of cooling/byproduct electricity to that for cooling/heating/byproduct electricity is that the plant size (and the magnitude of all inputs and outputs) is scaled up by a constant factor. Hence, no variation between these cases, for any given cogeneration plant, exists in any of its efficiency measures (i.e., the η_{CHP} and ψ_{CHP} values in Table 7.2, and the COP_{CHP} values and their exergy equivalents in Tables 7.3 and 7.4).

7.6 Performance and efficiency of integrated cogeneration/district energy

For the integrated systems, values of the energy-based measure of efficiency f_{sys} in Table 7.2 are higher for the electrical chiller-based systems (varying approximately from 1.7 to 2.3) than for the absorption chiller-based systems (for which f_{sys} values vary approximately between 0.5 and 1.0). This difference is, in part, attributable to the fact that the electrical chiller-based systems generate no byproduct electricity, and so the contribution to f_{sys} of the chiller COP, which has a high numerical value, is significant. If the electrical chiller-based systems produced mainly electricity and little cooling, the values of f_{sys} would be nearer to the energy efficiencies of the electrical generating stations (i.e., 0.4–0.5). In addition, for the electrical chiller-based systems, f_{sys} increases with increasing η_{CHP} (which here is just the electrical generation efficiency); this effect is expected since COP_{ch}, which provides the chiller contribution to f_{sys}, is fixed.

The f_{sys} values for the cases utilizing two-stage absorption chillers are generally higher than those for the cases involving one-stage absorption chillers. Also, for the absorption chiller systems, f_{sys} values generally decrease as the cogeneration unit is changed from steam turbine to combined cycle to gas turbine to diesel.

It is noted that f_{sys} values cannot be directly compared for the different cases considered, since each case produces different quantities of products (cool and/or heat and/or electricity). Values of f_{sys} are comparable for systems generating the same products. To avoid being misleading, f_{sys} values must be considered along with information on the quantities and types of products for the cases.

Values of the system exergy efficiency ψ_{sys} are comparable between different cases because, although different product energy forms may be present, they are weighted on an equivalent basis using the exergy method.

The insights provided by the system exergy efficiencies in Table 7.2 are different from those provided by the energy-based f_{sys} measures, for several reasons:

- The values of ψ_{sys} are generally lower than those for f_{sys}, with ψ_{sys} values being approximately 50 per cent of the f_{sys} values for the absorption chiller-based cases, and 5 per cent of the f_{sys} values for the electrical chiller-based cases. This is expected because values of exergy efficiencies for chillers and cogeneration units are generally lower than (at most equal to) values of the corresponding energy-based measures of performance.
- The values of ψ_{sys} for the electrical chiller-based cases (which range approximately from 0.08 to 0.10) are less than those for the absorption chiller-based systems (which range approximately from 0.3 to 0.5). This observation is attributable to the lack of byproduct electrical generation (with an exergy efficiency higher than that for cooling) in the electrical chiller cases. The exergy efficiencies for the systems utilizing two-stage absorption chillers are generally within a few per cent of those for the systems using one-stage units.
- For the electrical chiller cases, ψ_{sys} decreases as the electrical generating efficiency of the cogeneration unit decreases, while for absorption chiller cases, ψ_{sys} generally decreases as the cogeneration unit is changed from combined cycle to diesel to gas turbine to steam turbine.

The COP-based measures of performance in Table 7.3 provide additional insights into the integrated systems, including the following:

- For the electrical chiller systems, COP_{sys} is the same for all cogeneration units, and fixed at the values of the chiller COP of 4.5. This observation is expected as COP_{sys} represents the export electrical energy sacrificed to provide cooling.
- For the absorption chiller systems, COP_{sys} values are higher for the diesel and gas turbine cases (approximately 9–19) than those for the steam turbine and combined cycle cases (for which COP_{sys} ranges in value approximately from 4 to 5). This is expected since little electricity is curtailed in the former cases to generate the heat needed to drive the absorption chillers. In addition, COP_{sys} values are higher for the systems having two-stage absorption chillers, relative to those with one-stage units.
- For the integrated systems utilizing steam turbine and combined cycle cogeneration units, COP_{sys} values are similar (ranging from 4.3 to 5.4) for all types of chillers. Steam turbine and combined cycle cogeneration plants are those for which, most notably, continually increasing quantities of electrical output are curtailed as increasing product-heat temperatures are produced. That is, differences in chiller performance (as measured by COP_{ch} values) are approximately offset by the efficiencies of the cogeneration (or electrical generation) plants, which form the other portion of the integrated systems.

The exergy-based equivalents to the COP_{sys} values in Table 7.3 are lower than the COP_{sys} values, but generally lead to the same observations as discussed above for the COP_{sys} values.

7.7 Closure

Systems for integrated cogeneration and district energy are contrasted, accounting for performance and efficiency. The material demonstrates how the separate technologies can together operate mutually beneficially, but highlights the need for proper measures of merit. The comparisons are demonstrated to help inform decision-making by clarifying competing options, and to demonstrate the merits of different integrated cogeneration and district energy systems.

Chapter 8
Economics of cogeneration and district energy

Overview

The economics of cogeneration and district energy are described and examined, starting with fundamentals and general economic considerations such as methods for estimating TCI, performing economic evaluations and calculating revenue requirements. A detailed case study on the economics of cogeneration is provided. For the scenario considered in the case study, the fixed and TCI, as well as fuel costs, start-up costs, and working capital (WC), are explained. The cogeneration project economics are provided, and the facility investment, modified accelerated cost recovery system and product costs are described. Modern and historical applications of the case study are examined. Various economic considerations involved in integrated cogeneration-based district energy systems are considered, including economic considerations for generation and distribution as well as economic considerations for consumers. An analogy is presented between allocating wastes for cogeneration among products, and allocating economic costs among products. Optimization schemes and procedures involving economics are discussed, to provide a foundation for detailed routines and computer codes for assessing, comparing and optimizing the benefits of such systems. Various investigations of economics of cogeneration and district energy are described.

8.1 Introduction

Decisions related to the design on systems and components for cogeneration, for heating and district heating, and for cooling and district cooling – as well as integrated cogeneration-based district energy systems – are strongly dependent on economic considerations, in addition to technical, environmental and other factors. Hence, it is important that a sound understanding of the economics of these technologies and systems be attained by practitioners and decision-makers.

This chapter builds on the technical topics covered thus far in this book, throughout Chapters 2–7. Environmental factors, and particularly GHG emissions, are covered subsequently in Chapters 9 and 10. But these environmental factors can be incorporated into economic assessments in various ways.

8.2 Fundamentals

Each component, or group of components, of an integrated system for cogeneration and district energy, is in itself a complex system:

- Cogeneration systems (described in Chapter 3) consist of the electrical power generating cycle and can be based on steam turbines, gas turbines, combined cycles or reciprocating engines. The cogeneration unit provides electrical energy and heat for heating and/or cooling.
- District heating systems (described in Chapter 4) consist of a central heating plant that delivers heat in different forms (e.g., hot water or steam) to heat users, which can be from the residential, commercial, institutional or industrial sectors. The distribution system is an integral part of the district heating system. In general, a two-pipe system is used: one pipe delivers the hot fluid and the other returns the spent (cooler) fluid or the condensate.
- District cooling systems (described in Chapter 5) consist of a central chiller plant that provides a chilled fluid to users, which are often from the residential, institutional or commercial sectors. The distribution system is an integral part of the district cooling system. In general, a two-pipe system is used: one pipe delivers the chilled fluid and the second one returns the spent (warmer) fluid. If the district cooling system is integrated with a district heating system, a four-pipe system is often used: one pipe each for the delivery of the hot and cold fluids, and two pipes for the return of the used fluids. Some significant variations can occur in the design of the district cooling system depending on the type of chiller used: electric or absorption.

A general cogeneration/district energy system can incorporate all of the above systems. For instance, a thermal power plant-based cogeneration system can produce electricity and heat. Then, a distribution system can be used to deliver the hot fluid to consumers and, in the case of district cooling, either a distribution system can be used to deliver centrally produced cold fluid to consumers from the cogeneration plant, or steam can be delivered to individual consumers to drive absorption chillers.

8.3 General economic considerations

The economic analysis of an integrated energy system incorporating electricity production and hot and/or cold fluid generation and distribution is extremely complicated because of the large number of systems and subsystems involved, as well as the large differences in prices of fuels, electricity, labor and equipment.

The economic principles that apply to thermal systems design apply also to the design of integrated energy systems. For a relatively small-scale project such as a campus-integrated energy system or a hospital energy system, the economic considerations are similar to the economic principles described below. For a larger scale project involving utility generation as well as large distribution systems, the economic considerations become more involved at a macro-economic level, requiring a longer term approach and probably municipal or even provincial (state) incentives.

In section 8.3, we describe in brief these economic principles as they apply to a small power plant-based cogeneration project, following closely the excellent presentation provided elsewhere [203]. The method described is the "Revenue Requirements Method" [203]. The successful completion of a thermal design project requires the estimation of the major costs involved in the project. The Revenue Requirement Method permits the calculation of the cost of the main product using the following four steps [203]:

- estimation of the TCI,
- determination of the economic, financial, operating and market input parameters for detailed cost calculation,
- calculation of the total revenue requirement and
- calculation of the levelized product cost.

A fundamental distinction in costs is noted here: fuel costs and operation and maintenance costs are incurred continuously, while the initial investment is a one-time cost.

8.3.1 Estimation of TCI

The TCI is the sum of the fixed capital investment (FCI) and other financial outlays. The FCI is composed of the capital needed to purchase or lease the required land, to build the facilities and to buy and install the equipment. The cost estimated for the FCI consists of direct and indirect costs, details for which follow:

- **Direct costs**. Direct costs include the costs of permanent equipment, materials, labor and other resources required for the installation of the facility.
- **Indirect costs**. Indirect costs include engineering and supervision and other indirect resource needs, and do not become part of the facility but are required for the completion of the project.

Table 8.1 provides a rough breakdown of the items contributing to the TCI, in terms of FCI (direct and indirect) and other outlays for a typical process-like industrial facility. Note that onsite costs correspond to the installed cost for all items' process equipment.

Table 8.2 provides a rough breakdown of the TCI as a function of FCI and other outlays for typical power generation or chemical facilities, showing approximate relative values of the costs. In Table 8.1, the onsite costs correspond to the installed equipment cost for all items shown on the process flow diagram.

Note that all cost data used in an economic analysis must be brought to the same reference year, normally the year used as a basis for the cost calculations. For cost data based on conditions at different times, this normalization is accomplished using a cost-index, which is an inflation indicator that corrects the costs at one time for equipment, material, labor, supplies, etc. with the costs at the reference year or

Table 8.1 Breakdown of TCI for a typical process-like industrial facility (adapted from [203])

Fixed capital investment (FCI)			Other outlays
Direct costs (DC)		Indirect costs (IC)	
Onsite costs (ONSC)	**Offsite costs (OFSC)**		
Purchased equipment costs (PEC)	Land	Engineering	Start-up costs (SUC)
Purchased equipment installation	Civil work	Supervision	Working capital (WC)
Piping	Structural work	Construction	Licensing
Instrumentation and control	Architectural work	Contingencies	Research and development
Electrical equipment	Service facilities		Allowance for funds used during construction (AFUDC)

evaluation date. Cost-indices are commonly needed and are often monitored and supplied by governments.

The first step in determining the TCI is an estimation of the cost of the purchased equipment. The most accurate value is the vendor quotation for the equipment, although cost charts are often useful for cost estimates.

Using information from Tables 8.1 and 8.2, a simplified relationship can be expressed for the TCI required for a new system as the sum of the FCI, the start-up costs (SUC), the WC, the costs of licensing, research and development (LRD) and an allowance for the funds used during construction (AFUDC):

$$\mathrm{TCI} = \mathrm{FCI} + \mathrm{SUC} + \mathrm{WC} + \mathrm{LRD} + \mathrm{AFUDC} \tag{8.1}$$

This expression can also be applied to an expansion of an existing system.

The FCI can be expressed as the sum of direct costs (DC) and indirect costs (IC):

$$\mathrm{FCI} = \mathrm{DC} + \mathrm{IC} \tag{8.2}$$

where the direct costs include the on-site costs (ONSC) and offsite costs (OFSC), i.e.:

$$\mathrm{DC} = \mathrm{ONSC} + \mathrm{OFSC} \tag{8.3}$$

The onsite costs can be estimated directly from correlations given in Table 8.2. The offsite costs are often estimatable in two ways, depending on whether a new facility or expansion is being considered, as follows:

- 100–200 per cent (average value 120 per cent) of the onsite costs for a new facility or
- 40–50 per cent (average value 45 per cent) of the onsite costs for an expansion or modification of an existing facility.

Table 8.2 Estimated breakdown of costs contributing to TCI for a typical power generation or chemical facility (adapted from [203])

Cost breakdown			Estimated value (% of identified cost)			
			FCI	PEC	DC	TCI
Fixed capital investment (FCI)	Direct costs (DC)					
	• Onsite costs (ONSC)	Purchased equipment costs (PEC)	15–40			
		Purchased equipment installation	6–14			
		Piping	3–20			
		Instrumentation and control	2–8			
		Electrical equipment	2–10			
	• Offsite costs (OFSC)	Land	0–2	0–10		
		Civil, structural, architectural work	5–23	15–90		
		Service facilities	8–20	30–100		
	Indirect costs (IC)	Engineering and supervision	4–21	25–75	6–15	
		Construction	6–22		15	
		Contingencies[1]	5–20			
Other outlays		Start-up costs (SUC)	5–12			
		Working capital (WC)				10–20
		Licensing, research and development (LRD)				
		Allowance for funds used during construction				

[1]Contingencies also can be estimated as 8–25% of the other indirect costs.

That is, the OFSC can be expressed as a function of the ONSC as follows:

$$\text{OFSC} = \begin{cases} 1.2\,(\text{ONSC}) \text{ for a new system} \\ 0.45\,(\text{ONSC}) \text{ for a system expansion} \end{cases} \tag{8.4}$$

It can be further assumed from Table 8.2 that WC, the start-up costs, the licensing, research and development costs plus an AFUDC, and the indirect costs, respectively, can be expressed as follows:

$$\mathrm{WC} = 0.15\,(\mathrm{TCI}) \tag{8.5}$$

$$\mathrm{SUC} = 0.10\,(\mathrm{FCI}) \tag{8.6}$$

$$\mathrm{LRD} + \mathrm{AFUDC} = 0.15\,(\mathrm{FCI}) \tag{8.7}$$

$$\mathrm{IC} = 0.25\,(\mathrm{DC}) \tag{8.8}$$

Substituting (8.5)–(8.7) into (8.1) yields and expression for the TCI as a function of the FCI:

$$\mathrm{TCI} = 1.47\,(\mathrm{FCI}) \tag{8.9}$$

Combining (8.9) with (8.2), (8.3) and (8.8) yields an expanded expression for the TCI:

$$\mathrm{TCI} = 1.84\,(\mathrm{DC}) = 1.84\,(\mathrm{ONSC} + \mathrm{OFSC}) \tag{8.10}$$

which can be modified with (8.4) as follows:

$$\mathrm{TCI} = \begin{cases} 4.05\,(\mathrm{ONSC}) \text{ for a new system} \\ 2.67\,(\mathrm{ONSC}) \text{ for a system expansion} \end{cases} \tag{8.11}$$

The FCI for a new system usually:

- averages about 430 per cent of the purchased-equipment cost (PEC), although this value can vary significantly and normally lies in the range of 280–550 per cent,
- averages at about 283 per cent of PEC for a system expansion.

In other words:

$$\mathrm{FCI} = \begin{cases} 4.3\,(\mathrm{PEC}) \text{ for a new system} \\ 2.83\,(\mathrm{PEC}) \text{ for a system expansion} \end{cases} \tag{8.12}$$

Combining this result with (8.9) gives an alternative expression the total capital investment:

$$\mathrm{TCI} = \begin{cases} 6.32\,(\mathrm{PEC}) \text{ for a new system} \\ 4.16\,(\mathrm{PEC}) \text{ for a system expansion} \end{cases} \tag{8.13}$$

Thus, in order to estimate the total equipment investment for a new system, (8.13) can be used after the purchased-equipment costs have been estimated, or (8.11) can be used after the ONSC have been estimated.

Of course, manufacturer data can be used when available, and other data from the literature can be used to estimate the average cost of complete plants or portions of plants, often as a function of the plant size or capacity.

8.3.2 Economic evaluation

After the capital investment cost of equipment has been determined or estimated, an economic evaluation in normally carried out. Some of the engineering-economics principles that must be accounted for in such an evaluation are as follows:

- **Time value of money**. Future values are often converted to present values to permit consistent economic comparisons.
- **Inflation, escalation and levelization**. Inflation and cost escalations must be accounted for. The real interest rate does not include an adjustment for general price inflation, but this adjustment is included in the nominal interest rate. Since cost escalation applied to an expenditure over a period of years results in a non-uniform cost schedule, the concept of levelization is often used to convert a series of varying quantities to a financially constant quantity over a specified time interval.
- **Depreciation, economic life and salvage value**. Economic principles cover numerous methods for depreciating the value of an asset like a piece of process equipment. The economic life (or book life) of an asset is the best estimate of the time duration for which the asset can be used. The salvage value of an asset is its estimated value at the end of its economic life.
- **Financing**. The money to cover the total capital requirement of a project investment can come from one or more of borrowing capital (debt financing), and/or the sale of common or preferred stocks (equity financing) and/or existing funds of the company (self-financing).
- **Fuel, operating and maintenance costs**. Fuel costs are often treated as part of operating and maintenance (O&M) costs. Because fuel costs are normally significant in cogeneration, district energy and other integrated energy projects, fuel costs are often considered separately from O&M costs. Operating and maintenance costs can be divided into fixed and variable costs. The fixed costs include operating labor, maintenance labor, maintenance materials, overhead, administration and support, distribution and marketing. The variable operating costs depend on the average annual system capacity factor, which determines the equivalent number of hours of operation per year at full load. The variable operating costs are often of importance in many integrated energy projects because of the high variations in the heating and cooling loads during the year.
- **Taxes and insurance**. Income taxes are calculated by multiplying the income tax rate, which varies by jurisdiction, by the taxable income. Tax-deductible expenditures include fuel costs, O&M charges, return on debt and investment cost recovery. Often income, and thus income tax, is deferred to the latter years of a project. Annual property taxes also have to be taken into account. The annual insurance costs are typically between 0.5 and 1.5 per cent of the plant facility investment (PFI).

8.3.3 Revenue requirements

The annual total revenue requirements (total product costs) for a system are the revenues that must be collected in a given year through the sale of all products to compensate the system operating company for all expenditures incurred in the same year and to ensure sound economic plant operation. The annual total revenue requirements include:

- **Expenses.** Expenses include fuel costs and O&M costs.
- **Carrying charges.** These charges include taxes and insurance, income taxes, capital recovery (depreciation) and minimum acceptable returns (depending on financing method).

A year-by-year analysis must be undertaken to calculate the levelized revenue requirement and the levelized costs of the main products. Such a detailed analysis of a cogeneration system allows the levelized costs of electricity and steam to be determined.

8.3.4 Profitability evaluation methods

The profitability of a project is generally assessed using one or more of the following criteria:

- average rate of return,
- payback period,
- net present value,
- benefit–cost ratio and
- internal rate of return (IRR).

Which of the above criteria are used in practice depends on company preferences and the project to be evaluated. The first two criteria are commonly applied and straightforward to implement, but do not consider the time value of money; hence, these criteria should be avoided in projects with long economic lives. The methods using discounted cash flows are useful for evaluating alternative investments.

8.4 Cogeneration economics case study

8.4.1 Case-study scenario

This case study is based closely on the very informative case study in Reference 203. The case study considers a simple cogeneration plant containing a single-stage gas-turbine cycle with a net electrical output of 30 MW. The flue gases from the gas turbine preheat the delivery air from the compressor and then are used in a heat recovery steam generator to produce saturated steam.

A schematic of the cycle is shown in Figure 8.1 and the characteristics of each flow are described in Table 8.3. Values used for parameters in the case study are selected so as to be typical of industrial applications.

8.4.2 Fixed and total capital investment

The FCI for this project is \$39.5 million and the TCI is \$46.0 million. These values are obtained in a tabular fashion in Tables 8.4a–8.4d.

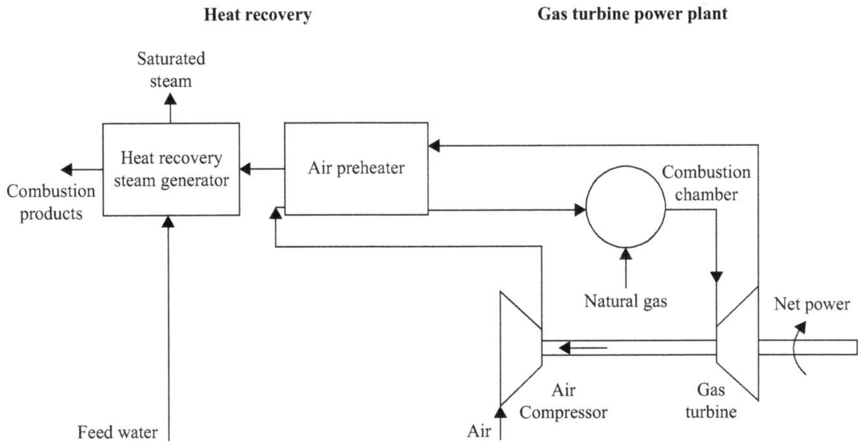

Figure 8.1 Cogeneration system based on a gas turbine power plant and heat recovery to raise steam from the turbine exhaust. This system is assessed in the case study.

Table 8.3 Characteristics of flows for the cogeneration system considered in the case study

Flow information			Flow properties	
Location	**Commodity**	**Mass flow rate (kg/s)**	**Temperature (K)**	**Pressure (bar)**
Entering compressor	Air	91.3	298.0	1.01
Exiting compressor	Air	91.3	603.7	10.1
Entering combustor	Air	91.3	850.0	9.62
Exiting combustor	Combustion gases	92.9	1,520.0	9.14
Exiting turbine	Combustion gases	92.9	1,006.1	1.10
Exiting air preheater	Combustion gases	92.9	779.8	1.07
Exiting HRSG	Combustion gases	92.9	426.9	1.01
Entering HRSG	Liquid water	14.0	298.0	20.0
Exiting HRSG	Saturated steam	14.0	485.5	20.0
Entering combustor	Natural gas (mainly methane)	1.64	298.0	12.0

HRSG, heat recovery steam generator.
Adapted from Reference 203.

By comparison, the simplified formulas shown earlier, i.e., (8.11) and (8.13), respectively, yield:

$$\text{TCI} = 2.67\ (\text{ONSC}) = 2.67 \times 21{,}230{,}000 = \$56{,}700{,}000$$

$$\text{TCI} = 4.16\ (\text{PEC}) = 4.16 \times 11{,}000{,}000 = \$45{,}800{,}000$$

The latter approximation appears reasonably accurate, whereas the former overestimates the value.

Table 8.4a Estimate of portion of TCI associated with FCI due to direct cost for the cogeneration system case study (costs in million dollars)

Onsite costs		
Purchased equipment costs (PEC)		
Air compressor	3.74	
Air preheater	0.94	
Combustion chamber	0.34	
Gas turbine	3.74	
Heat recovery steam generator	1.30	
Other plant equipment	0.94	
Total purchased equipment costs		11.00
Purchased equipment installation (33% of PEC)	3.63	
Piping (35% of PEC)	3.85	
Instrumentation and control (12% of PEC)	1.32	
Electrical equipment and materials (13% of PEC)	1.43	
Total onsite costs		21.2
Offsite costs		
Land	0.50	
Civil	2.31	
Service	3.85	
Total offsite costs		6.66
Total		27.9

Table 8.4b Estimate of portion of TCI associated with FCI due to indirect cost for the cogeneration system case study (costs in million dollars)

Engineering and supervision	2.23	
Construction costs and contractor profit	4.18	
Contingency (15% of the above sum)	5.15	
Total		11.6

Table 8.4c Estimate of portion of TCI associated with other outlays for the cogeneration system case study (costs in million dollars)

Start-up costs	1.25	
Working capital	1.92	
Costs of licensing, research and development	0.00	
Allowance for funds used during construction	3.35	
Total		6.5

Table 8.4d Estimate of TCI for the cogeneration system considered in the case study (costs in million dollars)

Fixed capital investment		
Direct costs (Table 8.4a)	27.9	
Indirect costs (Table 8.4b)	11.6	
Total fixed capital investment		39.5
Other outlays (Table 8.4c)		6.5
Total capital investment		46.0

8.4.3 Cogeneration project economics

The cogeneration economics are evaluated based on the details provided in Table 8.5. It is assumed that:

- The annual inflation rate is 5 per cent.
- The annual increase in the price of natural gas is 6 per cent.
- The design lasts two years.
- Construction begins immediately after the design ends and lasts for two years.

The economic life of the project is assumed to be 20 years, with the system life for tax purposes taken to be 15 years. The salvage value at the end of the project is considered to be zero.

The average capacity factor for the cogeneration system is estimated as 85 per cent, i.e., the system operates for 7,446 hours of the 8,760 hours in a year.

Based on 30 labor positions and an average cost of \$28/hr, the annual direct labor costs are \$1,750,000. Based on these costs, the annual fixed O&M costs and the annual variable O&M costs are estimated at full capacity as \$3,800,000 and \$350,000, respectively. At 85 per cent capacity, the annual variable O&M costs are \$298,000.

To determine the corresponding costs for the first year of operation, these costs are escalated, at a nominal escalation rate of 5 per cent per year, to the middle of the first year of operation. The escalated annual values for fixed and variable O&M costs, respectively, are \$4,620,000 and \$360,000.

8.4.4 Fuel costs, start-up costs and WC

The cost of natural gas is taken to be \$3/GJ based on lower heating value (LHV), which is taken to be 50.01 MJ/kg. The fuel mass flow rate is 1.64 kg/s (Table 8.3), so the annual fuel cost (FC) is:

$$\begin{aligned} \text{FC} &= \$0.003/\text{MJ} \times 50.01\ \text{MJ/kg} \times 1.64\ \text{kg/s} \times 7{,}446\ \text{hr/yr} \times 3{,}600\ \text{s/hr} \\ &= \$6{,}600{,}000/\text{yr} \end{aligned}$$

To calculate the annual fuel costs for the first year of commercial operation, this value is escalated by 6 per cent per year to that year:

$$\text{FC} = 6{,}600{,}000 \times 1.06 = \$8{,}340{,}000/\text{yr}$$

Table 8.5 Parameters and assumptions for the cogeneration case study economic evaluation of total revenue requirement

Quantity type	Parameter	Value
Inflation and escalation rates (%)	Average annual inflation rate (from start of design to end of economic life of plant)	5
	Average nominal escalation rate of all costs (except fuel) (from start of design to end of economic life of plant)	5
	Average nominal escalation rate of natural gas costs (from start of design to end of economic life of plant)	6
Pre-operation periods (yr)	Duration of design period	2
	Duration of construction period (after which commercial operation commences)	2
Plant life (yr)	Plant economic life	20
	Plant life for tax purposes	15
Financing	Plant financing fractions and required returns on capital:	

	Type of financing		
	Common equity	Preferred stock	Debt
Financing fraction (%)	35	15	50
Required annual return (%)	15	11.7	10
Resulting average cost of money (%)			12

Quantity type	Parameter	Value
Tax rates (%)	Average combined income tax rate (from start of design to end of economic life of plant)	38
	Average property tax rate (from start of design to end of economic life of plant) (% of plant facility investment)	1.5
	Average insurance tax rate (from start of design to end of economic life of plant) (% of plant facility investment)	0.5
Other data	Average capacity factor (%)	85
	Number of labor positions for operating and maintenance	30
	Average labor rate ($/hr)	28
	Annual fixed operating and maintenance costs ($)	3,800,000
	Annual variable operating and maintenance costs ($)	350,000
	Unit cost of fuel, based on lower heating value ($/GJ)	3
	Allocation of PFI by calendar year for design and construction (%)	
	• At end of design period	40
	• Midway through construction period	60
Other assumed values	Net salvage value ($)	0
	Investment tax credit ($)	0
	Grants in aid of construction ($)	0
Depreciation and leveling	MACRS depreciation used for company purposes	
	Straight-line depreciation used for company purposes	
	Revenue levelizing occurs with before-tax average cost of money	

After calculating the fuel and O&M costs, the start-up costs and the WC (shown as other outlays in Table 8.4c) can be estimated.

The start-up costs for a power plant are the sum of the following unescalated costs:

- one month of fixed O&M costs,
- one month of variable operating costs calculated at full load,
- one week of full-load fuel cost and
- 2 per cent of the PFI.

That is:

$$\begin{aligned}\text{SUC} = {} & 3{,}800{,}000/12 + 350{,}000 \times 0.85/12 + 6{,}600{,}000/52 + 0.02 \\ & \times 38{,}950{,}000 = \$1{,}247{,}000\end{aligned}$$

After escalation to middle of the construction period, SUC = \$1,400,000. Similarly, the WC is the sum of the unescalated expenses representing two months of fuel and variable operating costs at full load, and three months of labor costs plus a contingency of 25 per cent of the total of the above three items. Neglecting the variable operating costs for the cogeneration system:

$$\text{WC} = (6{,}600{,}000/6 + 1{,}750{,}000/4) \times 1.25 = \$1{,}920{,}000$$

After escalation to the end of the construction period, WC = \$2,280,000.

8.4.5 Plant facility investment and modified accelerated cost recovery system

The PFI, evaluated as FCI less land cost, is estimated from the data in Table 8.4 as \$38,950,000 at the start of the design period. Forty per cent of this amount must be escalated at an annual rate of 5 per cent to the start of the construction period when it is expended, and the remainder of the PFI is escalated to the middle of the construction period. The system financing fractions are given in Table 8.5. The AFUDC has to be calculated separately for each type of financing (debt, equity, etc.). Following the financing fractions, escalation rates and rate of return required for each type of financing:

$$\text{AFUDC} = \$4{,}985{,}000 \text{ (at end of the construction period)}$$

or

$$\text{AFUDC} = \$3{,}353{,}000 \text{ (at start of the design period)}$$

The contribution of AFUDC to the total investment increases with increasing rate of return on investment and increasing length of the design and construction period.

Table 8.6 concludes the calculations for the net capital investment. After calculating the total depreciable capital investment, the year-end book value of the project can be calculated using MACRS (modified accelerated cost recovery system) depreciation factors for 15 years.

Table 8.6 Calculation of the net capital investment for the cogeneration case study (in million dollars)

Total capital investment			
Net outlay			
Cost of land (start of construction period)	0.54		
Escalated PFI (middle of construction period)	44.23		
Start-up costs (middle of construction period)	1.44		
Working capital (end of construction period)	2.28		
Costs of licensing, research and development	0.00		
Grants in aid of construction	0.00		
Total net outlay		48.50	
AFUDC			
Common equity AFUDC	2.19		
Preferred equity AFUDC	0.73		
Debt AFUDC	2.07		
Total AFUDC		4.99	
Total capital investment (Total net outlay + Total AFUDC)			53.48
Non depreciable capital investment			
Cost of land		0.54	
Working capital		2.28	
Common equity AFUDC		2.19	
Total nondepreciable capital investment			5.01
Total depreciable capital investment (Total capital investment – Total nondepreciable capital investment)			48.5

8.4.6 Product costs

The total product costs are evaluated based on the annual total revenue requirements for the cogeneration system. The total product costs represent the revenues needed in a given year through the sale of all products:

- to compensate for all expenditures incurred in operating the system in the same year and
- to ensure sound economic plant operation.

The expenditures in the cogeneration case study include expenses related to fuel as well as operation and maintenance, and carrying charges related to taxes and insurance, income taxes, capital recovery (depreciation) and minimum acceptable returns.

The levelized costs of the main products, which together represent the levelized revenue requirement, are calculated in this case study with a year-by-year analysis. Based on the detailed analysis of the cogeneration system considered in this case study for 20 years, the following key results are obtained:

- The levelized cost of steam is \$0.035/kg.
- The levelized cost of electricity is 7.01 cents/kWh.

8.5 Applications of cogeneration economics case study

8.5.1 Historical application

This case study presented in the last section mirrors a cogeneration case study performed in the 1990s based on a single-stage gas-turbine cycle with a net electrical output of 30 MW [203]. The design and construction period extended over four years, from 1994 to 1998, after which commercial operation began.

Considering costs in 1994 US dollars, the FCI for the project was just under $40 million, and the TCI was about $46 million. The unit cost of fuel (based on LHV) at the time of construction was approximately 3 $/GJ. Based on conditions in the USA in the mid-1990s, an annual inflation rate of 5 per cent, an annual increase in the natural gas price of 6 per cent, an average escalation rate of all costs (except natural gas) of 5 per cent and an average labor cost of $28/hr are applied. An average combined income tax rate of 38 per cent was assumed based over the 20-year plant economic life from 1994 to 2017.

The total FCI for the plant was found to be $39.5 million and, after accounting for other outlays, the TCI was found to be $46 million. For the first year of commercial operation, the annual fuel cost was determined to be $8.3 million, and the start-up costs $1.4 million and the WC $2.3 million. The total product costs, representing the annual total revenue requirements for the system, on a levelized basis, were found to be 7 cents/kWh for electricity and 3.5 cents/kg for steam.

8.5.2 Modern application

The historical application for the case study presented in the previous subsection is updated in a simplified manner here to approximate a cogeneration plant constructed in 2016, again based on a single-stage gas-turbine cycle with a net electrical output of 30 MW. A similar construction period of four years and economic life of 20 years are considered.

The approximate manner used involves escalating costs based on price indexes, which are a measure of inflation effects. Here, we use price indexes published by the U.S. Bureau of Labor Statistics in Washington, DC (www.bls.gov), which is part of the U.S. Department of Labor.

A price index for a cogeneration plants is not available. But a specific price index for a large part of a cogeneration plant can be obtained by considering the manufacturing of turbine and turbine generator set units. The Producer Price Index for turbine and turbine generator set units manufacturing industries, which is based on industry data, indicates as of April 2016 that the $1.00 in 1994 has the same buying power $1.50 in 2016. A broader industry measure is provided by considering all manufacturing industries. In that case, the Producer Price Index for total manufacturing industries, which is based on industry data, indicates as of April 2016 that the $1.00 in 1994 has the same buying power $1.48 in 2016. Much more broadly, we can contrast and compare these values with the general consumer price index, which is based on average price data for all urban consumers. For the same period, the Consumer Price Index indicates as of April 2016 that $1.00 in

1994 has the same buying power $1.61 in 2016. In this analysis, a price index of $1.50 is considered for the period 1994–2016.

Escalating costs in 1994 U.S. dollars to 2016 U.S., dollars, the total product costs, representing the annual total revenue requirements for the system, on a levelized basis, are estimated to be 11 cents/kWh for electricity and 5 cents/kg for steam. These values are based on the FCI for the cogeneration plant in 2016 estimated as $60 million, as well as the total FCI in 2016 for the plant estimated as $59 million. Consequently, after accounting for other outlays, the TCI is estimated as $69 million. For the first year of commercial operation, the annual fuel cost is estimated to be $12 million, and the start-up costs $2.1 million and the WC $3.5 million.

8.6 Economic considerations for cogeneration/district energy systems

The economic considerations to be taken into account in evaluating integrated cogeneration/district energy systems are extremely complex [109,205,207–208]. Such systems are designed to produce simultaneously two or three products: electricity, a heating agent and a cooling agent. The system can be designed to serve a relatively small load such as a university campus or a hospital. Alternatively, a system can be designed to serve a large area of a downtown core such as Minneapolis-St. Paul or Edmonton, or can involve an entire city such as Stockholm, or an entire area of country such as eastern Europe.

8.6.1 Economic considerations for generation and distribution

The techno-economic considerations for a cogeneration plant and district-energy distribution system are a function of the magnitude of the project, and local prices, regulations and policies. The selection of a cogeneration system [204], that will provide electricity, heat (steam or hot water) and/or cold fluid for cooling to a relatively small consumer such as a university campus, hospital or a small commercial facility, often considers one or more of the following sizing criteria:

- peak electrical demand,
- minimum or base-load electrical demand,
- peak thermal demand,
- minimum or base-load thermal demand,
- maximum cost/benefit ratio and
- marginal cost/benefit ratio.

Planning a cogeneration system is considerably more involved than planning a heating, refrigerating and air conditioning (HVAC) system. One "rule of thumb" [204] to determine quickly if a cogeneration feasibility study should be undertaken suggests that cogeneration should be considered if the cost of electricity (expressed in $/kWh) is more than a factor of 0.012 times the cost of fuel (expressed in $/MJ). If the ratio is twice the factor or more, a cogeneration system is likely to have a simple payback period of three years or less. Numerous small cogeneration projects are very attractive economically, having simple payback periods of less than two years.

Some cogeneration projects make use of technologies such as thermal storage for hot fluids or ice storage for cold fluids. With storages, the cold or hot fluid can be produced when electricity prices are low or demands for heating and cooling are low, and then stored for later usage.

Thermal storage can make the economics of some cogeneration projects more favorable [109,163,205–208]. At the Chrysler Technology Centre in Auburn, MI, a chilled water storage has been installed that provided cost savings of $3.6 million. The central plant supplies hot water, chilled water and compressed air to the entire facility. Many societies consider cogeneration to be of importance. For example, ASHRAE (the American Society for Heating, Refrigeration and Air Conditioning Engineering) is conducting research in the field of cogeneration and has a standing committee dealing with cogeneration Issues.

If integrated cogeneration and district energy is provided from a utility power plant, the economic analysis becomes more complicated and has to take into account more factors than it would for a smaller facility. European experience indicates that a major advantage of such integrated systems is the fuel economy that can be achieved by generating electricity and a heating medium in a central facility. Also, fuel, such as waste industrial gases, that would have not been utilizable in a residential building for heating, often can be used in a central cogeneration facility. Experiences in countries such as Sweden and Denmark for such integrated energy projects are positive. Of major importance in the success of these types of projects is the government legislation (as is the case in Denmark) that encourages implementation of central district heating.

Different economic issues usually need to be addressed for new facilities compared to retrofit facilities. A new central cogeneration/district heating project is normally assessed based on the costs and prices of the electricity, heating and cooling provided. The split of costs between the cost of electricity and the cost of the heating/cooling provided can be based on a variety of criteria (e.g., peak electrical load, avoided cost). The economics of such a project are complicated, being dependent on the selection of fuels, state and municipal incentives, government bonds for financing, costs of infrastructure in delivering the products, etc. It has been observed that these types of projects often have a high degree of public acceptance. The economics for the modification and retrofit of an existing facility for central cogeneration and district heating/cooling are usually assessed based on some of the above criteria, as well as on additional factors such as the impact on electricity price, the cost of the heating/cooling provided vs. the existing scenario and fuel savings. Numerous studies have reported of the advantages of these types of retrofit projects.

For comparative purposes, costs and prices from the mid-1990s are considered. Then, the price of steam in district heating projects based on about 40 district heating companies in the U.S. at the time ranged from about $5 to $12 per thousand pounds of steam. Toronto District Heating Corporation in Canada charged between $6.50 and $8.50 per thousand pounds of steam. Consolidated Edison of New York operated one of the largest district heating systems built at the time (apart from those in centrally planned economies), which serves a large part of Manhattan Island. In addition to providing space and water heating, steam from the system is used in numerous restaurants for food preparation as well as for driving absorption chillers and steam turbine-driven chillers for air conditioning. The peak steam delivery in winter 1994 was 12.2 million pounds/hr, or 3.5 GW. The 1994 revenues of the company were $5,140,472,000 from electricity sales, $890,107,000 from gas sales and $342,507,000 from steam sales. The average steam unit cost in 1994 was $11.16 per thousand pounds.

8.6.2 Economic considerations for consumers

The following economic benefits should be taken into account by a potential consumer when considering linking to a district heating and cooling system:

- Operating personnel for the consumer's HVAC system can be reduced or eliminated.
- Both property and liability insurance is significantly reduced with the elimination of the boiler in the consumer's mechanical room.
- Usable space is increased when a boiler or chiller and its ancillary equipment is no longer necessary, and can be removed.
- The noise level in the facility is reduced.
- Equipment maintenance needs are reduced with a corresponding reduction in maintenance staff.
- A large, central plant can usually achieve a higher thermal efficiency, and utilize advanced pollution-control equipment more economically.
- A central cogeneration can often operate on a wider range of fuels.

Of utmost significance in today's environment is the positive environmental contributions that can be achieved with a central district heating/cooling system. Some of these contributions, which often can be transformed in economic advantages, are

- Equipment with higher efficiency is used in central plants, leading to reduced emission levels of pollutants such as carbon dioxide and nitrogen oxides. Also, avoiding the need for delivery of fuel to some individual users can reduce the number of vehicles on roads.
- Delivering steam for absorption chillers, or centrally produced cold water, to individual consumers can reduce the usage of CFC refrigerants, which are in part considered responsible for depletion of the earth's atmospheric ozone layer.

Many studies report a variety of methods used for calculating the system costs and then distributing them to the consumers. Commonly, at least three different tariff elements are used:

- a connection charge consisting of a facility investment and service-line contributions; this charge is billed once the customer is connected to the network,
- a fixed charge consisting of subscription and system-capacity contributions and
- a variable-consumption charge.

8.7 Analogy between allocating wastes and economic costs for cogeneration

Many researchers have investigated the relations between economics and thermodynamics. In particular, a growing field of study linking exergy and economics, called thermoeconomics or exergoeconomics, has evolved.

One of the objectives of exergoeconomics, when it is applied to cogeneration processes, is determining the appropriate allocations of costs associated with the system with the co-products. The types of costs considered include both fixed capital costs and operating costs such as fuel costs. A good understanding of the proper allocations of costs is important because it allows individual product prices to be established that cover the costs of producing the products and allow for a margin or profit. In addition, such an understanding identifies when the product prices are such that the products are being sold below cost.

One of the outcomes of many exergy and economic studies is that the most appropriate way to allocate costs among the products of a cogeneration system may be to do so based on the exergy contents of the products. Other cost allocation methods, particularly those based on energy, are inadequate in that they divide costs in ways that radically differ from market prices.

Interestingly, there appears to be an analogy between the exergy-based methods for allocating costs and both the exergy-based method for allocating wastes (described later in Chapter 9) and the exergy-based method for allocating carbon dioxide emissions (described later in Chapter 10).

The analogy between allocating wastes such as carbon dioxide emissions and economic costs for cogeneration is illustrated in Figure 8.2. Although thought

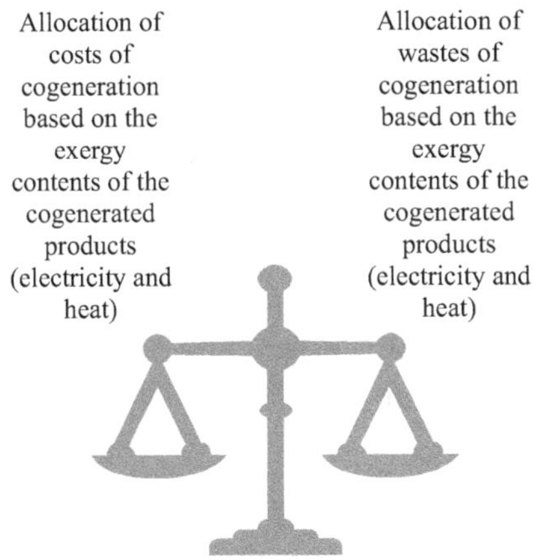

Figure 8.2 Conceptual illustration of the analogy between allocating wastes such as carbon dioxide emissions and economic costs for cogeneration. An equivalency is suggested between the allocation of costs and the allocation of wastes, and it is argued in this book that the allocations are best made based on the exergy associated with the products of cogeneration

provoking and intriguing, the analogy is in need of development. The insights this analogy brings forth can assist in permitting the allocation of costs to be more appropriately carried out and better understood. This topic is beyond the scope of the present book, as it is only in its infancy. Nevertheless, it would almost certainly be worthwhile to investigate further this analogy, as it may provide useful information for allocating of costs more appropriately.

8.8 Investigations of economics of cogeneration and district energy

Many economic investigations, often exergy-based, have been undertaken of processes for the cogeneration of electricity and heat as well as district energy.

8.8.1 Heating and district heating

Numerous assessments and/or optimizations have been reported of heating processes and district heating based on techno-economic factors including exergy.

8.8.1.1 Heating

Exergoeconomic analyses have been reported of ground-source heat pump systems for building applications [209].

A greenhouse heating system using a solar-assisted ground-source heat pump was assessed techno-economically by examining the relation between capital costs and thermodynamic losses for devices [210]. Ratios of thermodynamic loss rate to capital cost showed that devices in successful air conditioning are configured so as to achieve an overall optimal design by appropriately balancing the exergy-based and economic characteristics of the overall system and its devices.

A solar-assisted heating system for residential buildings has been assessed and optimized using exergoeconomics [211].

8.8.1.2 District heating

A ground-coupled heat pump with vertical underground heat exchangers has been investigated using exergoeconomic methods [212].

A more sophisticated multi-objective thermoeconomic optimization with an artificial intelligence method (evolutionary algorithm) was performed for a vertical ground-source heat pump [213]. The multi-objective optimization considers thermodynamic and thermoeconomic objectives simultaneously, with the assessment minimizing the total levelized cost of the system product or its exergy destruction.

Hot water piping systems have been optimized with thermoeconomics in work applicable to district heating [214], using several approaches:

- Simultaneous optimization of pipe diameter and insulation thickness based on the maximization of exergy efficiency and cost minimization.
- Sequential optimization of pipe diameter based on the minimization of total cost excluding heat losses.
- Sequential optimization of insulation thickness based on the minimization of cost of insulation and heat losses.

Exergoeconomic analyses have been reported of geothermal district heating systems [215].

8.8.2 Cooling and district cooling

Numerous assessments and/or optimizations have been reported based on techno-economic factors including exergy. Less work is reported for district cooling, although much of the work being undertaken for cooling can be applied to district cooling.

8.8.2.1 Vapor-compression chilling

Exergoeconomic techniques have been utilized to optimization a refrigeration evaporator coil with continuous fins [216].

The geometry of a precooling air reheater for air conditioning was optimized thermoeconomically [217]. The annualized total cost function was expressed as the precooler capital cost plus the operating cost attributable to the precooler irreversibility, and optimized based on the optimum heat transfer area and total irreversibilities.

An exergoeconomic optimization was reported of the geometry of continuous fins on a tube array on a refrigeration air-cooled condenser [218]. The total cost was based on annual capital cost and compensation cost for irreversibilities. For a given condenser capacity, system behavior was investigated considering changes in condenser tube diameter. An optimum heat exchanger area achieved an optimum balance among the exergy destruction components.

An exergoeconomic optimization was undertaken to determine the effect of matrix thermal conductivity on an air-conditioning rotary regenerator [219]. The operating cost was determined using different unit costs for the pressure and thermal components of exergy, and varying regenerator geometry revealed the effect of thermal conductivity on the minimum exergoeconomic objective function to be about 1 per cent for mild steel and 3 per cent for aluminum.

A steady-flow endoreversible refrigerator was optimized with finite-time exergoeconomic optimization for Carnot, Diesel, Otto, Atkinson and Brayton refrigeration cycles, utilizing profit rate as the objective [220]. The trade-off was evaluated between profit and COP.

8.8.2.2 Absorption chilling

An exergoeconomic optimization was reported of an aqua-ammonia absorption refrigeration system for air conditioning, by minimizing the overall operation and amortization cost and utilized exergetic costs [221]. A capital cost increase of only 2 per cent was seen to yield a 15 per cent product cost reduction.

Thermoeconomic analyses have been reported of single- and double-effect absorption refrigeration systems using a lithium bromide and water solution as the working fluid [222]. Direct-fired and hot-water heating are considered for single-effect refrigeration and a steam heating – with cogeneration – for double-effect refrigeration. The exergetic cost of the cool product is seen to be dependent on the heat source exergy and higher for the direct-fired system.

An industrial vapor-compression chiller of 1.5 MW cooling capacity was assessed with a scope-oriented thermoeconomic method [223] and subject to a related design optimization [224]. The roles of the condenser and the throttling valve are considered and marginal cost analyses are performed for key component attributes (condensation pressure, expander isentropic efficiency) to rationally allocate cost.

A finite-time exergoeconomic optimization of a combined refrigeration cycle was carried out with a profit objective [225] and demonstrated the balance between optimal profit and optimal COP.

8.8.3 Cogeneration and district energy

Economic investigations are common for the cogeneration of electricity and heat, often exergy-based, and oftentimes applicable to district energy.

For instance, a general approach was developed for assigning costs to the products of cogeneration [226].

Methods and applications for exergoeconomic evaluations and optimizations of cogeneration systems have been reported [227–230]. One of these included an exergy cost balance for each component and the overall system and an assessment of exergy cost generation within the system [230].

A simple cogeneration system was assessed with an iterative exergy-aided cost minimization [231].

8.8.3.1 Diesel engine cogeneration

Exergy and thermoeconomic analyses of diesel engine cogeneration have been performed [232,233]. Optimal exergoeconomic values (in US dollars) for total cost flow rate, unit electricity cost and unit steam cost, respectively, are determined to be 2,320 \$/h, 0.0670 \$/kWh and 0.0450 \$/kg with an iterative optimization procedure is used and 2,930 \$/h, 0.0890 \$/kWh and 0.0522 \$/kg for the non-optimized base case.

8.8.3.2 Gas turbine cogeneration

An optimization of simple gas turbine cogeneration showed the optimum turbine inlet temperature for exergy cost minimization to be 1,200°C [234], and that the gas turbine cost increased greatly at turbine-inlet temperatures above 1,300°C.

A 1,000-kW gas turbine cogeneration unit was assessed with an exergy-costing method [235]. An entropy-production flow was applied to system while material flows were decomposed into thermal, mechanical and chemical exergy.

Exergy and exergoeconomic analyses of a cogeneration system with a micro-gas turbine have been reported, considering exergy cost balances for the overall system and its components as well as the trade-off among exergy consumption and cost generation within the system [236].

A thermoeconomic analysis of an gas turbine cogeneration system containing a waste boiler identified the system using the Kawasaki Heavy Industries M1T-06 gas turbine as the best based on payback period, the system using the Hitachi Zosen CCS7 gas turbine to be the best based on maximum savings over 10 years [237]. The cogeneration system provides from a single fuel source one-third of an institutional facility's demand for 9-MW of electrical power and 1.8 kg/s of saturated steam at 0.86 MPa.

8.8.3.3 Combined cycle cogeneration

A thermoeconomic analysis of the Bilkent combined cycle cogeneration plant in Turkey was reported in [238].

Comparisons of several combined-cycle cogeneration configurations for varying steam demands were carried out using the specific-exergy costing method by applying cost balances to subsystems and unit exergy costs [239]. Configurations with condensing steam turbines that permit steam extraction were deemed advantageous thermoeconomically to those having back-pressure steam turbines.

The thermoeconomic potential for trigeneration of a biomass-driven steam-turbine plant showed that the overall production cost decreases with steam pressure and increases with steam temperature [240]. Four configurations were considered. Two involve dedicated steam production:

- Low-pressure saturated steam is generated to meet the process heat demand and electricity is purchased from the utility, and an electric vapor-compression chiller.
- High-pressure superheated steam is generated to meet the process heat demand and the absorption-chiller heat demand to produce cold water.

The other two configurations involve extraction steam:

- Steam is extracted at appropriate points on the turbine to meet the heat demands for process heating and cold water production in an absorption chiller and to generate electricity for export.
- Steam is extracted at appropriate points on the turbine to meet the heat demands for process heating and cold water production in an absorption chiller, without generating electricity for export.

8.9 Closure

The economics of cogeneration and district energy are described, as are relevant economic methods. A detailed case study is provided, and various economic considerations involved in integrated cogeneration-based district energy systems are considered. An analogy is presented between allocating wastes for cogeneration among products, and allocating economic costs among products. Numerous methods exist for the simulation and economic optimization of general energy systems, many of which are applicable to integrated cogeneration and district energy systems. A detailed methodology and computer code for such simulation and economic optimization, which can account for constraints such as environmental emission limits, is included in Chapter 11.

Chapter 9

Environmental impact of cogeneration systems: wastes and their allocation

Overview

The environmental impact of cogeneration systems is considered, with a focus on wastes from cogeneration and their allocations to products. The question of how to allocate wastes for an energy system with multiple products such as cogeneration has long provoked much debate and disagreement, in part because proposed methods lack consistency, simplicity, sound reasoning, ease of use, and universal – or at least widespread – acceptability. Various allocation methods for cogeneration wastes are described, including methods based on techno-economic measures such as energy content of products, exergy content of products, economic value of products, incremental fuel consumption to electrical production, incremental fuel consumption to thermal energy production and shared waste savings between electrical and thermal energy, as well as less rigorous waste-allocation methods. The rationale for the methods for allocating cogeneration wastes are detailed, with a particular focus on energy and exergy factors. The advantages of utilizing exergy methods for allocating cogeneration wastes and the balance they provide among cogenerated products are described. The authors in fact propose that exergy methods can form the basis of rational and meaningful allocation methods for wastes that are superior to other such allocation methods. Three detailed case studies are considered. Steam- and hot water-based cogeneration systems are considered in the first two case studies. In the third, a comparison is presented of the waste allocations for cogeneration and equivalent independent plants. The material in this chapter is particularly important because, by permitting wastes to be allocated more appropriately among the commodities of cogeneration, its environmental benefits can be better understood and exploited. These benefits should accrue to society through the better design and utilization of cogeneration technologies based on environmental considerations, and through better decision and policy-making by companies and government.

9.1 Introduction

Like other energy systems, systems for cogeneration or CHP release wastes that impact the environment and ecosystems. The types of wastes include solid, liquid and gaseous emissions, and vary with the specific technology applied and the energy source(s) that drive it. The overall waste releases to the environment are relatively straightforward to evaluate for a cogeneration system.

But the breakdown or allocation of the wastes to the multiple product outputs of cogeneration is not straightforward. In fact, many companies, government agencies and researchers have struggled with the question of how to allocate wastes for an energy system that has multiple products and multiple inputs. Some work has been carried out in this area, especially for cogeneration. For example, several attempts have been made to determine how to allocate emissions among the products of cogeneration systems [241–244].

However, the results obtained as of yet are not universally accepted and, in the view of the authors, are often not based on sound reasoning. Furthermore, the results obtained as of yet often conflicting. Further, the methods developed often are overly complex, rendering it difficult to use them and to convince decision and policy-makers of their potential benefits.

A relatively clear example of the problems in this area can be seen by examining the work on cogeneration systems. The several existing methods of allocating gaseous emissions among outputs include "efficiency methods," "work potential methods" and "heat content methods." The results obtained with each method are generally different. Further, the reasoning behind each is often suspect and/or lacking. The situation illustrated by this example for cogeneration systems becomes even more difficult to address for more complex systems, such as those involving trigeneration (the simultaneous production of electrical, heating and cooling services).

9.1.1 Motivation

The issues raised in the preceding paragraphs are difficult to manage and likely require advances. To provide insights to practitioners and researchers, they are described and discussed in this chapter. In particular, it is shown that the direct use of exergy methods can form the basis of sound and meaningful allocation methods for cogeneration wastes, although research is needed in this area.

Many governments have launched a number of initiatives involving air issues and the energy sector. Two examples follow:

- Much work on defining cogeneration and assessing the performance of systems and their emissions has been carried out by European government bodies and agencies. For instance, the European Parliament [245] issued a directive in February 2004 on the promotion of cogeneration based on a useful heat demand in the internal energy market. Also, the European Committee for Standardization (CEN) and European Committee for Electrotechnical Standardization (CENELEC) put forth in 2004 a workshop agreement manual for determination of CHP [246]. Although these two documents have their critics [247,248], they are widely accepted for primary energy savings for cogeneration. However, these documents ignore exergy, and thus suffer from the same difficulties of allocation methods mentioned earlier.

- Environment Canada has pursued the Ozone Annex, NOx/VOC Plan and acid rain initiatives, the Strategic Options Process for air toxics and the National Plan for Climate Change. Environment Canada is interested in emissions in industry and their implications for air quality across Canada.

With a move toward output-based standards or guidelines for wastes and comparative evaluations of wastes and pollution for all energy sources, better information is needed on estimating the equivalence along heat, cooling and electricity. Both shaft (mechanical) power and electricity are much more valuable than most if not all forms of industrial heat, but this depends on the temperature and quality of the steam or hot water extracted from the process. The normal definition of efficiency (fuel utilization) does not reflect appropriately this relationship.

Addressing cogeneration wastes and pollution, and consequent environmental impact, requires an appropriate allocation of credits or penalties for these among energy producers and users.

9.1.2 Outline of chapter

In this chapter, we discuss the environmental wastes associated with cogeneration systems. In addition, we investigate methods, based on exergy and other indicators, for allocating wastes for complex energy systems having multiple inputs and products, such as cogeneration systems. This method is compared with the other allocation methods, in as simple a manner as possible.

Exergy and energy methods of analysis are considered in these evaluations. Calculations are made, using actual and hypothetical case studies, of how much output power is lost, and thermal energy gained, when heat is captured or extracted for cogeneration purposes. Discussion is provided on an overall approach with regard to energy efficiency and shared allocations of cogeneration wastes.

The scope of this chapter includes the following:

- An examination of directly pertinent information in the literature on cogeneration wastes, along with an assessment of methods for allocating emissions discussed in relevant publications and correspondences (e.g., letters between the National Council for Air and Stream Improvement and World Resources Institute [243]).
- An investigation of allocation methods for wastes of cogeneration and a comparison of the methods for allocating wastes, for various energy commodities.
- Applications to several cogeneration system case studies of the waste-allocation methods based on various approaches, and a comparison of the results.

9.2 Wastes from cogeneration

The wastes from cogeneration plants are normally a combination of the wastes associated with the electricity generation process upon which the cogeneration plant is based, as well as the wastes associated with typical heating processes. The latter are again based on the process upon which the cogeneration plant is premised.

In simple terms, the wastes from cogeneration plants are basically dependent on the energy resource used. For fossil fuel-based cogeneration, the wastes are typically those associated with fossil fuel-based processes in general. For nuclear-based cogeneration, the main wastes are those typical of nuclear power plants, such as spent nuclear fuel and waste heat. For solar thermal-based cogeneration, there are few wastes.

The wastes, in general, include gaseous, liquid and solid wastes. For combustion-based cogeneration plants, for instance, the gaseous wastes are mainly the environmental emissions that exit the combustion chamber and are released through the stack. The liquid wastes are largely related to the spent cooling water used in some cogeneration facilities. The solid wastes are largely related to the ash and other solids left over from the combustion process.

As pointed out earlier, numerous governments have implemented various initiatives involving air, water and land quality and the energy sector as well as energy-intensive sectors. Environment Canada has pursued the Ozone Annex, NO_x/VOC Plan and acid rain initiatives, the Strategic Options Process for air toxics and the National Plan for Climate Change, and is interested in emissions in industry and their implications for air quality across the country. Efforts to define cogeneration and assess the performance of systems and their emissions have been carried out in Europe. For instance, the European Parliament [245] issued a directive promoting cogeneration based on a useful heat demand in the internal energy market, while the CEN and CENELEC developed a manual for determination of CHP [246]. These documents are widely accepted for primary energy savings for cogeneration, yet they ignore exergy and thus suffer from the same difficulties as mentioned earlier of allocation methods.

9.3 Allocation methods for cogeneration wastes

Limited work has been done on methods for allocating the wastes from energy systems that produce multiple products, such as cogeneration plants.

Given recent concerns about climate change, some methods have been developed for allocating carbon dioxide emissions from cogeneration systems to the electrical and thermal energy products, as is described in the next chapter. But not much effort has been put forth on extending these efforts to all wastes. The need for these methods is premised on the fact that when the owner of the cogeneration plant, the thermal energy user and the electrical energy user are not the same, a method for allocating the wastes is needed to ensure that each party is attributed with their appropriate share of the wastes from the system. In addition, having a meaningful allocation method allows the sources of wastes to be better understood and thereby reduced where and when appropriate.

Strickland and Nyboer [241,242] list seven methods of calculating the fuel allocation to the thermal and electrical energy products of a cogeneration system. The bases for these allocation seven methods are as follows

- Energy content of the products
- Exergy content of the products
- Economic value of the products

- Incremental fuel consumption to electrical production
- Incremental fuel consumption to the heat production
- Shared waste savings between heat and electricity
- Agreement

In their work, Strickland and Nyboer [241,242] adapt the calculational methods introduced earlier by Phylipsen *et al.* [244]. These methods can be generalized by multiplying the fuel allocation by the appropriate waste factor to evaluate the share of wastes allocated to each product. A generalized presentation is provided in the next section.

9.3.1 Selected methods for allocating cogeneration wastes

The allocation methods presented by Strickland and Nyboer [241,242] based on the work by Phylipsen *et al.* [244] are further adapted and simplified so that they present the fractions, rather than the total, wastes allocated to each product. The categorizations follow for convenience those used by Strickland and Nyboer [241,242].

9.3.1.1 Allocation of wastes based on energy content of products

The waste allocations of cogeneration are evaluated in proportion to the energy contents of the products as follows:

$$f_E = E/(E+Q) \tag{9.1}$$

$$f_Q = Q/(E+Q) \tag{9.2}$$

where f_E and f_Q denote, respectively, the fractions of the wastes allocated to the electrical and thermal products, and E and Q denote, respectively, the net outputs of electrical energy and thermal energy from the cogeneration system. Note that the term Q can represent an actual transfer of thermal energy, or the net thermal energy transferred via a material flow in and out of a heat exchange device.

Although straightforward and simple, this allocation method ignores or is silent on the quality of energy and focuses on the quantities involved only. Consequently, it can be argued that this method underestimates the share of wastes allocated to the electrical product and, by corollary, overestimates the share of wastes allocated to the thermal product.

9.3.1.2 Allocation of wastes based on exergy content of products

The allocations of the fraction of the wastes allocated to the electrical product f_E and the fraction of the wastes allocated to the thermal product f_Q are evaluated in proportion to the exergy contents of the products as follows:

$$f_E = Ex_E/(Ex_E + Ex_Q) \tag{9.3}$$

$$f_Q = Ex_Q/(Ex_E + Ex_Q) \tag{9.4}$$

where Ex_E and Ex_Q denote, respectively, the net outputs of electrical exergy and thermal exergy from the cogeneration system. It is noted that electrical energy and electrical exergy are equivalent, so that $Ex_E = E$.

In this allocation method, one can treat the thermal product in two ways:

- simply thermal energy and
- thermal energy transfer via heat transport materials.

Both these ways of treating thermal products are described below.

For simple thermal energy, the corresponding thermal exergy can be written as

$$Ex_Q = Q\tau \tag{9.5}$$

where τ denotes the exergetic temperature factor and can be evaluated as

$$\tau = 1 - T_o/T \tag{9.6}$$

Here, T denotes the temperature at which heat Q crosses the system boundary, and T_o denotes the temperature of the reference environment. For a reference environment at a temperature of 300 K (27°C), the value of the exergetic temperature factor τ is 0.25 for heat transfers at 400 K, 0.5 for heat transfers at 600 K and 0.75 for heat transfers at 1,200 K.

If, on the other hand, the thermal energy is delivered via a heat transport fluid, i.e., a material flow into and out of a heat exchange device, then the term Ex_Q is evaluated as the difference between the exergy of the incoming flow and the outgoing flow. The flowing commodity can be steam, hot water, cold water, etc. The factors that must be taken into account in determining the corresponding exergy flow rates include mass flow rate, temperature, pressure and sometimes vapor fraction.

An additional important point regarding the exergy-based allocation method is that the choice of reference environment for determining exergy quantities is significant, and can affect the results. Normal practice involves selecting a reference environment that is similar to the actual environment. But other reference environments can be used. For example, Upton [243], in an exercise to allocate emissions, evaluates exergy values using a reference-environment temperature of 100°C, which is well in excess of the actual environment annual mean temperature (perhaps 10°C to 20°C, depending on location). Upton [243] uses this value because it is related to the process; by inference it considers thermal energy below this temperature to be non-useful. Although this choice of a reference-environment temperature is permitted when using exergy methods, it is important to note that care must be exercised to ensure that a consistent reference environment is used throughout an analysis, or across multiple analyses if they are to be compared or contrasted.

It is noted that the type of exergy-based allocation method presented by Strickland and Nyboer [241,242] is not general, in that it assumes that the thermal product of cogeneration can be modeled only as pure thermal energy, rather than also considering the transfer of heat via flowing materials.

This allocation method accounts for the quality and quantity of the commodities involved. Consequently, compared to the energy allocation method, the exergy method avoids underestimating the share of the wastes allocated to the electrical product, and allocates a lower portion of the wastes to the thermal product.

9.3.1.3 Allocation of wastes based on economic value of products

The waste allocations are evaluated in proportion to the economic values of the products as follows:

$$f_E = c_E E/(c_E E + c_Q Q) \tag{9.7}$$

$$f_Q = c_E Q/(c_E E + c_Q Q) \tag{9.8}$$

where c_E and c_Q denote, respectively, the unit economic values of the electrical product and the thermal product of the cogeneration system.

Two important points are noted:

- The unit economic values presented here are on an energy basis (i.e., the economic value of a type of energy commodity per unit quantity of energy of that commodity), but the unit economic values based on other quantities – such as exergy – could also be used.
- The economic values can be determined using several economic measures in several ways. For instance, they can reflect as the costs to produce the commodities, or their prices, or some other economic measure for them).

This method is sometimes considered to have advantages to owners of cogeneration systems that sell electrical and thermal products separately.

It is not necessary to know the numerical values of both unit economic values, c_E and c_Q, when applying this method of waste allocation. Instead, it is the ratio of the unit economic values that is significant. This phenomenon can be seen by modifying the expressions for f_E and f_Q, respectively, for this method of allocation as follows:

$$f_E = E/\left[E + Q(c_E/c_Q)^{-1}\right] \tag{9.9}$$

$$f_Q = Q/\left[E(c_E/c_Q) + Q\right] \tag{9.10}$$

Here, c_E/c_Q denotes the ratio of the unit economic value for electricity to that for thermal energy. Since the unit economic value for electricity normally exceeds that for thermal energy, it is usually found that $c_E/c_Q > 1$.

9.3.1.4 Allocation of wastes based on incremental fuel consumption to electrical production

The allocations of wastes are evaluated here by dividing the total fuel consumed in the cogeneration among the electrical and thermal products, while considering electricity generation to be a byproduct of the thermal energy production process. Then, the wastes are allocated in proportion to the fuel division. This allocation method is consistent with the "Fuel Charged to Power" (FCP) method used by many cogeneration consulting firms.

Two steps are used to divide the fuel consumption.

First, the fuel consumption attributed to thermal energy production is evaluated as the hypothetical amount of fuel that would be consumed by an independent device for providing the same thermal energy as the cogeneration system (e.g., a reference steam boiler if the thermal energy is in the form of steam). That is,

$$F_Q = Q/\eta_b \tag{9.11}$$

where F_Q denotes the fuel consumption attributed to the production of thermal energy Q, and η_b denotes the energy efficiency of the independent device for providing the same thermal energy as the cogeneration system (e.g., a reference steam boiler).

Second, the fuel consumption attributed to electricity generation, F_E, is evaluated by subtracting this hypothetical amount of fuel from the total primary fuel energy consumed by the cogeneration system, F. That is,

$$F_E = F - F_Q \tag{9.12}$$

Then, the wastes allocation fractions f_E and f_Q are determined as the ratios F_E/F and F_Q/F, respectively. That is,

$$f_Q = Q/(F\eta_b) \tag{9.13}$$

$$f_E = 1 - f_Q \tag{9.14}$$

9.3.1.5 Allocation of wastes based on incremental fuel consumption to thermal energy production

This wastes allocation method is similar to the previous one, except the waste allocations are evaluated here by dividing the total fuel consumed in the cogeneration among the electrical and thermal products, while considering thermal energy production to be a byproduct of the electricity generation process. Again, the wastes are then allocated in proportion to the fuel division, using the following two steps to divide the fuel consumption.

First, the fuel consumption attributed to electricity generation is evaluated as the hypothetical amount of fuel that would be consumed by an independent device for providing the same electrical energy as the cogeneration system (e.g., a reference power plant). That is,

$$F_E = E/\eta_{pp} \tag{9.15}$$

where η_{pp} denotes the energy efficiency of the independent device for providing the same electrical energy as the cogeneration system (e.g., a reference power plant).

Second, the fuel consumption attributed to thermal energy production, F_Q, is evaluated by subtracting this hypothetical amount of fuel from the total primary fuel energy consumed by the cogeneration system. That is,

$$F_Q = F - F_E \tag{9.16}$$

Again, the waste allocation fractions f_E and f_Q are then determined as the ratios F_E/F and F_Q/F, respectively, as follows:

$$f_E = E/(F\eta_{pp}) \tag{9.17}$$

$$f_Q = 1 - f_E \tag{9.18}$$

9.3.1.6 Allocation of wastes based on shared waste savings between electrical and thermal energy

The allocations are evaluated for each product in proportion to the hypothetical fuel that would be used to produce that product independently, relative to the total hypothetical fuel that would be used to produce both products independently.

Using the terms previously defined, the hypothetical fuel consumption attributed to an independent process for thermal energy production (e.g., a reference steam boiler if the thermal energy is in the form of steam) is evaluated as

$$F_Q = Q/\eta_b \tag{9.19}$$

and the hypothetical fuel consumption attributed to an independent process for electricity generation is evaluated as

$$F_E = E/\eta_{pp} \tag{9.20}$$

Then, the waste allocation fractions f_E and f_Q are determined as the ratios F_E/F and F_Q/F, respectively, as follows:

$$f_E = (E/\eta_{pp})/(E/\eta_{pp} + Q/\eta_b) \tag{9.21}$$

$$f_Q = (Q/\eta_b)/(E/\eta_{pp} + Q/\eta_b) \tag{9.22}$$

This allocation method therefore shares the wastes among the products in a particular format. This method somewhat extends the concepts used in the previous two waste allocation methods, but is more of a compromise in terms of treating one or the other product as the primary one.

9.3.1.7 Allocation of wastes by agreement

Allocation of wastes to each product of cogeneration can be determined purely based on an agreement between the various parties involved in a project.

9.3.1.8 Allocation of wastes based on other factors

Wastes can also be allocated according to other formulas and based on other factors.

9.3.2 Rationale for allocating cogeneration wastes

The various waste allocation methods for cogeneration processes discussed in the previous subsection are examined and compared in subsequent sections. First, however, it is useful to understand the basic intentions and considerations in allocating wastes, and that is the topic of this subsection.

9.3.2.1 Fundamentals of allocating cogeneration wastes

Considering a general waste for a cogeneration processes, the total waste release, C, for a cogeneration (or multi-product production) process can be expressed as

$$C = C_E + C_Q \tag{9.23}$$

where C_E and C_Q denote, respectively, the waste associated with the electrical and thermal energy products, when they are produced in the cogeneration process. We can also express the total waste as

$$C = F\varphi \tag{9.24}$$

where F denotes the total fuel use in the process and φ a waste coefficient for the fuel. The terms F and φ must be on consistent bases (e.g., if F is in energy units, then φ must be the waste per unit fuel energy consumed). The total waste, C, can also be written as

$$C = (F_E + F_Q)\varphi \tag{9.25}$$

where F_E and F_Q denote, respectively, the fuel consumption associated with the electrical and thermal energy products, when they are produced in the cogeneration process. Clearly, $F_E + F_Q = F$. Furthermore, we can write:

$$C = (f_E + f_Q)F\varphi \tag{9.26}$$

where the fractions of fuel consumption associated with the electrical and thermal energy products, respectively, can be expressed as

$$f_E = F_E/F \tag{9.27}$$

$$f_Q = F_Q/F \tag{9.28}$$

Clearly, the fractions of fuel consumption associated with the electrical and thermal energy products are related as follows:

$$f_E + f_Q = 1 \tag{9.29}$$

9.3.2.2 Energy and exergy factors in allocating cogeneration wastes

From an energy basis, the fuel consumption associated with generating the electricity in the cogeneration process can be written as

$$F_E = E/\eta_E \tag{9.30}$$

where η_E denotes the energy efficiency of generating the electrical energy product within a cogeneration process.

Correspondingly, the fuel consumption associated with producing the thermal energy in the cogeneration process can be expressed as

$$F_Q = Q/\eta_Q \tag{9.31}$$

where η_Q denotes the energy efficiency of producing the thermal energy product within a cogeneration process.

Combining the above equations, we can write the following expressions for the fractions f_E and f_Q, respectively:

$$f_E = (E/\eta_E)/(E/\eta_E + Q/\eta_Q) \tag{9.32}$$

$$f_Q = (Q/\eta_Q)/(E/\eta_E + Q/\eta_Q) \tag{9.33}$$

Alternatively, we can use an exergy basis rather than an energy basis in establishing the above equations. Then, the fuel exergy consumption, Ex_{FE}, associated with generating the electrical exergy, Ex_E, in the cogeneration process can be expressed as

$$Ex_{FE} = Ex_E/\psi_E \tag{9.34}$$

where ψ_E denotes the exergy efficiency of generating the electrical energy product within the cogeneration process. Correspondingly, the fuel exergy consumption, Ex_{FQ}, associated with producing the thermal exergy, Ex_Q, in the cogeneration process can be expressed as

$$Ex_{FQ} = Ex_Q/\psi_Q \tag{9.35}$$

where ψ_Q denotes the exergy efficiency of producing the thermal energy product within the cogeneration process. Combining the above equations, we can write the following expressions for the fractions f_E and f_Q, respectively, using exergy terms:

$$f_E = (Ex_E/\psi_E)/(Ex_E/\psi_E + Ex_Q/\psi_Q) \tag{9.36}$$

$$f_Q = (Ex_Q/\psi_Q)/(Ex_E/\psi_E + Ex_Q/\psi_Q) \tag{9.37}$$

9.3.2.3 Exergy vs. energy in allocating cogeneration wastes: balance between cogenerated products

In this subsection, we discuss the trade-off between thermal and electrical products of cogeneration as well as the implications for waste allocations.

When an electrical generation process is modified so that it cogenerates, i.e., becomes a cogeneration process, often some of the electrical product is sacrificed for a gain in thermal output. When considering energy quantities, it is usually seen that the thermal energy gain is usually very great, even for a minor decrease in electrical energy output. Further, there is often no dependence on the temperature at which the thermal energy is delivered. When considering exergy quantities, however, the trade-off between electrical and thermal exergy products is more balanced. That is, a small decrease in electrical exergy output usually leads to a relatively small and similar in magnitude increase in thermal exergy output, while a large decrease in electrical exergy output usually leads to a correspondingly large increase in thermal exergy output. Also, the increase in thermal exergy is directly dependent on the temperature at which the thermal energy is delivered; generally, the greater is the temperature, the greater is the thermal exergy.

As an example, we consider a previous study a coal-fired electrical generating station by one of the authors [249–251]. There, the effects of modifying the electrical generating station for cogeneration demonstrated that the overall variation in exergy efficiency is relatively small, while the corresponding variation for the energy efficiency is large. These results are listed in Table 9.1, where the exergy efficiencies are observed to vary between 35 and 39 per cent, while the

Table 9.1 Energy and exergy efficiencies for a cogeneration plant based on various modifications to an electrical generating station[1]

Operating mode	**Product heat temperature (°C)**	**Efficiency (%)**	
		Exergy	**Energy**
Electricity generation only	–	37	37
Low-temperature cogeneration[2]	36	39	69
Medium-temperature cogeneration[2]	243	37	60
High-temperature cogeneration[2]	383	35	55

[1]Based on data in Rosen [249–251].
[2]For cogeneration cases, 50% of the resulting process heat is assumed to be useful product.

energy efficiencies vary between 37 and 69 per cent. Some interesting observations can be drawn from Table 9.1:

- The exergy results demonstrate that the benefits of cogeneration are not really due to the shift from electricity generation to heat production, since there is a balanced trade-off between the exergy of the two product commodities and the overall exergy efficiency remains relatively fixed.
- The main benefits of cogeneration are in reality due to the fact that the thermal energy produced offsets the need for a separate heat production process that uses additional fuel and that – on an exergy basis – is inefficient.
- The energy results present an entirely different perspective, one that is skewed due to the fact that energy analysis values electrical and thermal energy equally.

The observation that a decrease in electrical exergy output of a cogeneration plant usually leads to a relatively similar magnitude increase in thermal exergy output, but that a decrease in electrical energy output of a cogeneration plant usually leads to a dissimilar magnitude increase in thermal energy output, suggests the following:

- The exergy efficiency of generating the electrical product within the cogeneration process, ψ_E, is similar to the exergy efficiency of generating the thermal product within the cogeneration process, ψ_Q.
- The energy efficiency of generating the electrical product within the cogeneration process, η_E, is not similar to the energy efficiency of generating the thermal product within the cogeneration process, η_Q.

9.3.2.4 Advantages of exergy methods for allocating cogeneration wastes

As a consequence of the above two bullets, it can be seen from the analyses presented earlier that

- The allocation method based on exergy contents of the products (see section 9.3.1.2) most closely approximates the allocation expressions presented in sections 9.3.2.1–9.3.2.3.
- The other waste allocation methods in section 9.3.1 are significantly inaccurate relative to the objective of allocating wastes fairly and accurately.

These two bullets are discussed further in the next section, where the different waste allocation methods from section 9.3.1 are compared, bearing in mind the information presented in this section.

As a consequence of the analysis presented in this section, it can be seen that the exergy-based allocation method provides a rational means to determine the more productive modifications for a plant, when the objective is to reduce wastes.

In applying the exergy-based method for allocating wastes, there is a need to know the exergy contents of the various commodities that may be encountered in cogeneration. Such information is presented in Chapter 2 (see especially section 2.12).

Another advantage of the method of allocating wastes for cogeneration processes based on the exergy content of the products is that the allocation method is generalizable to any number and type of products. For instance, the exergy-based method can accommodate:

- cogeneration processes with multiple electricity and heat outputs,
- trigeneration processes (i.e., cogeneration processes in which, in addition to electricity and heat outputs, cooling capacity is also a product) and
- other processes producing two or more products (e.g., a desalination process that yields potable water and electricity, a fuel production process to produce hydrogen which also yields pure oxygen as a product or byproduct, or a chemical process yielding two different chemical commodities).

Most of the other allocation methods described in section 9.3.1 are much less flexible.

9.3.3 Further discussion and comparison of allocation methods for cogeneration wastes

Based on the information presented to this point in this section, the authors propose that the most rational and meaningful method of allocating wastes for cogeneration processes is to do so based on the exergy content of products. To justify this view, the different waste allocation methods discussed for cogeneration processes in section 9.3.1 are compared. In particular, the problems inherent in the other waste allocation methods for cogeneration processes are discussed.

The allocation method based on energy contents (section 9.3.1.1) leads to inaccurate breakdowns of the wastes, essentially because such a method presumes that the energy efficiency of generating the electrical product within a cogeneration process, η_E, is approximately similar to the energy efficiency of generating the thermal product within the cogeneration process, η_Q. As discussed earlier (section 9.3.2.3), this presumption is not valid, as values for η_E and η_Q can vary widely.

The allocation method based on shared savings between electrical and thermal energy (section 9.3.1.6) leads to inaccurate breakdowns of the wastes, essentially because such a method presumes that the energy efficiency of generating the electrical product within a cogeneration process, η_E, is approximately similar to the energy efficiency of generating the electrical product via a separate process, η_p. This presumption is not valid, as values for η_E and η_{pp} normally vary widely.

Similarly, this allocation method presumes that the energy efficiency of generating the thermal product within a cogeneration process, η_Q, is approximately similar to the energy efficiency of generating the thermal product via a separate process, η_b, again a presumption that is not valid, as values for η_Q and η_b normally vary widely. It makes sense that these efficiencies vary since one of the main reasons to consider cogeneration is that it allows one to generate two products simultaneously with a higher efficiency than would be the case if each product were produced in a separate and independent process.

Note that one could determine the shared allocations (section 9.3.1.6) based on exergy, rather than energy. Doing so would in fact overcome many of the problems associated with the shared waste allocation method based on energy. This observation is attributable to the fact that the exergy-based efficiencies for electricity generation in the part of a cogeneration system responsible for electricity generation and in a pure electricity generation process are similar (i.e., $\psi_E \approx \psi_{pp}$), while the exergy-based efficiencies for thermal energy production in the part of a cogeneration system responsible for thermal energy production and in a pure thermal energy production process are similar (i.e., $\psi_Q \approx \psi_b$). Thus, the shared allocation method based on exergy reduces approximately to the allocation method based on the exergy contents of the products.

The allocation methods based on incremental fuel consumption to either electrical production (section 9.3.1.4) or thermal energy production (section 9.3.1.5) both lead to inaccurate breakdowns of the wastes. The reasons are similar and follow below:

- The allocation method based on incremental fuel consumption to electrical production basically presumes erroneously that the energy efficiency of generating the thermal product within a cogeneration process, η_Q, is approximately similar to the energy efficiency of generating the thermal product via a separate process, η_b. Even more problematic, the method then presumes that the value of the energy efficiency for generating the electrical product within the cogeneration process, η_E, can simply be selected so that the overall wastes total correctly. The ensuing values of η_E can as a result vary radically and for the most part arbitrarily.
- Similarly, the allocation method based on incremental fuel consumption to thermal energy production essentially presumes erroneously that the energy efficiency of generating the electrical product within a cogeneration process, η_E, is approximately similar to the energy efficiency of generating the electrical product via a separate process, η_p. Further, the method then presumes that the value of the energy efficiency for generating the thermal product within the cogeneration process, η_Q, can simply be selected so that the overall wastes total correctly. As for the values of η_E in the preceding bullet, the ensuing values of η_Q can as a result vary radically and for the most part arbitrarily.

In general, the effect of the incremental-based allocations is that they arbitrarily underestimate the wastes from one of the products of a cogeneration process at the expense of the other. Both incremental-based allocations methods are thus unfair, since we seek the true and fair distribution of wastes among products – based on the efficiency of production for each within the cogeneration process.

It is noted that one could determine the incremental-based allocations (sections 9.3.1.4 and 9.3.1.5) based on exergy. However, this determination is not carried out here since the incremental allocation method is itself somewhat arbitrary and therefore not rational.

A common problem shared by the two incremental-based allocation methods (sections 9.3.1.4 and 9.3.1.5) and the shared allocation method (section 9.3.1.6) is that they introduce independent devices for providing thermal energy (e.g., a reference steam boiler) and electrical energy (e.g., a reference power plant). The results obtained using these allocation methods are dependent on the energy efficiencies of these independent devices (η_b for the reference steam boiler and η_{pp} for the reference power plant). But, the values of η_b and η_{pp} can vary relatively widely depending on the specific devices chosen (e.g., high- vs. medium- vs. low-efficiency models), and these variations cause the waste allocations evaluated with these methods to vary over correspondingly wide ranges.

The allocation method based on economic values of products (section 9.3.1.3) leads to inaccurate breakdowns of the wastes because that method allows economic parameters to skew the allocations. The proper allocations of wastes for a cogeneration process should be based entirely on principles of thermodynamics or physics. Economic parameters such as costs and prices vary with time and location, but proper waste allocations do not as they are dependent on characteristics of the technology involved. If one nevertheless chooses to modify the appropriate waste allocations by penalizing certain products in terms of their wastes, through economic or other means, then it must be recognized that the resulting waste allocations deviate arbitrarily from the appropriate waste allocations.

Similarly, the allocation method based an agreement between the various stakeholders in a project (section 9.3.1.7) leads to inaccurate breakdowns of the wastes because that method allows arbitrary factors that generally are not based entirely on principles of thermodynamics or physics to affect the allocations. If one nevertheless chooses to modify the appropriate wastes allocations by penalizing certain products in terms of their wastes, through factors such as agreements between various stakeholders, then it must be recognized that the resulting waste allocations almost certainly deviate arbitrarily from the appropriate waste allocations.

9.3.4 Section closure

In summary, it is pointed out that all of the allocation methods described in this section, except the exergy-based one, assign some arbitrary and/or subjective values to the differences between the product commodities. We need, instead, a rigorous scientific method, to help get the correct allocation and to remove the arbitrariness, and the exergy approach provides such a method. If, after determining the exergy-based allocations of wastes, we nevertheless choose to allocate wastes differently – for economic, political or other reasons – we can do so, but at least we do so knowing the appropriate unbiased allocation.

Note that the material covered in this section is a subset of the more general question of how to allocate a type of waste for a multi-product production process. The aim certainly is to allocate the waste to each product according to the actual waste that is in fact attributable to that product, accounting for all thermodynamic losses, when it is produced in the multi-product production process. Usually, the waste allocation breakdown is directly proportional to the breakdown of fuel use that is attributable to each product, when it is produced in the multi-product production process.

The methods for allocating wastes for cogeneration described in this section are demonstrated and illustrated for three case studies in subsequent sections. In the first two case studies, the methods of allocating wastes are illustrated for actual cogeneration plants. In the third case study, a hypothetical cogeneration scenario is compared to equivalent separate processes for producing thermal and electrical energy, and the implications of the different allocation methods are discussed. The third case study is extended in the next chapter to illustrate carbon dioxide emissions how the results could be used to determine carbon dioxide emission credits that might be used in emissions trading.

9.4 Case study 1: steam-based cogeneration

9.4.1 System description

This case study considers the University of Toronto Cogeneration System, an actual system in Toronto, Ontario, Canada.

Data for this system are drawn mainly from information sheets by Consumers Gas [252], as compiled by Klein [121].

The basic technical parameters for the cogeneration system are summarized in Table 9.2, including data on the cogeneration engine and cogenerated heat. Some general thermodynamic parameters for the cogeneration system from the literature [252] are presented in the top part of Table 9.3.

Table 9.2 Technical parameters for University of Toronto cogeneration system[1]

Parameter	Value
Engine type	Gas turbine
Heat use	Heating of campus using 6 km steam tunnel system
Heat quantity and type (base load)	13,607 kg/hr (30,000 lb/hr) of 1,379 kPa (200 psi) steam
Supplemental firing	Heat recovery steam generator can be supplementary-fired to 40,823 kg/hr (90,000 lb/hr) steam at 1,379 kPa (200 psi)
Environmental controls	Water injection to control nitrogen oxide emissions (to 42 ppm)
Installation date	1993

[1]Primary data source: Consumers Gas [252].

Table 9.3 Specified and evaluated thermodynamic parameters for University of Toronto cogeneration system

Parameter	Value
Fuel	
Type/composition	Natural gas
Supply rate (kg/s)	0.3949
Thermal product	
Form	Steam (dry saturated)
Mass flow rate (kg/s)	3.78
Temperature (°C)	197.6
Pressure (absolute) (bar)	14.8
Energy	
Fuel energy supply rate (MW)	19.75
Product electrical rate (MW)	6
Product thermal energy rate (MW)	10.393
Energy efficiency (%)	83
Exergy	
Fuel exergy supply rate (MW)	20.35
Product electrical exergy rate (MW)	6
Product thermal exergy rate (MW)	3.654
Exergy efficiency (%)	46.0

9.4.2 Energy and exergy values

Energy and exergy data for the University of Toronto Cogeneration System are presented in Table 9.3, including efficiencies and flow rates for products and inputs. The values are based on data in the literature [252] and evaluations by the present authors.

The energy efficiencies provided in sources are taken to be based on lower heating value. The input fuel is natural gas, which is modeled as methane in calculations. Furthermore, for calculating the chemical exergy of methane, the ratio of chemical exergy to lower heating value for methane is used, which is 1.03 based on data in Reference 26.

The reference-environment temperature and pressure are taken to be constant at 10°C and 1 bar, respectively. The conditions represent approximate annual means.

It can be seen that the energy and exergy flow rates for the fuel are similar, as are the energy and exergy flow rates for the electrical products. However, the product thermal energy and thermal exergy rates differ markedly for both processes, as do the energy and exergy efficiencies.

9.4.3 Waste allocation results

The results of applying the methods for allocating wastes for the University of Toronto cogeneration plant are presented in Table 9.4. These values are compared by illustrating them in Figure 9.1, which is based on the data in Table 9.4.

The allocations of wastes are seen to vary markedly depending on the allocation method employed. The authors contend that the exergy-based allocations are

Table 9.4 Allocation of wastes for University of Toronto cogeneration plant

Waste allocation basis	Waste allocation fraction	
	To electrical product	To thermal product
Energy content of products	0.366	0.634
Exergy content of products	0.621	0.379
Incremental fuel consumption allocated to electrical production[1]	0.415	0.585
Incremental fuel consumption allocated to heat production[2]	0.868	0.132
Shared waste savings between electricity and heat[1,2]	0.598	0.402
Economic value of products[3]	$5/[5+7(c_E/c_Q)^{-1}]$	$7/(5c_E/c_Q+7)$

[1]The heater/boiler that would have been used in the production of the same amount of heat as produced by cogeneration is assumed to have an energy efficiency of 90%.
[2]The power plant that would have been used in the production of the same amount of electricity as produced by cogeneration is assumed to have an energy efficiency of 35%.
[3]c_E/c_Q is the ratio of the economic value of the product electricity to that of the product heat.

the most appropriate and that using the other waste allocation methods can be misleading, since the resulting wastes may deviate widely from those obtained using the exergy-based method. For the allocation method based on economic value of products, in particular, the results depend on the value of the ratio of the economic value of the electricity produced c_E to the economic value of the thermal energy produced c_Q. The value of this ratio, even considering only the present time, varies with location. In Table 9.4, therefore, the values of the waste allocations for this method are left variable. In Figure 9.1, a range of c_E/c_Q values are considered.

So as to appreciate the wide range of possible waste allocations possible when using the method based on economic value of products, the waste allocation values are plotted in Figure 9.2 for a wide range of c_E/c_Q values, for the University of Toronto cogeneration plant. At a cost ratio of 1 (i.e., $c_E = c_Q$), electrical and thermal energy have the same economic value, while electricity is the more valuable commodity when $c_E/c_Q > 1$ and heat is more valuable when $c_E/c_Q < 1$. It is observed in Figure 9.2 that all wastes are attributable to heat for a value ratio $c_E/c_Q = 0$. As the value of the ratio increases, more wastes are shifted from heat to electricity. As the ratio approaches infinity, the wastes approach being entirely attributable to electricity.

9.5 Case study 2: hot water-based cogeneration with district energy

9.5.1 System description

This case study considers the Cornwall Cogeneration and District Energy System, an actual system in Cornwall, Ontario, Canada. Data for this system are drawn mainly from information sheets by Wiggin [253], as compiled by Klein [121]. Basic technical parameters and details for the cogeneration system are summarized

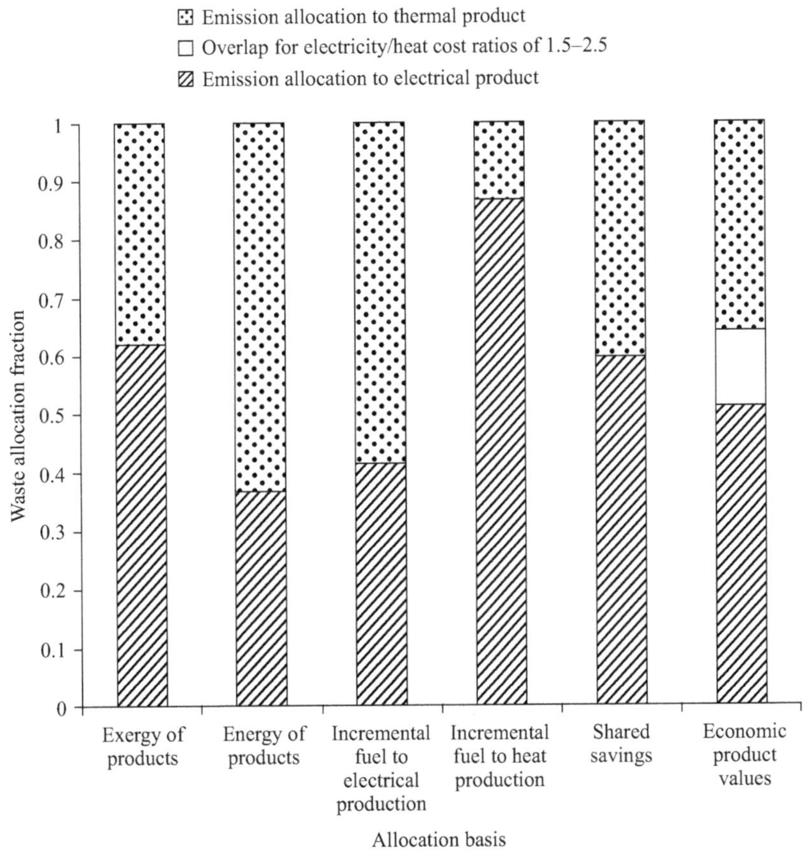

Figure 9.1 Allocation of wastes for University of Toronto cogeneration plant for various waste allocation bases

in Table 9.5. Some general thermodynamic parameters for the cogeneration system from Consumers Gas [252] are presented in the top part of Table 9.6.

9.5.2 Energy and exergy values

Energy and exergy data for the Cornwall Cogeneration and District Energy System are presented in Table 9.6, including efficiencies and flow rates for products and inputs. The values are based on data in the literature [253] and evaluations by the present authors.

As for the case study in the previous section, the energy efficiencies are based on lower heating value; the input fuel, natural gas, is modeled as methane and the ratio of chemical exergy to lower heating value for methane (1.03, based on data in Moran *et al.* [26]) is used in evaluating the chemical exergy of methane; and the reference-environment temperature and pressure are taken to be constant at approximate annual mean values: 10°C and 1 bar, respectively. Again, the energy and exergy flow rates for the fuel are similar, as are the energy and exergy flow

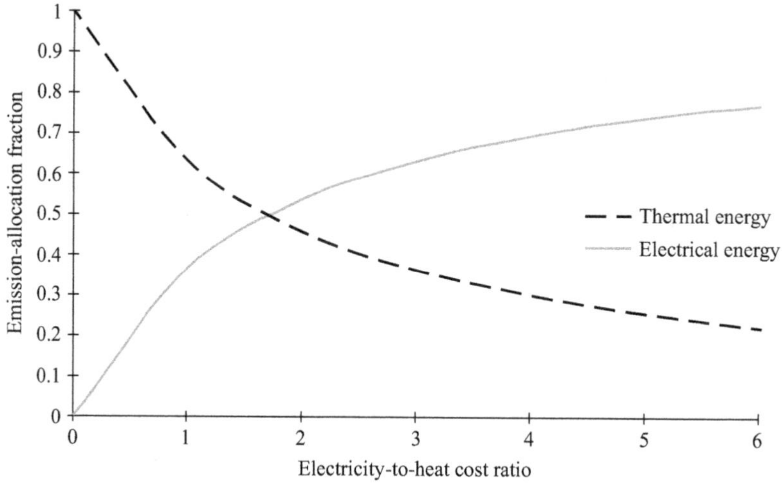

Figure 9.2 Variation of the emission-allocation fraction with the ratio of the cost of electricity to the cost of heat, when waste allocations are based on the economic values of the electrical and thermal products, for the University of Toronto cogeneration plant

Table 9.5 Technical parameters for Cornwall Cogeneration and District Energy System[1]

Parameter	Value
Engine type	Two reciprocating engines
Heat use	Municipal district heating using 4.5 km hot-water distribution network
Heat quantity and type (base load)	7 MW via 120°C and 1,585 kPa hot water
Supplemental firing	None
Environmental controls	Engines use lean-burn technology
Installation date	1995

[1]Primary data source: Wiggin [253].

rates for the electrical product, but the product thermal energy and thermal exergy rates as well as the energy and exergy efficiencies differ markedly.

9.5.3 Waste allocation results

The results of applying the methods for allocating wastes for the Cornwall Cogeneration and District Energy System are presented in Table 9.7. These values are compared by illustrating them in Figure 9.3, which is based on the data in Table 9.7.

For both case studies, it is clear that the allocations of wastes vary markedly, depending on the allocation method used. The authors contend, as discussed

Table 9.6 Specified and evaluated thermodynamic parameters for Cornwall Cogeneration and District Energy System

Parameter	**Value**
Fuel	
Type/composition	Natural gas
Supply rate (kg/s)	0.2660
Thermal product	
Form	Hot water
Mass flow rate (kg/s)	15.12
Temperature (°C)	120
Pressure (absolute) (bar)	16.85
Energy	
Fuel energy supply rate (MW)	13.3
Product electrical rate (MW)	5
Product thermal energy rate (MW)	7
Energy efficiency (%)	90
Exergy	
Fuel exergy supply rate (MW)	13.70
Product electrical exergy rate (MW)	5
Product thermal exergy rate (MW)	1.134
Exergy efficiency (%)	44.8

Table 9.7 Allocation of wastes for Cornwall Cogeneration and District Energy System

Waste allocation basis	**Waste allocation fraction**	
	To electrical product	**To thermal product**
Energy content of products	0.417	0.583
Exergy content of products	0.815	0.185
Incremental fuel consumption allocated to electrical production[1]	0.415	0.585
Incremental fuel consumption allocated to heat production[2]	1.074	−0.074
Shared waste savings between electricity and heat[1,2]	0.647	0.353
Economic value of products[3]	$6/[6+10.4\,(c_E/c_Q)^{-1}]$	$10.4/(6c_E/c_Q+10.4)$

[1]The heater/boiler that would have been used in the production of the same amount of heat as produced by cogeneration is assumed to have an energy efficiency of 90%.
[2]The power plant that would have been used in the production of the same amount of electricity as produced by cogeneration is assumed to have an energy efficiency of 35%.
[3]c_E/c_Q is the ratio of the economic value of the product electricity to that of the product heat.

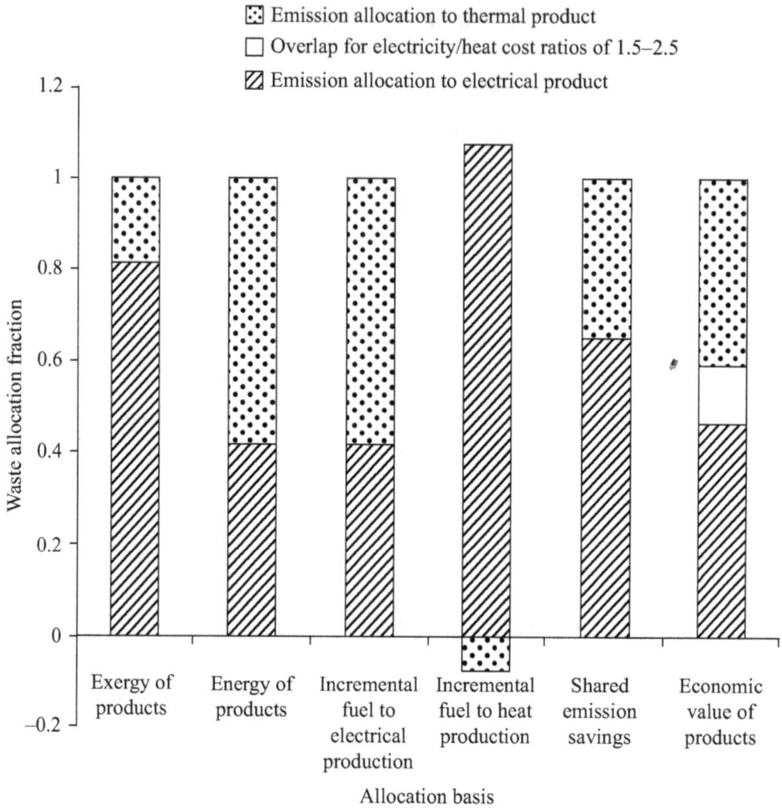

Figure 9.3 Allocation of wastes for Cornwall Cogeneration and District Energy System for various waste allocation bases

throughout this chapter, that the exergy-based allocations are the most appropriate. Thus, using the other waste allocation methods can be very misleading, since the resulting waste allocations may deviate widely from those obtained using the exergy-based method.

Some of the problems with the other waste allocation methods are illustrated in Table 9.7 and Figure 9.3. Some examples follow:

- Where the waste allocation is based on incremental fuel consumption to heat production for the Cornwall system, the illogical situation exists in which the allocations of wastes to the thermal product are evaluated to be negative and the allocations to the electrical product to exceed 100 per cent. This result is simply a consequence of the flaws in that allocation method and its use of an energy efficiency η_{pp} for an independent device for providing the same electrical energy as the cogeneration system (e.g., a reference power plant). Here, a value of $\eta_{pp} = 35\%$ is used. If, instead, the value of η_{pp} is selected such that $\eta_{pp} = 38\%$, then the allocations of wastes to both products would be positive and less than 100 per cent.

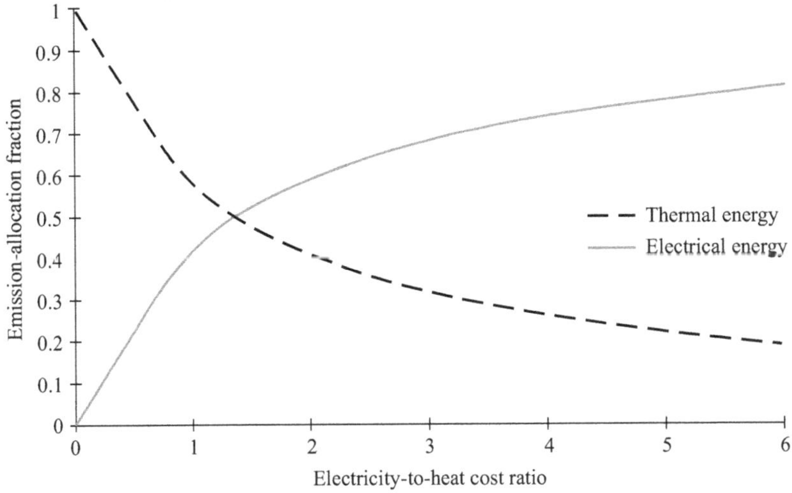

Figure 9.4 Variation of the emission-allocation fraction with the ratio of the cost of electricity to the cost of heat, when waste allocations are based on the economic values of the electrical and thermal products, for the Cornwall Cogeneration and District Energy System

- For the allocation method based on economic value of products, the results depend on the value of the ratio of the economic value of the electricity produced c_E to the economic value of the thermal energy produced c_Q. The value of this ratio, even considering only the present time, varies with location. In Table 9.7, therefore, the values of the waste allocations for this method are left variable. In Figure 9.3, a range of c_E/c_Q values are considered.

So as to appreciate the wide range of possible waste allocations possible when using the method based on economic value of products, the waste allocation values are plotted in Figure 9.4 for a wide range of c_E/c_Q values, for the Cornwall Cogeneration and District Energy System. At a cost ratio of 1 (i.e., $c_E = c_Q$), electrical and thermal energy have the same economic value, while electricity is the more valuable commodity when $c_E/c_Q > 1$ and heat is more valuable when $c_E/c_Q < 1$. It is observed in Figure 9.4 that all wastes are attributable to heat for a value ratio $c_E/c_Q = 0$. As the value of the ratio increases, more wastes are shifted from heat to electricity. As the ratio approaches infinity, the wastes approach being entirely attributable to electricity.

9.6 Case study 3: comparison of waste allocations for cogeneration and equivalent independent plants

9.6.1 Scenario description

A hypothetical cogeneration scenario is compared to equivalent separate processes for producing the same thermal and electrical energy, and the implications of the different allocation methods are described. The main characteristics of the processes being compared are as follows:

- The input fuel is natural gas in all cases.
- The hypothetical cogeneration system produces 4 MW of electrical power and 4 MW of thermal power from a fuel energy input rate of 10 MW.
- The overall energy efficiency of the hypothetical cogeneration system is therefore 80 per cent, and its electrical generation energy efficiency is 40 per cent as is its thermal energy production energy efficiency. The energy efficiencies are taken to be based on lower heating value.
- The separate processes consist of (i) an electricity generation system which produces 4 MW of electrical power from a fuel energy input rate of 10 MW and (ii) a heating system which produces 4 MW of thermal power from a fuel energy input rate of 5 MW.
- The energy efficiency of the overall (combined) process is 53 per cent (i.e., $8/15 \times 100\%$). The energy efficiency of the separate electricity generation system 40 per cent, while the energy efficiency of the separate heating system is 80 per cent.
- The reference-environment temperature and pressure are taken for all cases to be 10°C and 1 bar, respectively.

The 4 MW of thermal energy produced in each case is taken to be made up of 2 MW of steam and 2 MW of hot water. For simplicity, the state of the steam is taken to be the same as for the steam produced in the University of Toronto Cogeneration Plant described in the previous section (dry saturated steam at a pressure of 1,379 kPa (200 psi) gauge), while the state of the hot water is taken to be the same as for the hot water produced in the Cornwall Cogeneration and District Energy System described in the previous section (hot water at a temperature of 120°C and a pressure of 1,585 kPa gauge).

9.6.2 Energy and exergy values

Energy and exergy data for the cogeneration and independent processes for heating and electricity generation are presented in Table 9.8. Those data include efficiencies and flow rates for products and inputs. Details are presented in Table 9.9 on the two thermal products: steam and hot water.

The results in Table 9.8 demonstrate that the energy and exergy flow rates for the fuel are similar, as are the energy and exergy flow rates for the electrical products. However, the product thermal energy and thermal exergy rates differ markedly for both processes, as do the energy and exergy efficiencies.

Two particular results are observed in Table 9.8 regarding efficiency:

- Cogeneration is much more efficient (on energy or exergy bases) than the independent processes for producing the same thermal and electrical products.
- The exergy efficiencies are lower than the energy efficiencies, reflecting the fact that the thermal energy products are both of lower usefulness (or quality) than electricity. Of the two thermal products, the exergy values indicate that the usefulness of the steam is greater than that for the hot water.

Table 9.8 Specified and evaluated energy and exergy parameters for the cogeneration and independent processes

Parameter	**Independent processes**			**Cogeneration**
	Heating	**Electricity generation**	**Overall (combined)**	
Energy				
Fuel[1] energy input rate (MW)	5	10	15	10
Product electricity generation rate (MW)	–	4	4	4
Product thermal energy rate (MW)				
Steam	2	–	2	2
Hot water	2	–	2	2
Total	4	–	4	4
Energy efficiency (%)	80	40	53.3	80
Exergy				
Fuel[1] exergy input rate[2] (MW)	5.15	10.3	15.45	10.3
Product electrical exergy generation rate (MW)	–	4	4	4
Product thermal exergy rate (MW)				
Steam	0.777	–	0.777	0.777
Hot water	0.323	–	0.323	0.323
Total	1.100	–	1.100	1.100
Exergy efficiency (%)	21.4	38.8	33.0	49.5

[1]In all cases, the fuel is natural gas.
[2]Ratio of chemical exergy to lower heating value for methane is taken to be 1.03, based on data in Moran *et al.* [26].

Table 9.9 Specified and evaluated thermodynamic parameters for the thermal products[1] of the cogeneration and independent processes

Thermal product parameter	**Independent processes**			**Cogeneration**
	Heating	**Electricity generation**	**Overall (combined)**	
Product steam				
Steam mass flow rate (kg/s)	0.727	–	0.727	0.727
Steam temperature (°C)	197.6	–	197.6	197.6
Steam absolute pressure (bar)	14.8	–	14.8	14.8
Product hot water				
Hot water mass flow rate (kg/s)	4.32	–	4.32	4.32
Hot water temperature (°C)	120	–	120	120
Hot water absolute pressure (bar)	16.85	–	16.85	16.85

[1]In all cases, there are two types of thermal products: steam (dry saturated) and hot water.

9.6.3 Waste allocation results

The results of applying the methods for allocating wastes for the cogeneration and independent processes are presented in Table 9.10 and 9.11, and comparatively illustrated in Figure 9.5 (based on data in Table 9.10 and 9.11). Fractional breakdowns are shown for the allocation of wastes.

For the independent processes, there exists no ambiguity regarding the waste allocations to the thermal and electrical products. For the cogeneration process, the

Table 9.10 Waste allocations for the independent systems for producing heat and electricity

System	Waste-allocation basis	Waste-allocation fraction	
		To electrical product	To thermal product
Heating	All to single product	0.000	1.000
Electricity generation	All to single product	1.000	0.000
Overall (combined)	All electrical generation wastes to electricity generation and all thermal production wastes to thermal energy production	0.667	0.333

Table 9.11 Waste allocations for the cogeneration system

Waste-allocation basis	Waste-allocation fraction	
	To electrical product	To thermal product
Exergy content of products	0.784	0.216
Energy content of products	0.500	0.500
Incremental fuel consumption allocated to electrical production[1]	0.556	0.444
Incremental fuel consumption allocated to heat production[2]		
Reference power plant efficiency $\eta_{pp} = 35\%$	1.143	–0.143
Reference power plant efficiency $\eta_{pp} = 40\%$	1.000	0.000
Shared waste savings between electricity and heat[1,2]		
Reference power plant efficiency $\eta_{pp} = 35\%$	0.720	0.280
Reference power plant efficiency $\eta_{pp} = 40\%$	0.692	0.308
Economic value of products		
Electrical-to-thermal cost ratio $c_E/c_Q = 1.5$	0.600	0.400
Electrical-to-thermal cost ratio $c_E/c_Q = 1.8$	0.643	0.357
Electrical-to-thermal cost ratio $c_E/c_Q = 2.1$	0.677	0.323

[1]An efficiency of 90% is assumed for the boiler that would have been used in the production of the same amount of heat as produced by the cogeneration system.

[2]Efficiencies of 35% and 40% are considered for the power plant that would have been used in the production of the same amount of electricity as produced by the cogeneration system.

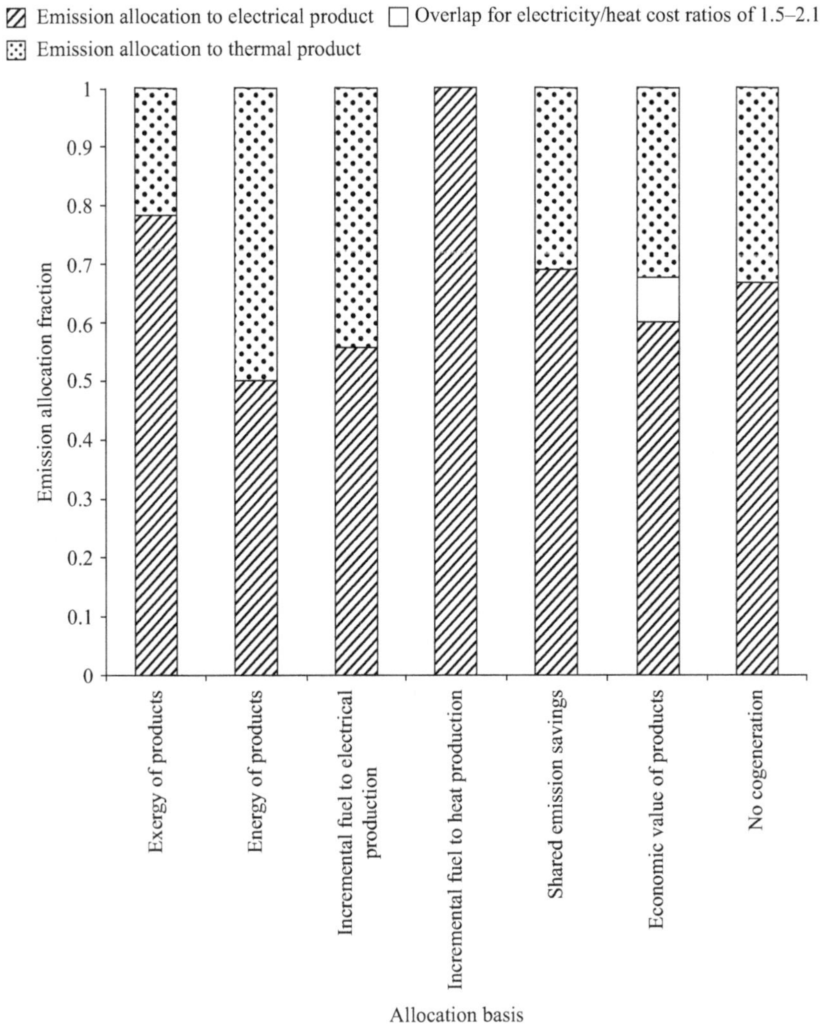

Figure 9.5 Fractional allocation of wastes for independent and cogeneration processes considered (based on a reference power plant efficiency of 40 per cent). The case at the right labeled "no cogeneration" represents independent processes for equivalent electricity generation and heat production

appropriate method to allocate wastes is not clear, so the allocation methods discussed earlier are applied. It is observed that the allocations of wastes vary markedly, depending on the allocation method used.

We contend, as discussed throughout this chapter, that the exergy-based allocations are the most appropriate. Thus, using the other allocation methods for wastes can be misleading, since the resulting waste allocations may deviate widely from those obtained using the exergy-based method.

Some of the problems with the other allocation methods are illustrated in Table 9.11 and Figure 9.5. Some examples follow:

- In one case (where the allocation is based on incremental fuel consumption to heat production), the absurd situation exists in which the allocations of wastes to the thermal product are evaluated to be negative and the allocations to the electrical product to exceed 100 per cent. This result is simply a consequence of the flaws in that allocation method and its use of an energy efficiency η_{pp} for an independent device for providing the same electrical energy as the cogeneration system (e.g., a reference power plant). Here, a value of $\eta_{pp} = 35\%$ is used. For comparison, an alternate value of η_{pp} is selected (40%); then, the waste allocations to both products do not exceed 100 per cent.
- For the allocation method based on the economic value of products, the result allocations depend on the value of the ratio of the economic value of the electricity produced c_E to the economic value of the thermal energy produced c_Q. The value of this ratio, even considering only the present time, varies with location. Here, for illustration only, economic parameter values from a U.S. report [254] are used. With these values, the cost of electricity is approximated as US$0.060/kWh, and the cost of steam as US$0.0334. The electrical-to-thermal cost ratio is then $c_E/c_Q = 1.8$. In Table 9.11, therefore, this value is used. Also, values of c_E/c_Q of 1.5 and 2.1 are used to indicate the sensitivity of the allocations to the cost ratio.

9.7 Closure

Cogeneration systems in general emit wastes that impact the environment and ecosystems. The wastes are dependent on the type of cogeneration technology and the energy source(s) that drive it. Although the overall waste releases to the environment are relatively straightforward to evaluate for a cogeneration system, the allocation of the wastes to cogenerated products is not straightforward. Of the many waste allocation methodologies, the exergy-based method allows for a rational and meaningful allocation of cogeneration wastes for systems. By permitting wastes to be allocated more appropriately among the different commodities generated in cogeneration systems, the results allow the environmental benefits of technologies that cogenerate to be better understood and exploited. These results should therefore allow the more beneficial among competing technologies to be identified in a rational and meaningful manner.

The results consequently indicate that the exergy-based method should be used in allocating wastes for cogeneration devices, as it can help ensure proper decision-making regarding issues such as

- how and where cogeneration technologies should be used,
- how wastes should be reduced in a given device,

- what effect introducing cogeneration will have on overall wastes and
- a fair way to establish detailed trading schemes for wastes where trading schemes are envisioned or appropriate (the case of the GHG carbon dioxide is considered specifically in the next chapter).

The results presented in this chapter should be of benefit to designers of energy systems, and to decision and policy-makers in companies and government. If the results are used appropriately, they should allow benefits to accrue to society through the selection and design of better energy technologies, based on environmental considerations.

The results may also be useful as input to the regulations developed in Europe for defining and assessing the benefits of cogeneration [245,246]. In particular, the incorporation of exergy into these regulations would likely improve them markedly, both in general and in particular for allocating wastes.

Indirectly, due to the analogy between cost and waste allocations, the results may also lead to economic benefits, as the results should permit the costs associated with cogeneration technologies to be more appropriately allocated among the different commodities generated.

Continuing research is merited to advance this work. In particular, the work should be expanded from cogeneration to trigeneration and other multigeneration systems. Also, the exergy-based allocation method needs to be applied at a more detailed level (e.g., using hourly plant data over a year rather than mean annual data).

Chapter 10

Climate change and cogeneration: addressing carbon dioxide emissions

Overview

Climate change issues related to cogeneration are examined, with a focus on carbon dioxide emissions and how to address them. The allocation of carbon dioxide emissions for cogeneration is highly contentious, in part because of the importance of mitigating them to address and also because proposed methods are often viewed as unsound and inconsistent, difficult of use because of their complexity, and lacking in widespread acceptability. After detailing carbon dioxide emissions from cogeneration, allocation methods for these emissions are described. This includes descriptions of selected methods for allocating carbon dioxide emissions for cogeneration, basic considerations in allocating such emissions, and a comparison of exergy- and energy-based methods for allocating such emissions. The authors contend that exergy methods provide the underpinnings of logical and practical allocation methods for carbon dioxide emissions, unlike other such allocation methods. A comprehensive case study is presented that compares carbon dioxide emissions allocations for cogeneration and independent plants and explains how to use this information for determining and trading carbon dioxide emissions credits. The latter includes discussions of carbon dioxide emissions credits for trading purposes from switching to cogeneration from equivalent independent plants, carbon dioxide emissions credits when an electricity user switches to cogeneration, carbon dioxide emissions credits when a heat user switches to cogeneration, and carbon dioxide emissions credits for other cases. The importance of the material in this chapter is noteworthy because it can permit the environmental benefits of cogeneration to be better understood and exploited – by allocating carbon dioxide emissions more appropriately among the electrical and thermal products of cogeneration. These benefits can improve cogeneration design and utilization, and enhance related decision-making by companies and policy-making by governments. In addition, exergy-based carbon dioxide emissions allocations provide a sensible, meaningful fair way to establish schemes for emissions trading.

10.1 Introduction

Climate change has been demonstrated by many to have a wide range of potential effects on human activities around the world and to be a significant risk to humanity. For instance, many predict that, if historical fossil fuel use trends continue and atmospheric concentrations of GHGs continue to increase, the Earth's mean surface temperature will increase by 2–6°C and the sea level will rise by 30–60 cm by 2100, and various disruptions will ensue such as sea-level rises and flooding of coastal settlements, reduced availability of fresh water, and displacement of fertile zones for agriculture and food production toward higher latitudes. Climate change has as a consequence become a significant concern to policy-makers and the public, and motivated studies and activities on climate change mitigation and adaptation.

Global climate change is believed to be attributable to the emissions of GHGs, the most significant of which is carbon dioxide. Cogeneration systems that operate using fossil fuels emit such gases, and many government bodies and industries need to know how to allocate GHG emissions for an energy system with multiple products like cogeneration. Some work has been carried out to determine how to allocate emissions among the products of cogeneration systems, as noted in Chapter 9. But methods developed to date are neither based on sound principles nor widely accepted, sometimes yielding conflicting results. The complexity of the methods makes them challenging to employ, making decision-makers in industry and government unlikely to use them. The complexity is evidenced by the fact that, to allocate cogeneration emissions among outputs, the methods proposed vary widely (e.g., efficiency, work potential, and heat content methods) provide quite varying results. This problem is even more pronounced for more complex systems like trigeneration.

Many initiatives have been instituted in recent years to address climate change and the emissions that contribute to it, with a focus on the energy sector or sectors that use energy heavily. For instance, Environment Canada has pursued a National Plan for Climate Change, whereby it seeks to understand the potential for emissions trading and energy quality in the industry and its implications. Also, European governments have been carrying out work to define cogeneration systems and to evaluate their performance and GHG emissions. Indicators for carbon dioxide emissions allocation from cogeneration are contained in a directive of the European Parliament [245] promoting cogeneration based on a thermal demand and a manual of the European Committee for Standardization and European Committee for Electrotechnical Standardization [246]. These documents are widely accepted for primary energy savings for cogeneration, although they have critics.

Activities that lead to climate change (sometimes referred to as global warming) are shown in Figure 10.1, along with the cascade of effects that ensue. A prime driver is shown to be emissions of GHGs. These are largely from anthropogenic activity, indicating that humanity is contributing through many of its economic and other activities. Carbon dioxide, the most significant GHG, mainly results from the combustion of fossil fuels, which account for the great majority of energy use globally. Other sources of GHG emissions include methane emissions from human activity (e.g., natural gas leaks, coal mining), CFC releases and deforestation. Energy-related activities are clearly major direct and indirect emitters of GHGs. Options for stabilizing climate change include restraining emissions of GHGs, improving efficiency and fuel substitution.

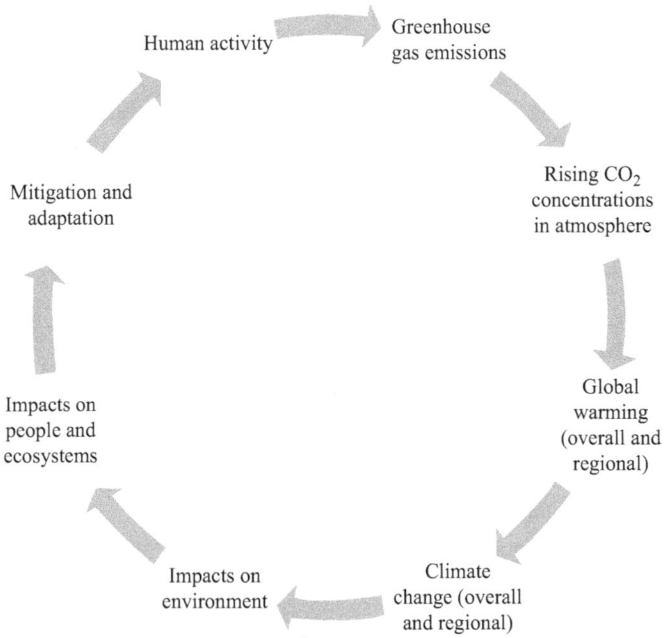

Figure 10.1 Activities that lead to climate change (sometimes referred to as global warming) and the cascade of effects that ensue, including efforts to mitigate or adapt to it. The cycle starts with human activity, which leads to GHG emissions

The need for standards for GHG emission guidelines, GHG emissions trading mechanisms, and comparative evaluations of GHGs, the equivalence between heat, cooling and electricity needs better quantification. The concept of comparing emission mass per unit energy output (in kg/MWh, for example) of carbon dioxide and other emissions must consider the trade-off between electricity and heat for cogeneration, if suitable comparisons are to be made. This has been approximated in the 1992 CCME Gas Turbine emission guideline. More definitive and universally acceptable work is needed to address this issue for energy products such as high- and low-pressure steam, and hot and cold water. Such information can also be useful for the conversion of emission factors from mass per unit energy (e.g., kg/MWh), into cost per unit mass (e.g., $/tonne) externalities and cost per unit energy (e.g., $/MWh) of reduction measures. Emissions trading of GHGs requires a shared allocation of credits/allowances for carbon dioxide and other relevant gases among industrial, commercial and other energy producers and users.

Methods for allocating GHG emissions for cogeneration systems are described in this chapter. The methods follow closely in many instances the numerous allocation methods for waste from cogeneration, covered in Chapter 9. Simplicity in methodology is sought where possible, although the need for accuracy is maintained. Discussions are included regarding how the methods can be used for GHG emissions trading. Note, however, that this chapter focuses on carbon dioxide; this is done for simplicity and because it is the primary GHG. But, the material is extendable to CO_2 equivalent emissions in terms of GHG potential.

This chapter covers existing methods for allocating carbon dioxide emissions reported in the literature, an assessment of the allocation methods for emissions based on various indicators and a comparison of the methods, applications to several example cogeneration systems of the carbon dioxide emissions allocation methods and a comparison of the results, and explanations of how the results can be used in future emissions-trading concepts.

10.2 Carbon dioxide emissions from cogeneration

General wastes from cogeneration plants are described in section 9.2, and include carbon dioxide. A more focused discussion on the carbon dioxide emissions is presented here, since it is a special case in that it is considered the primary contributor to climate change.

Carbon dioxide is a special type of emission in that it is not an avoidable waste when a hydrocarbon or other carbon-based fuel is utilized. The natural expended form of the carbon, after its chemical energy has been extracted, is carbon dioxide. This carbon dioxide either exits the plant or must be stored. The only real way to avoid carbon dioxide emissions is to not use carbon-based fuels, which essentially changes the basis of the process. By contrast, consider other types of pollutants such as sulfur and nitrogen oxides; these are avoidable via a wide range of engineering techniques, and their emission is not inherent to the use of the carbon-based fuel. Avoiding these emissions does not involve a fundamental change to the process in which they are used.

As for wastes in general, the carbon dioxide emissions for a cogeneration plant usually are a combination of the carbon dioxide emissions associated with the electricity generation process upon which the cogeneration plant is based, as well as the carbon dioxide emissions associated with the heating process(es) upon which the cogeneration plant is premised.

In simple terms, the carbon dioxide emissions from cogeneration plants are basically dependent on the energy resource or fuel used. Some specific cases follow:

- **Hydrocarbon-based cogeneration**. For hydrocarbon-based cogeneration, the carbon dioxide emissions normally are those associated with hydrocarbon-based processes in general. As these are well documented elsewhere, they are not discussed in detail here.
- **Biomass cogeneration**. The main carbon dioxide emissions typical of biomass cogeneration plants depend on the nature of the biomass fuel. For biomass grown using sunlight, there are usually no or very few net emissions of carbon dioxide. This is because the biomass growth is based on photosynthesis, which involves the input of carbon dioxde – typically from the atmosphere.
- **Nuclear- and renewable energy-based cogeneration**. For nuclear- and renewable energy-based cogeneration, hydrocarbon fuels are not in general used in

significant quantities during operation so there are no carbon dioxide emissions during operation (although there may be some over the life cycle of the plants).

Carbon capture and sequestration (CCS) is proposed as a means of avoiding the emissions of carbon dioxide, via techniques that separate and capture the carbon dioxide before it is emitted and then storing it for a long period of time (ideally permanently). The storage may involve holding the carbon dioxide in gaseous form, or it may involve converting it to a liquid and solid form that is more amenable to long-term storage. For fossil fuel-based plants, CCS is being examined and demonstrated for typical combustion processes that would normally release carbon dioxide through a stack.

10.3 Allocation methods for carbon dioxide emissions from cogeneration

Some work has been done on general methods for allocating carbon dioxide emissions from multiple-product energy systems, although some methods have been developed for allocating carbon dioxide emissions from cogeneration to its electrical and thermal energy products. These methods are important when the owner of the cogeneration system, the thermal energy user and the electrical energy user differ, since each party needs to be credited with its appropriate share of the CO_2 emissions, especially if carbon trading or caps are in place or envisioned. Meaningful allocation methods also permit reductions in the sources of carbon dioxide emissions by improving understanding.

As pointed out in Chapter 9, Strickland and Nyboer [241,242] list seven allocation methods of calculating the fuel allocation to the thermal and electrical energy products of a cogeneration system, including allocation based on product energy content, product exergy content, economic value of the products, incremental fuel consumption to electrical production, incremental fuel consumption to the heat production, shared emission savings between heat and electricity, and agreement. These methods were intended for use with carbon dioxide emissions and, in fact, are an adaption of the methods of Phylipsen *et al.* [244] in which the product of the fuel allocation and an appropriate carbon dioxide emission factor determines the share of carbon dioxide emissions allocated to each product.

Others have also investigated methods for allocating GHG emissions associated with industry, for example, the World Resources Institute, Washington, DC and the National Council for Air and Stream Improvement, Inc., Corvallis, OR [243]. In general, the allocation methods discussed by Upton [243] are variations on those discussed by Strickland and Nyboer [241,242] and Phylipsen *et al.* [244].

10.3.1 Simplified selected methods for allocating carbon dioxide emissions for cogeneration

The allocation methods presented in Chapter 9 for wastes, based on the work of Strickland and Nyboer [241,242] and Phylipsen *et al.* [244], are further adapted in this section to provide fractions of carbon dioxide emissions allocated to each product. In this chapter, the terms f_E and f_Q in Chapter 9 become the fractions of the carbon dioxide emissions allocated to the electrical and thermal products for a cogeneration system, respectively.

The allocation of carbon dioxide based on energy content of products is carried out by allocating in proportion to the product energy contents, as done in section 9.3.1.1 for wastes, as follows:

$$f_E = E/(E+Q) \tag{10.1}$$

$$f_Q = Q/(E+Q) \tag{10.2}$$

where E and Q remain, respectively, the net outputs of electrical energy and thermal energy from the cogeneration system. As pointed out earlier, this simple method ignores energy quality, and so likely underestimates the share of the emissions allocated to the electrical product.

The allocation of carbon dioxide based on exergy content of products contents, as in section 9.3.1.2 for wastes, involves allocating in proportion to the exergy contents. That is,

$$f_E = Ex_E/(Ex_E + Ex_Q) \tag{10.3}$$

$$f_Q = Ex_Q/(Ex_E + Ex_Q) \tag{10.4}$$

where Ex_E and Ex_Q remain, respectively, the net outputs of electrical exergy and thermal exergy from the cogeneration system, as explained in section 9.3.1.2 and, more generally, in Chapter 2. Since this allocation method accounts for the quality and quantity of the commodities involved, it avoids underestimating the share of the carbon dioxide emissions allocated to the electrical product as done with the energy method, and allocates a lower portion of the carbon dioxide emissions to the thermal product.

The allocation of carbon dioxide based on economic value of products is performed by allocating in proportion to the economic values of the products, following in section 9.3.1.3 for wastes, as follows:

$$f_E = c_E E/(c_E E + c_Q Q) = E/\left[E + Q(c_E/c_Q)^{-1}\right] \tag{10.5}$$

$$f_Q = c_E Q/(c_E E + c_Q Q) = Q/\left[E(c_E/c_Q) + Q\right] \tag{10.6}$$

where c_E and c_Q remain, respectively, the unit economic values of the electrical product and the thermal product of the cogeneration system, and c_E/c_Q denotes the ratio of the unit economic value for electricity to that for thermal energy. The unit economic value for electricity usually exceeds that for thermal energy, so normally $c_E/c_Q > 1$. The unit economic values are on an energy basis, but the method can be altered if desired to accommodate unit economic values based on exergy or other quantities. The economic values are general here, but can represent commodity production costs or prices or other economic measures.

The allocation of carbon dioxide based on incremental fuel consumption to electrical energy production is implemented, as done in section 9.3.1.4 for wastes, by allocating carbon dioxide emissions in proportion to a specific fuel division, in which the total fuel consumed in the cogeneration system is divided among the electrical and thermal products while considering electricity generation to be a byproduct of the thermal energy production process. That is,

$$f_Q = F_Q/F = Q/(F\eta_b) \tag{10.7}$$

$$f_E = F_E/F = 1 - f_Q \tag{10.8}$$

where F denotes the total primary fuel energy consumed by the cogeneration system, F_Q denotes the fuel consumption attributed to the production of thermal energy Q based on a hypothetical independent device for providing the same thermal energy as the cogeneration system, η_b denotes the energy efficiency of the independent device for providing the same thermal energy as the cogeneration system, and F_E denotes the fuel consumption attributed to electricity generation E.

The allocation of carbon dioxide based on incremental fuel consumption to thermal energy production, conversely, is implemented, as done in section 9.3.1.5 for wastes, by allocating carbon dioxide emissions in proportion to a specific fuel division, in which the total fuel consumed in the cogeneration system is divided among the electrical and thermal products while considering thermal energy production to be a byproduct of the electricity generation process. That is,

$$f_E = F_E/F = E/(F\eta_{pp}) \tag{10.9}$$

$$f_Q = F_Q/F = 1 - f_E \tag{10.10}$$

where F_E denotes the fuel consumption attributed to electricity generation E based on a hypothetical independent device for providing the same electrical energy as the cogeneration system, η_{pp} denotes the energy efficiency of the independent device for providing the same electrical energy as the cogeneration system, and F_Q denotes the fuel consumption attributed to the production of thermal energy Q.

The allocation of carbon dioxide based on shared emission savings between electrical and thermal energy is performed, as done in section 9.3.1.6 for wastes, by allocating carbon dioxide emissions in proportion to the hypothetical fuel that would be used to produce each product independently, relative to the total hypothetical fuel that would be used to produce both products independently, using the terms for the previous two methods. That is, the emission allocation fractions are determined in a manner that shares the emissions among the products as follows:

$$f_E = F_E/F = (E/\eta_{pp})/(E/\eta_{pp} + Q/\eta_b) \tag{10.11}$$

$$f_Q = F_Q/F = (Q/\eta_b)/(E/\eta_{pp} + Q/\eta_b) \tag{10.12}$$

By extending the concepts in the previous two carbon dioxide emissions allocation methods, this method is somewhat of a compromise in terms of treating one product or the other as primary.

The allocation of carbon dioxide based agreement, as in section 9.3.1.7 for wastes, involves allocating solely based on an agreement between the various parties involved in a project in proportion to the exergy contents. Although this method can seem and is relatively arbitrary, it can be important for carbon dioxide due to agreements that are being made in many jurisdictions and countries, and that are being sought globally, to combat climate change.

Finally, it is noted that, as for wastes in general, carbon dioxide emissions can be allocated according to other formulas and based on other factors. Such other methods may prove necessary, given the complexities of treaties and other instruments being implemented to limit or reduce emissions of carbon dioxide and other GHGs, in many jurisdictions and countries and even globally, as part of efforts to mitigate global warming and climate change.

10.3.2 Basic considerations in allocating carbon dioxide emissions for cogeneration

Following the description wastes in the previous chapter but considering the waste to be gaseous carbon dioxide, the total carbon dioxide emissions for a cogeneration process, again denoted C, here and throughout this chapter, can be written as

$$C = C_E + C_Q \tag{10.13}$$

where C_E and C_Q denote, respectively, the CO_2 emissions associated with the electrical and thermal energy products of a cogeneration process. We can also express the total CO_2 emissions as

$$C = F\varphi = (F_E + F_Q)\varphi = (f_E + f_Q)F\varphi \tag{10.14}$$

where F_E and F_Q, respectively, denote the fuel consumption associated with the electrical and thermal energy products of cogeneration, F $(=F_E + F_Q)$ denotes the total fuel use in the process, and φ denotes a CO_2 emission coefficient for the fuel. Also, f_E $(=F_E/F)$ is the fraction of fuel consumption associated with the electrical product and f_Q $(=F_Q/F)$ is the fraction of fuel consumption associated with the thermal product. These terms are related as follows: $f_E + f_Q = 1$.

As described in the previous chapter for a general waste, we can write the following expressions for the fractions f_E and f_Q on an energy basis:

$$f_E = (E/\eta_E)/(E/\eta_E + Q/\eta_Q) \tag{10.15}$$

$$f_Q = (Q/\eta_Q)/(E/\eta_E + Q/\eta_Q) \tag{10.16}$$

Here, F_E $(=E/\eta_E)$ denotes the fuel consumption associated with generating the electricity via cogeneration, F_Q $(=Q/\eta_Q)$ denotes the fuel consumption associated with producing the thermal energy via cogeneration, η_E denotes the energy efficiency of generating the electrical energy product within a cogeneration process, and η_Q denotes the energy efficiency of producing the thermal energy product within a cogeneration process.

Analogously, we can write the following expressions for the fractions f_E and f_Q on an exergy basis:

$$f_E = (Ex_E/\psi_E)/(Ex_E/\psi_E + Ex_Q/\psi_Q) \tag{10.17}$$

$$f_Q = (Ex_Q/\psi_Q)/(Ex_E/\psi_E + Ex_Q/\psi_Q) \tag{10.18}$$

Here, Ex_{FE} ($=Ex_E/\psi_E$) denotes the fuel exergy consumption associated with generating the electrical exergy Ex_E via cogeneration, Ex_{FQ} ($=Ex_Q/\psi_Q$) denotes the fuel exergy consumption associated with producing the thermal exergy Ex_Q via cogeneration, ψ_E denotes the exergy efficiency of generating the electrical energy product within the cogeneration process, and ψ_Q denotes the exergy efficiency of producing the thermal energy product within the cogeneration process.

10.3.3 Exergy vs. energy in allocating cogeneration CO_2 emissions

Since a decrease in the electrical exergy generated via cogeneration normally is associated with a relatively similar magnitude increase in thermal exergy output, but a decrease in electrical energy output of cogeneration normally is associated with a dissimilar magnitude increase in thermal energy output, we noted in the previous chapter that the exergy efficiency of generating the electrical product of cogeneration, ψ_E, is similar to the exergy efficiency of generating the thermal product within cogeneration, ψ_Q, while the energy efficiency of generating the electrical product of cogeneration, η_E, is not similar to the energy efficiency of generating the thermal product within the cogeneration process, η_Q. This suggests that the allocation method for carbon dioxide based on exergy contents of the products (see section 10.3.1) most closely approximates the allocation expressions presented in section 10.3.2, while the other emissions allocation methods for carbon dioxide in section 10.3.1 are significantly inaccurate relative to the objective of allocating emissions fairly and accurately. Thus, the exergy-based allocation method provides a rational means to determine the more productive modifications for a plant, when the objective is to reduce CO_2 emissions.

As a consequence, the most rational and meaningful method of allocating carbon dioxide emissions for cogeneration can be argued to be based on the exergy content of products. To expand, the carbon dioxide emissions allocation methods for cogeneration from section 10.3.1 are compared, pointing out problems inherent in the other CO_2 emission allocation methods for cogeneration.

The allocation method based on energy contents (section 10.3.1) leads to inaccurate breakdowns of the carbon dioxide emissions since it is premised on the energy efficiency of generating the electrical product via cogeneration η_E is approximately similar to the corresponding energy efficiency η_Q. Since η_E and η_Q can vary significantly, this premise is not valid (see section 10.3.2).

The allocation method based on shared emission savings between electrical and thermal energy (section 10.3.1) leads to inaccurate breakdowns of the carbon dioxide emissions because it assumes that the energy efficiency of generating the electrical product via cogeneration η_E is approximately similar to the energy efficiency of generating the electrical product via a separate process η_{pp} and because it assumes that the energy efficiency of generating the thermal product via cogeneration η_Q is approximately similar to the energy efficiency of generating the thermal product via a separate process η_b. These assumptions are not valid since η_E and η_{pp} normally vary significantly, as do η_Q and η_b. But, as noted in the previous chapter for a general waste, the shared-emissions allocations (section 10.3.1) could for carbon dioxide be based on exergy rather than energy, and this could overcome many of the problems associated with the shared-emissions allocation method based on energy. This is because the exergy-based efficiencies for electricity generation in the part of a cogeneration system responsible for electricity generation and in a pure electricity generation process are similar (i.e., $\psi_E \approx \psi_{pp}$), as are the exergy-based efficiencies for thermal energy production in the part of a cogeneration system responsible for thermal energy production and in a pure thermal energy production process (i.e., $\psi_Q \approx \psi_b$), approximating the allocation method for carbon dioxide based on the exergy contents of the products.

The allocation methods based on incremental fuel consumption to either electrical production or to thermal energy production (section 10.3.1) both lead to inaccurate breakdowns of the carbon dioxide emissions. This can be explained by noting that the allocation method based on incremental fuel consumption to electrical production assumes incorrectly that the energy efficiency of generating the thermal product within a cogeneration process η_Q is approximately similar to the energy efficiency of generating the thermal product via a separate process η_b, and that the energy efficiency for generating the electrical product within the cogeneration process η_E can be selected just so the overall emissions total. An analogous explanation relates the allocation method based on incremental fuel consumption to thermal energy production, as pointed out in the previous chapter for a general waste. The impact of the incremental-based allocations is significant, in that they somewhat arbitrarily underestimate the emissions from one cogenerated product at the expense of the other, making both incremental-based allocations methods unfair. This is particularly so because of a problem shared by both incremental-based allocation methods (section 10.3.1), as well as the shared-emissions allocation method: they introduce independent devices for providing thermal energy and electrical energy which can vary widely depending on the specific devices assumed, causing the emissions allocations evaluated with these methods to vary widely correspondingly.

The allocation method based on economic values of products (section 10.3.1) provides inaccurate divisions of the carbon dioxide emissions because the results vary as economic conditions (costs, prices, interest rates, etc.) change, whereas allocations of carbon dioxide emissions in reality ought to be based on technical principles and the characteristics of cogeneration technologies. However, given efforts to develop carbon taxes and cap-and-trade schemes for GHG emissions, an allocation method based on economic principles could be useful from a practical viewpoint. But if carbon dioxide emissions allocations are penalized through economic means, the resulting emissions allocations likely deviate from the appropriate emissions allocations. Similarly, the allocation method based an agreement between the various stakeholders (section 10.3.1) leads to inaccurate breakdowns of the carbon dioxide emissions because it allows non-technical factors to affect the allocations. As with the allocation method based on economic values of products, this method could intentionally include factors that are aimed at reducing carbon dioxide emissions, and thus could have practical use. But, again, if carbon dioxide emissions allocations by skewed through agreements among stakeholders, the resulting emissions allocations likely deviate from the appropriate emissions allocations for carbon dioxide.

Hence, all of the allocation methods for carbon dioxide emissions described in section 10.3.1, except the exergy-based one, introduce some arbitrary values, whereas a rigorous technical method is needed to determine the correct allocation. Only the exergy-based allocation approach provides this. Given political efforts to reduce carbon dioxide emissions, allocations may be established for carbon dioxide emissions that incorporate economic, political or other factors, even if the resulting allocations are not sound from a purely technical basis.

The methods for allocating carbon dioxide emissions for cogeneration plants that are described in section 10.3.1 are illustrated for a case study in the next section. There, a hypothetical cogeneration scenario is compared to equivalent separate processes for thermal and electrical energy, the implications of the carbon dioxide allocation methods are determined, and possible applications of the methods for determining carbon dioxide emission credits for emissions trading are examined.

10.4 Case study: comparing CO_2 emissions allocations for cogeneration and independent plants and use for determining and trading CO_2 emissions credits

In this section, the case study in section 9.6 of Chapter 9 is extended to determine the carbon dioxide emission for a hypothetical cogeneration scenario and to compare it to equivalent separate processes for producing the same thermal and electrical energy. The various allocation methods described in the previous section are utilized. Finally the manner in which the results can be used to determine carbon dioxide emission credits that might be used in emissions trading schemes is illustrated and discussed.

10.4.1 Recap of case study 3 of section 9.6: comparison of waste allocations for a cogeneration system and equivalent independent plants

In Case Study 3 of section 9.6, waste allocations for a cogeneration plant and equivalent independent plants, operating in a reference environment at a temperature of 10°C and a pressure of 1 bar, are evaluated and compared. The hypothetical cogeneration system is 80 per cent efficient on an energy basis, and produces 4 MW of electrical power and 4 MW of thermal power from an energy input rate of 10 MW in the form of natural gas fuel. The separate electricity generation system produces 4 MW of electrical power from a natural gas energy input rate of 10 MW. The separate heating system produces 4 MW of thermal power from a natural gas energy input rate of 5 MW. The energy efficiency of the combined separate systems is 53 per cent. In both cases, the 4 MW of thermal power produced is taken to be made up of 2 MW of steam (dry saturated at a pressure of 200 psi gauge) and 2 MW of hot water (at a temperature of 120°C and a pressure of 1,585 kPa gauge). Selected parameters for the cogeneration and independent processes, drawn from Tables 9.8 and 9.9, are shown in Table 10.1.

Energy and exergy analyses show that the energy and exergy flow rates for the fuel are similar, as are the energy and exergy flow rates for the electrical products. However, the product thermal energy and thermal exergy rates differ markedly for both processes, as do the energy and exergy efficiencies. The cogeneration system is observed to have a significantly higher energy efficiency and exergy efficiency than the combined independent processes for producing the same thermal and electrical products. But the exergy efficiencies are found to be lower than the energy efficiencies, reflecting the lower usefulness or quality or value of the thermal energy product compared to the electrical product.

Table 10.1 Selected parameters for cogeneration and independent processes

Parameter	Independent processes			Cogeneration
	Heating	**Electricity generation**	**Overall (combined)**	
General				
Fuel	Natural gas	Natural gas	Natural gas	Natural gas
Thermal products	Dry saturated steam (0.727 kg/s at 120°C and 14.8 bar) Hot water (4.32 kg/s at 197.6°C and 16.85 bar)	–	Dry saturated steam (0.727 kg/s at 120°C and 14.8 bar) Hot water (4.32 kg/s at 197.6°C and 16.85 bar)	Dry saturated steam (0.727 kg/s at 120°C and 14.8 bar) Hot water (4.32 kg/s at 197.6°C and 16.85 bar)
Energy				
Fuel energy input rate (MW)	5	10	15	10
Electrical energy generation rate (MW)	–	4	4	4
Product thermal energy rate (MW)	4	–	4	4
Energy efficiency (%)	80	40	53.3	80
Exergy				
Fuel exergy input rate (MW)	5.15	10.3	15.45	10.3
Electrical exergy generation rate (MW)	–	4	4	4
Product thermal exergy rate (MW)	1.100	–	1.100	1.100
Exergy efficiency (%)	21.4	38.8	33.0	49.5

10.4.2 Carbon dioxide allocation results

For the cogeneration and independent electrical generation and heating processes, we now determine the carbon dioxide emissions. In the evaluations, the carbon dioxide emissions for natural gas are taken to be 50 kg CO_2/GJ natural gas.

The results of applying the methods for allocating CO_2 emissions for the cogeneration and independent processes are presented in Tables 10.2 and 10.3, respectively, and comparatively illustrated in Figure 10.2 (based on data in Tables 10.2 and 10.3). In these presentations, absolute emission rates are given. The data in Tables 10.2 and 10.3 and in Figure 10.2 are based on the fractional breakdowns for the wastes of the cogeneration system provided in Table 9.10 and Figure 9.5, respectively.

As for wastes in general, there exists no ambiguity regarding the allocations of carbon dioxide emissions to the thermal and electrical products for the independent processes, but an appropriate carbon dioxide emissions allocation method is required for the cogeneration process. We apply the allocations methods described earlier in section 10.3.1.

The results demonstrate that, depending on the allocation method used, the allocations of CO_2 emissions vary markedly for the cogeneration system. The exergy-based allocations are viewed as the most appropriate, avoiding the misleading and somewhat arbitrary the resulting emissions breakdowns based on other allocation methods. For instance, when the allocation is based on incremental fuel consumption to heat production, an odd result occurs in which the carbon dioxide emissions allocations to the thermal product are negative and to the electrical product exceed 100 per cent. This odd results is due to the weakness in that allocation method's use of an energy efficiency η_{pp} for an independent device for providing the same electrical energy as the cogeneration system. A value of $\eta_{pp} = 35\%$ is used here, but when the value is raised to 40 per cent, the allocations of carbon dioxide emissions to both products are below 100 per cent. As another example, when the allocation method based on economic value of products is used,

Table 10.2 Allocation of carbon dioxide emission rate for the independent processes for producing heat and electricity

System	Emission-allocation basis	Emission allocation (g CO_2/s)		
		To electrical product	**To thermal product**	**Total**
Heating	All to single product	0	250	250
Electricity generation	All to single product	500	0	500
Overall (combined)	All electrical generation wastes to electricity generation and all thermal production wastes to thermal energy production	500	250	750

Table 10.3 Allocation of carbon dioxide emission rate for the cogeneration

Emission-allocation basis	Emission allocation (g CO_2/s)		
	To electrical product	**To thermal product**	**Total**
Exergy content of products	392	108	500
Energy content of products	250	250	500
Incremental fuel consumption allocated to electrical production[1]	278	222	500
Incremental fuel consumption allocated to heat production[2]			
Reference power plant efficiency $\eta_{pp} = 35\%$	572	–72	500
Reference power plant efficiency $\eta_{pp} = 40\%$	500	0	500
Shared emission savings between electricity and heat[1,2]			
Reference power plant efficiency $\eta_{pp} = 35\%$	360	140	500
Reference power plant efficiency $\eta_{pp} = 40\%$	346	154	500
Economic value of products			
Electrical-to-thermal cost ratio $c_E/c_Q = 1.5$	300	200	500
Electrical-to-thermal cost ratio $c_E/c_Q = 1.8$	322	178	500
Electrical-to-thermal cost ratio $c_E/c_Q = 2.1$	339	161	500

[1]An efficiency of 90% is assumed for the boiler that would have been used in the production of the same amount of heat as produced by the cogeneration system.
[2]Efficiencies of 35% and 40% are considered for the power plant that would have been used in the production of the same amount of electricity as produced by the cogeneration system.

the results depend on the ratio of the economic value of the electricity produced to the economic value of the thermal energy produced, a temporally and spatially varying ratio.

10.4.3 CO_2 emissions credits for trading purposes from switching to cogeneration from equivalent independent plants

In this subsection, the case study is extended to demonstrate how the results could be utilized to determine carbon dioxide emission credits that could be used in emissions trading. The characteristics of the cogeneration plant and the independent electrical power plant and heating plant are as already described in this case study. Also, the types of thermal energy used remain unchanged from those described in the case study.

In part of the case study, we consider two energy users, one of electricity and one of thermal energy. The decrease in CO_2 emissions attributable to the energy users are evaluated, when each switches from obtaining the energy required (electricity or thermal energy) from a producer of the just the required energy to a supplier using cogeneration. The decrease in CO_2 emissions for each energy user is its CO_2 emissions credits.

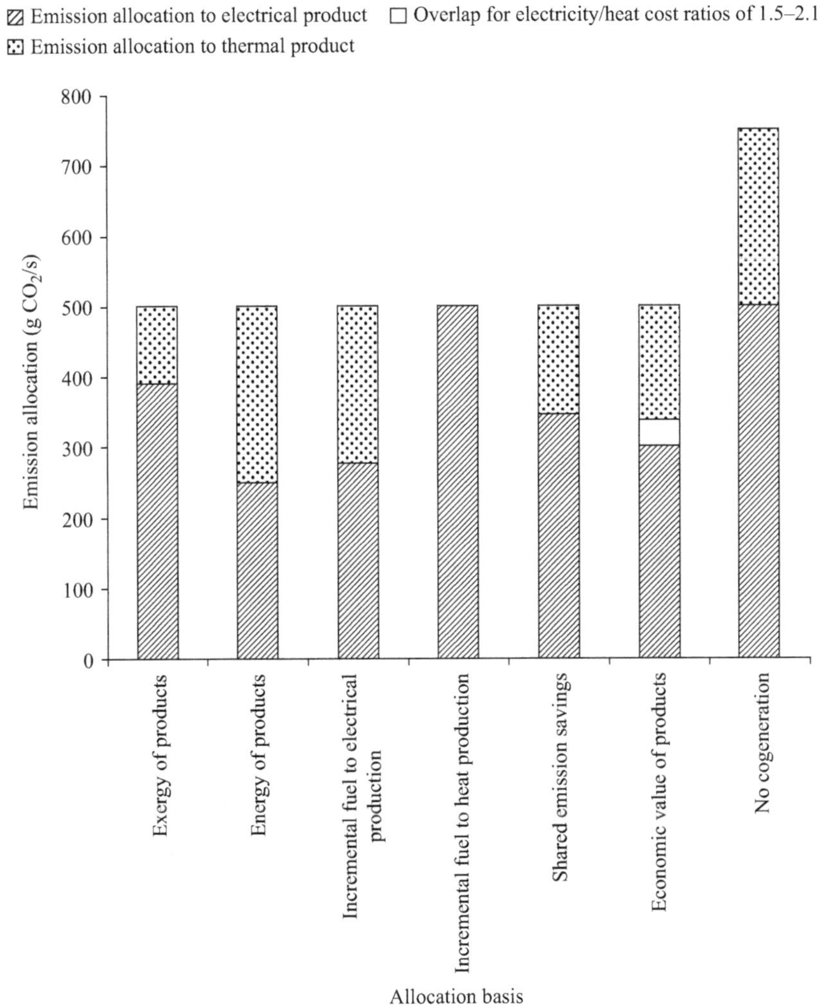

Figure 10.2 Allocation of emissions (in g CO_2/s) for independent and cogeneration processes considered (based on a reference power plant efficiency of 40 per cent). The case at the right labeled "no cogeneration" represents independent processes for equivalent electricity generation and heat production

10.4.3.1 CO_2 emissions credits when an electricity user switches to cogeneration

In this situation, we consider an electrical consumer which normally obtains electricity from a power plant. We determine the decrease in CO_2 emissions attributable to that consumer (i.e., the CO_2 emissions credit for the consumer) if the consumer switches to obtaining electricity from a supplier that uses a cogeneration plant.

We consider a multi-step calculation procedure. For simplicity, we consider a unit energy use by the consumer of 1 GJ of electricity. Parameter values that characterize the technologies considered are drawn from section 10.4.2.

The total CO_2 emissions of the cogeneration plant (to produce 1 GJ of electricity as well as thermal energy) can be evaluated as the product of the total fuel use in the cogeneration plant and the carbon dioxide emissions factor for the fuel. That is,

$$\text{CO}_2 \text{ emissions} = (\text{Fuel use in cogeneration plant})(\text{Fuel CO}_2 \text{ emissions factor}) \tag{10.19}$$

Since the fuel use in the cogeneration plant can be expressed as the ratio of the electricity use and the cogeneration efficiency for electricity, the above expression can be written as follows:

$$\begin{aligned} \text{CO}_2 \text{ emissions} = &(\text{Electricity use/Cogeneration efficiency for elec.}) \\ &\times (\text{Fuel CO}_2 \text{ emissions factor}) \end{aligned} \tag{10.20}$$

For a unit electricity use of 1 GJ and a cogeneration efficiency for electricity of 40 per cent, as well as a CO_2 emissions factor for the natural gas fuel of 50 kg CO_2/GJ fuel, the CO_2 emissions can be determined as follows:

$$\begin{aligned} \text{CO}_2 \text{ emissions} &= (1 \text{ GJ electricity}/0.40 \text{ GJ electricity}/\text{GJ fuel})(50 \text{ kg CO}_2/\text{GJ fuel}) \\ &= 125.0 \text{ kg CO}_2/\text{GJ electricity} \end{aligned} \tag{10.21}$$

The CO_2 emissions attributable to the consumer when obtaining electricity from a power plant can be evaluated as the product of the fuel use in the power plant and the carbon dioxide emissions factor for the fuel. That is,

$$\text{CO}_2 \text{ emissions} = (\text{Fuel use in power plant})(\text{Fuel CO}_2 \text{ emissions factor}) \tag{10.22}$$

Since the fuel use in the power plant can be expressed as the ratio of the electricity use and the power plant efficiency for electricity generation, the above expression can be written as follows:

$$\begin{aligned} \text{CO}_2 \text{ emissions} = &(\text{Electricity use/Power plant efficiency}) \\ &\times (\text{Fuel CO}_2 \text{ emissions factor}) \end{aligned} \tag{10.23}$$

Again for a unit electricity use of 1 GJ and a CO_2 emissions factor for the natural gas fuel of 50 kg CO_2/GJ fuel, but for a power plant efficiency for electricity generation of 40 per cent, the CO_2 emissions can be determined as follows:

$$\begin{aligned} \text{CO}_2 \text{ emissions} &= (1 \text{ GJ electricity}/0.40 \text{ GJ electricity}/\text{GJ fuel})(50 \text{ kg CO}_2/\text{GJ fuel}) \\ &= 125.0 \text{ kg CO}_2/\text{GJ electricity} \end{aligned} \tag{10.24}$$

The total CO_2 emissions of the cogeneration plant (to produce 1 GJ of electricity as well as thermal energy) is thus seen to be the same as the total CO_2 emissions attributable to the consumer when obtaining the same quantity of electricity from a power plant. Of the total CO_2 emissions of the cogeneration plant (to produce 1 GJ of electricity as well as thermal energy), the CO_2 emissions attributable to the consumer when obtaining electricity from a cogeneration plant can be evaluated as the fraction of the total CO_2 emissions of the cogeneration plant attributed to electricity production.

In this chapter, we have argued that the division of CO_2 emissions for cogeneration among electrical and thermal products should be based on the exergy contents of the products. For the present case, it was shown in section 9.6.3 that 78.4 per cent of the total CO_2 emissions for the cogeneration plant should be attributed to the electrical product based on exergy (see Table 9.10). Thus, the CO_2 emissions attributable to the consumer when obtaining electricity from a cogeneration plant can be evaluated as follows:

$$\begin{aligned} CO_2 \text{ emissions for user} &= (\text{Total } CO_2 \text{ emissions of cogeneration plant}) \\ &\quad \times (\text{Fraction for electricity}) \end{aligned} \tag{10.25}$$

Since the total CO_2 emissions of the cogeneration plant is 125.0 kg CO_2/GJ electricity and the allocation fraction for electricity based on exergy is 0.784, the CO_2 emissions attributable to the consumer when obtaining electricity from a cogeneration plant can be evaluated as

$$\begin{aligned} CO_2 \text{ emissions for user} &= (125.0 \text{ kg } CO_2/\text{GJ electricity})(0.784) \\ &= 98.0 \text{ kg } CO_2/\text{GJ electricity} \end{aligned} \tag{10.26}$$

Finally, the CO_2 emissions credit for switching to cogeneration, evaluated as the decrease in CO_2 emissions attributable to the electricity user, can be evaluated as the difference between the CO_2 emissions attributable to the consumer when obtaining electricity from a power plant and from a cogeneration plant. That is,

$$\begin{aligned} CO_2 \text{ emissions credit} &= (CO_2 \text{ emissions for elec. from power plant} \\ &\quad \text{attributed to consumer}) - (CO_2 \text{ emissions for} \\ &\quad \text{elec. from cogen. attributed to consumer}) \end{aligned} \tag{10.27}$$

Since the total CO_2 emissions of the cogeneration plant remains 125.0 kg CO_2/GJ electricity and the CO_2 emissions attributable to the consumer when obtaining electricity from a cogeneration plant is 98.0 kg CO_2/GJ electricity, the CO_2 emissions credit for switching to cogeneration can be evaluated as follows:

$$\begin{aligned} CO_2 \text{ emissions credit} &= (125 \text{ kg } CO_2/\text{GJ electricity}) - (98.0 \text{ kg } CO_2/\text{GJ electricity}) \\ &= 27.0 \text{ kg } CO_2/\text{GJ electricity} \end{aligned} \tag{10.28}$$

The results, presented in Table 10.4 and illustrated in Figure 10.3, can assist in evaluating CO_2 emissions credits for electricity users, for trading and other purposes.

Table 10.4 Carbon dioxide emissions and emissions credit for switching to cogeneration for electricity and heat users[1]

Energy user	CO_2 emissions (kg)
For user of electricity (1 GJ)	
CO_2 emissions for user of electricity from power plant	125.0
CO_2 emissions for user of electricity from cogeneration plant	98.0
CO_2 emissions credit for user of electricity for switching from electric power plant to cogeneration plant	27.0
For user of heat (1 GJ)	
CO_2 emissions for user of heat from heating plant	62.5
CO_2 emissions for user of heat from cogeneration plant	27.0
CO_2 emissions credit for user of heat for switching from heating plant to cogeneration plant	35.5

[1]Allocation of emissions from cogeneration to electrical and heat products is determined based on exergy contents of products.

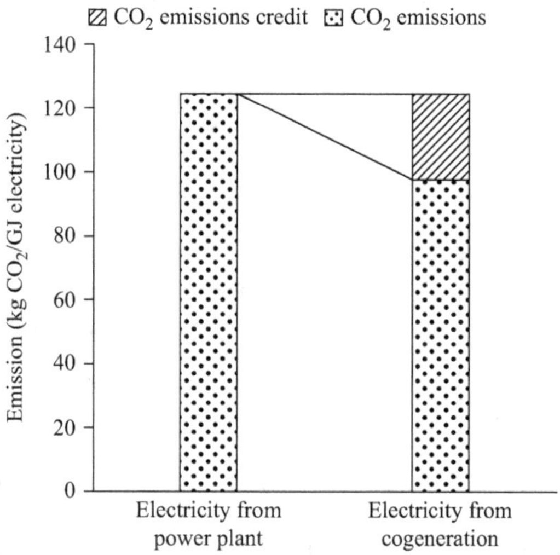

Figure 10.3 Normalized emissions of carbon dioxide for electricity generation via an electrical power plant and cogeneration, and the corresponding carbon dioxide emissions credit when an electricity user switches to a supplier using cogeneration from a supplier using independent electricity generation

10.4.3.2 CO_2 emissions credits when a heat user switches to cogeneration

In this situation, we consider a heat consumer who normally obtains heat from a heating plant. We wish to determine the decrease in CO_2 emissions attributable to that consumer (i.e., the CO_2 emissions credit for the consumer) if the consumer switches to obtaining heat from a supplier that uses a cogeneration plant.

We again consider a multi-step calculation procedure, for simplicity, a unit energy use by the consumer of 1 GJ of thermal energy. The thermal energy used by the consumer is of the type described in section 9.6, and parameter values that characterize the technologies considered are drawn from section 9.6.

As in the last subsection, the total CO_2 emissions of the cogeneration plant (to produce 1 GJ of heat as well as electrical energy), evaluated as the product of the total fuel use in the cogeneration plant and the fuel carbon dioxide emissions factor, is 125.0 kg CO_2/GJ heat. But, the CO_2 emissions attributable to the consumer when obtaining heat from a heating plant can be evaluated as the product of the fuel use in the heating plant and the carbon dioxide emissions factor for the fuel. The heating plant energy efficiency is 80 per cent, i.e., 0.80 GJ heat is produced for each 1 GJ of fuel utilized. So, following a similar approach as in the last subsection:

$$\begin{aligned} CO_2 \text{ emissions} &= (\text{Fuel use in heating plant})(\text{Fuel } CO_2 \text{ emissions factor}) \\ &= (\text{Heat use/Heating plant efficiency})(\text{Fuel } CO_2 \text{ emissions factor}) \\ &= (1\text{ GJ heat}/0.80\text{ GJ heat/GJ fuel})(50\text{ kg } CO_2/\text{GJ fuel}) \\ &= 62.5\text{ kg } CO_2/\text{GJ heat} \end{aligned} \tag{10.29}$$

Of the total CO_2 emissions of the cogeneration plant (to produce 1 GJ of heat as well as electrical energy), the CO_2 emissions attributable to the consumer when obtaining heat from a cogeneration plant can be evaluated as the fraction of the total CO_2 emissions of the cogeneration plant attributed to heat production. Again, since we have argued that the division of CO_2 emissions for cogeneration among electrical and thermal products should be based on product exergy contents and since for the present case 21.6 per cent of the total CO_2 emissions for the cogeneration plant should be attributed to the thermal product based on exergy (see Table 9.10), the CO_2 emissions attributable to the consumer when obtaining heat from a cogeneration plant can be evaluated as follows:

$$\begin{aligned} CO_2 \text{ emissions for user} = \; & (\text{Total } CO_2 \text{ emissions of cogeneration plant}) \\ & (\text{Fraction for heat}) \\ = \; & (125.0\text{ kg } CO_2/\text{GJ heat})(0.216) \\ = \; & 27.0\text{ kg } CO_2/\text{GJ heat} \end{aligned} \tag{10.30}$$

Finally, the CO_2 emissions credit for switching to cogeneration, evaluated as the decrease in CO_2 emissions attributable to the heat user, can be evaluated as the difference between the CO_2 emissions attributable to the consumer when obtaining heat from a power plant and from a cogeneration plant. That is,

$$\begin{aligned} CO_2 \text{ emissions credit} &= (CO_2 \text{ emissions for heat from heating plant attributed to consumer}) - (CO_2 \text{ emissions for heat from cogen. attributed to consumer}) \\ &= (62.5 \text{ kg } CO_2/\text{GJ heat}) - (27.0 \text{ kg } CO_2/\text{GJ heat}) \\ &= 35.5 \text{ kg } CO_2/\text{GJ heat} \end{aligned} \tag{10.31}$$

The results are presented in Table 10.4 and illustrated in Figure 10.4, and demonstrate the manner in which they can assist in evaluating CO_2 emissions credits for thermal energy users, for trading and other purposes.

10.4.3.3 CO_2 emissions credits for other cases

The procedures illustrated in the previous two subsections can be formalized for variations of the cases considered here, as well as for various other cases.

In particular, other GHGs than carbon dioxide can also be considered, so as to provide more comprehensive data for climate change mitigation efforts.

In addition, some of the other situations that could be considered include on- and offsite plants, different fuels, different thermal energy requirements and trigeneration systems.

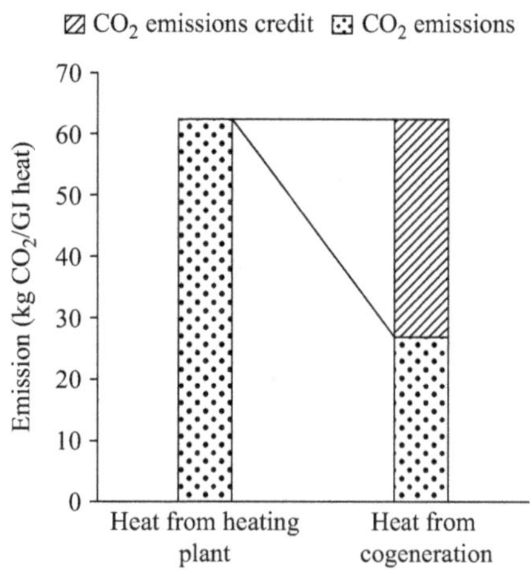

Figure 10.4 Normalized emissions of carbon dioxide for heat production via a heating plant and cogeneration, and the corresponding carbon dioxide emissions credit when a heat user switches to a supplier using cogeneration from a supplier using independent heat production

10.5 Closure

Cogeneration can assist efforts to combat global climate change, which is believed to be attributable to the emissions of GHGs, the most significant of which is carbon dioxide. The exergy-based method of carbon dioxide emissions allocation allows for a rational and meaningful allocation of such emissions for cogeneration systems, and is superior to several other methods that have been proposed for allocating such emissions. By permitting carbon dioxide emissions to be allocated more appropriately among the different commodities generated in cogeneration systems, the results allow the environmental benefits of cogeneration and other technologies that produce multiple products to be better understood and exploited, and superior technologies to be identified rationally and fairly. The exergy-based approach to carbon dioxide emissions allocations method provides a sensible basis for a meaningful overall approach for emissions trading. The exergy-based method helps with proper decision-making regarding how carbon dioxide emissions should be reduced in a process, the effect of cogeneration on overall carbon dioxide emissions, and a fair way to establish detailed schemes for carbon dioxide emissions trading. It would be useful to advance this approach by developing detailed methodologies for emissions trading that utilize the overall exergy-based emissions allocations approach and perhaps applying the exergy-based allocation method with more detailed data (e.g., hourly or daily rather than annual data).

The material in this chapter can benefit energy systems designers and decision-makers in industry and government, as well as to society through the design and utilization of better energy processes and systems based on climate change considerations. The results may also help inform regulations developed in Europe for defining and assessing the benefits of cogeneration [245,246]. In particular, the incorporation of exergy into these regulations would likely improve them markedly for allocating carbon dioxide emissions.

The material in this chapter on allocating carbon dioxide emissions for cogeneration can be extended to multi-product production processes like trigeneration (where the products are electricity, heat and cold), as the goal is to allocate the carbon dioxide emission to each product according to the actual emission that is in fact attributable to it. Often the carbon dioxide emission allocations are proportional to the fuel use breakdown by product, for a multi-product process. The method of allocating CO_2 emissions for cogeneration based on product exergy content is in fact generalizable to multiple products like trigeneration processes, whereas the other allocation methods are less flexible.

Chapter 11

Modeling and optimization of cogeneration-based district energy systems accounting for economics and environmental impact

Overview

The modeling and optimization of cogeneration-based district energy systems, especially while accounting for economics and environmental impact, are described. An energy equilibrium model is given and its mathematical formulation specified. Models and descriptions are presented for the principal relevant technologies: cogeneration production, heating and district heating, chilling and district cooling, and configurations for integrating cogeneration and district energy technologies. The economics of cogeneration and district energy technologies needed in optimization procedures are discussed. Methodologies for the analysis of economic impacts are outlined, including means of determining partial social welfare change and payback period. Various methodologies for the analysis of environmental impacts are explained, and an illustrative case study is presented. After describing the scenarios considered in the case study, parameter values for the illustrative model are presented as are the resulting partial social welfare changes and possible improvements. The modeling and optimization of cogeneration-based district energy systems help illustrate and demonstrate the comparative merits of different integrated systems for cogeneration and district energy. Often the optimization of cogeneration and district energy technologies accounts for environmental emission limits in terms of constraints, although the emissions can also be addressed as objectives.

11.1 Introduction

Since many energy policies, e.g., strategies of utilizing new energy technologies, may have long-term economic impacts, many energy-related economic models have been developed to aid in energy policy analysis and decision-making. In one class of energy-related economic models, the effects of energy policies are modeled as shifts in energy market equilibrium positions. Such models employ mathematical programming, and are often called energy equilibrium models for short (e.g., [255–260]).

An energy equilibrium model of a competitive energy market closely examines the interaction between energy supplies and demands, and determines the optimal levels of production (supply) and consumption (demand) that satisfy the equilibrium property that the prices consumers pay for each commodity should equal the marginal costs of production. In the energy equilibrium model, supply is represented by a cost-minimizing linear sub-model and demand by a smooth vector-valued function of prices. There are several algorithms for their solution (see, e.g., [261–264]), including the well-known Project Independence Evaluation System (PIES) algorithm of Ahn and Hogan [265].

In multi-period energy equilibrium, environmental impacts can be important factors, and the analyst may wish to introduce environmental measures into the model system (e.g., [266–269]). These environmental measures can be used to evaluate the environmental impacts over the time horizon of the model for a given energy policy, or can be constrained by applying upper bounds to achieve environmental impact control. In the latter case, the model can provide the optimal solution of the model system while accounting for the factors as the target objectives of environmental impact control.

Cogeneration-based district energy (DE) systems use central electricity-generation, heating and/or cooling facilities to provide integrated electricity, heating and/or cooling services for communities, and often have advantages over conventional heating and cooling systems. In this chapter, we employ an energy equilibrium model to study conventional heating, cooling and electricity-generation systems and cogeneration-based district energy DE systems in order to analyze and to compare the potential economic and environmental benefits of utilizing cogeneration-based DE systems for a range of scenarios, and to develop the optimal configuration of the model system considering factors such as economic and environmental impacts.

The energy equilibrium model is set up, formulated and solved within the software called the Waterloo Energy Modeling System (WATEMS), which employs sequential nonlinear programming to calculate a spatial inter-temporal equilibrium of energy supply and demands.

In section 11.2, we present the energy equilibrium model and its mathematical formulation. Sections 11.3 and 11.4 explore the methodologies for analysis and evaluations. In section 11.5, an illustrative case study is presented. Section 11.6 provides closing remarks.

11.2 Energy equilibrium model and its mathematical formulation

The energy equilibrium model integrates and utilizes the theoretical and analytical methods of several disciplines including engineering (especially process analysis), economics and operations research. Among these disciplines, models based on economic theory tend to emphasize customer behavioral description on the demand side, while models of engineering processes tend to be concerned with the production and technical aspects of the supply activities. These two approaches are jointly employed in the equilibrium model formulation. Through this combined approach, a comprehensive framework is provided in which the evolution of a market can be analyzed under various assumptions regarding production technologies and consumption behavior, and the formulation and evaluation of various policies and plans can be carried out with explicit recognition of both technical constraints and customer response.

In the energy equilibrium model, supply is represented by a cost-minimizing linear sub-model and demand by a smooth vector-valued function of prices. In the discussion that follows, superscripts in parentheses refer to time periods, while superscripts without parentheses are powers, e.g., $Z^{(t)}$ is the value of Z in period t but R^t is R raised to the power t.

Following Ahn and Hogan [265], the general type of energy equilibrium model is usually represented as follows:

Supply side:

$$\text{Min.} \sum_{t=1}^{T} R^t C^{(t)} Z^{(t)}$$

$$\text{s.t.} \quad D^{(t)} Z^{(t)} = Q^{(t)}, \qquad t = 1, 2, \ldots T$$

$$FZ \leq f$$

Demand side:

$$Q^{(t)} = g^{(t)}(P^{(t)}) \tag{11.1}$$

Equilibrium condition:

$$P^{(t)} = R^{-t} V^{(t)}$$

where

$Q^{(t)}$ denotes the vector of demands in period t,
$P^{(t)}$ denotes the vector of prices in period t,
R denotes the discount factor per period,
$C^{(t)}$ denotes the vector of unit costs associated with supply variables,
Z denotes the vector of all supply activities, composed of sub vectors $Z^{(t)}$,
$D^{(t)}$ denotes the matrix for demand requirement constraints relating $Q^{(t)}$ to $Z^{(t)}$,
F denotes the matrix for constraints involving only Z,
f denotes the vector for right hand side values for constraints,
$g^{(t)}(P^{(t)})$ denotes the demand function relating $Q^{(t)}$ to $P^{(t)}$ and
$V^{(t)}$ denotes the vector of dual variable for the demand requirement constraint.

Under the competitive market assumption, the elements of an optimal dual vector $V^{(t)}$ can be regarded as the supply prices, representing the marginal cost of meeting an additional unit of demand. Then, the solution V^* (or P^*) and Q^* become equilibrium price and demand vectors, respectively. An accompanying solution Z^* represents an equilibrium production profile of the supply activities.

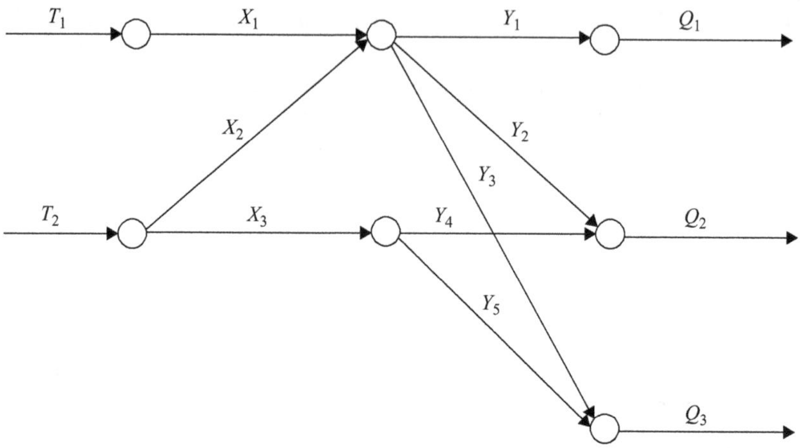

Figure 11.1 Network diagram for conventional and cogeneration-based DE systems

Consider now an energy equilibrium model for the provision of heating, cooling and electrical services using conventional technologies and cogeneration-based DE systems. The model is developed in order to examine or optimize the economic and environmental benefits of utilizing cogeneration-based DE systems for a range of scenarios. The flows of energy in the systems can been represented in a network diagram (Figure 11.1).

In the network diagram, each arc of energy flow represents a variable in the model. The model incorporates two types of energy supply (production) technologies, namely conventional electricity generation (T_1) and cogeneration (T_2), and only three end-use demands, electricity (Q_1), heating (Q_2) and cooling (Q_3). On the supply side, the conventional system produces only electricity (X_1), while the cogeneration system produces not only electricity (X_2) but also heating (X_3). The intermediate products, electricity and heat, can be converted into end-use demands at a cost through conversion and distribution activities (Y_1, Y_2, Y_3, Y_4 and Y_5). Exports and imports are not considered in the system.

The energy equilibrium model is set up, formulated and solved within the software WATEMS, which employs sequential nonlinear programming to calculate a spatial intertemporal equilibrium of energy supply and demand. The modeler specifies the structure of a WATEMS model and places the data for the model in an input spreadsheet of a special format. The solution is placed in another spreadsheet, where tables and graphs may be generated for a report. The solution is calculated by a GAMS program which repeatedly calls MINOS [270] for the sequential nonlinear programming of the PIES algorithm. Solutions to smaller models may be calculated on a microcomputer, but larger models may require a more powerful computer. References 264,271,272 provide more detailed descriptions of WATEMS than that provided here.

11.3 Methodologies for analysis of economic impacts

Many methods can be used to analyze the economic impacts of implementing cogeneration-based DE systems. In this book, two particular evaluation methods are considered and performed:

1. analysis of the present worth of partial social welfare change and
2. analysis of payback-period.

11.3.1 Partial social welfare change

The present worth of the total partial social welfare change, ΔW, from the business-as-usual (BAU) levels to those of scenario "*s*," is equal to the change in the consumers' total value or willingness to pay minus the change in the total costs incurred by the producers, including distribution and transportation costs. The difference in consumers' total value between the two scenarios is calculated for each commodity, region and period, and summed with discounting. The difference in consumption valued at the BAU price, plus one-half the product of price change and quantity change between the two scenarios acts as a suitable approximation to the difference in total value for each commodity in a period [273] and present worth can be annualized by multiplying by the appropriate discount rate (in this case, 5 per cent).

The present worth of partial social welfare change can be expressed as

$$\begin{aligned}\Delta W(s) = {} & \sum_i \sum_t (1+r)^{-t}[q_{\mathrm{BAU}}(i,t) - q_s(i,t)]P_{\mathrm{BAU}}(i,t) \\ & + 0.5 \sum_i \sum_t (1+r)^{-t}[q_{\mathrm{BAU}}(i,t) - q_s(i,t)][P_s(i,t) - P_{\mathrm{BAU}}(i,t)] \\ & - \sum_i \sum_t (1+r)^{-t}[Pcost(\mathrm{BAU}) - Pcost(s)] \end{aligned} \tag{11.2}$$

where

$\Delta W(s)$ denotes the total present worth of partial social welfare change for scenario *s*,
$q_s(i, t)$ denotes the quantity at time *t* of demand variable *i*, for scenario *s*,
$q_{\mathrm{BAU}}(i, t)$ denotes the quantity at time *t* of demand variable *i*, for scenario BAU,
$P_s(i, t)$ denotes the price at time *t* of demand variable *i*, for scenario *s*,
$P_{\mathrm{BAU}}(i, t)$ denotes the price at time *t* of demand variable *i*, for scenario BAU,
$Pcost(s)$ denotes the sum of the discounted cost incurred by producers, for scenario *s*,
$Pcost(\mathrm{BAU})$ denotes the sum of the discounted cost incurred by producers, for scenario BAU and
r denotes the discount rate.

If ΔW is negative, the model of the energy system has incurred a cost (loss) to carry out scenario "*s*." Otherwise, the model generates benefit (gain) from the scenario.

11.3.2 Payback-period method

The payback period is the number of time periods required for the accruing net present worth benefit from an investment to equal the cost of the investment. The payback period can be evaluated as follows:

$$\begin{aligned} &\text{Min.} \quad n \\ &\text{s.t.} \quad \sum_{t=1}^{n}\sum_{i}(1+r)^{-t}[Q_s(i,t)P_s(i,t) - Pcost(S)] = I \end{aligned} \tag{11.3}$$

where n denotes payback period and I cost of the investment.

11.4 Methodologies for analysis of environmental impacts

Since the nature and effects of environmental impacts can be a major concern in energy-policy analysis, and since policies to control environmental impacts may have long-term economic repercussions, environmental measures, e.g., CO_2 emission levels, have been introduced by the authors into the model system [269,274]. In the present work, we concentrate on CO_2 emissions; the methodology used and described below can be straightforwardly extended for other types of emissions.

Following Nordhaus [267], the following equations concerning annual emissions and accumulated emission levels are used to find CO_2 emission levels, as follows:

Annual CO_2 emission rate:

$$E(t) = \sum_{i} er_i X_i(t) \tag{11.4}$$

Accumulated CO_2 emission rate:

$$M(t) = (1-\sigma)E(t) + \delta M(t-1) \tag{11.5}$$

where

$E(t)$ denotes the annual rate of CO_2 emissions in year t,
$M(t)$ denotes the accumulation of CO_2 in the atmosphere in year t,
er_i denotes the CO_2 emission quantity per unit consumption of energy flow i,
σ denotes the rate of CO_2 absorption by oceans, plants and other absorbing media,
δ denotes the parameter indicating the natural reduction of CO_2 accumulated in the atmosphere and
$X_i(t)$ denotes the quantity of energy flow i, during period t (decision variables).

The parameters σ and δ in (11.5) have been estimated by Nordhaus [265] based on data of concentrations and emissions of CO_2 in 1985 and 1986. Using these values, (11.5) becomes:

$$M(t) = 0.5 \times E(t) + 0.995 \times M(t-1) \tag{11.6}$$

Equation (11.6) can be activated as a constraint in the model when limitations on the final accumulated CO_2 emission levels are desired. Of the two emission measures, the accumulated CO_2 emissions are more important, due to their direct link to global temperature increases, when examining climate changes to greenhouse effects.

Using (11.4) and (11.5), the CO_2 emission measures, $E(t)$ and $M(t)$, can be

(i) used to evaluate the CO_2 emission levels over the time horizon of the model for a given energy policy or scenario, or
(ii) constrained within upper limits (e.g., $E(t) \leq$ upper limit, or $M(t) \leq$ upper limit).

In the latter case where the constraints on upper bounds on CO_2 emissions are considered, the model can provide not only optimal forecasts of future energy supplies and demands, but also a view of the impact of CO_2 emissions on the model system over the long term.

The standard WATEMS has been updated to create and solve the energy equilibrium model with corresponding CO_2 emission-level constraints. The new version of WATEMS is different from the standard WATEMS since it is altered to let the modeler add the CO_2 emission control sub model, enter CO_2 emission data, and report on the solution for CO_2 emission-control analysis. References 269, 274 present details on the new version of WATEMS.

11.5 Case study

In order to clarify the structure of the energy equilibrium model, its solution technique and the methods described earlier of quantitative and comparative analysis, an illustrative use of the energy equilibrium model is presented in this section. The case study considers the provision of heating, cooling and electrical services via conventional and cogeneration-based DE systems. The economic impacts of utilizing cogeneration-based DE systems for a range of scenarios are estimated.

11.5.1 Description of scenarios

This illustrative model considers ten periods each of one year duration, spanning from year one to year ten. Other versions of the model that have different forecasting periods can be designed in order to assess the impacts over various time horizons.

Four scenarios are considered in this case study. In Scenario 0, the base-case scenario, conventional systems for heating, cooling and electricity generation are used over the next ten periods. The results of this scenario, termed the "business as usual" (BAU) case, provide key comparative parameters such as partial social welfare change. Subsequent improvements in partial social welfare change are expressed as percentage increases from the BAU value.

The other three scenarios are designed to study the impacts of various energy strategies for implementing cogeneration-based DE systems. Scenario 1 considers replacing conventional systems by cogeneration-based DE systems immediately. Scenario 1 is somewhat unrealistic since, in long-term energy planning, the reaction of supply method choice to a change in its technology is a process that requires time; hence, the time-lagged effect in new technology diffusion can play an important role. For this reason, Scenario 2 was developed to consider the replacement of the conventional systems by cogeneration-based DE systems gradually, in 10 per cent increments each period to simulate such a time-lag effect. Hence, Scenario 2 should be more realistic than Scenario 1. In order to perform a rough sensitivity analysis on the cost of delaying new technology implementation, Scenario 3 was developed (particularly for comparison with Scenario 2). In Scenario 3, the replacement of DE systems follows the same pattern as in Scenario 2 but we start implementing cogeneration-based DE systems in period 4 (i.e., with a three-period delay).

11.5.2 Parameter values for the illustrative model

Starting with the supply stage (left side of Figure 11.1), annual electricity production from conventional non-cogeneration production systems and annual electricity and heating production from cogeneration systems are represented in the diagram by arcs, $X_1^{(t)}$, $X_2^{(t)}$, $X_3^{(t)}$, respectively. Parameters C_1^X, C_2^X, C_3^X specify the respective unit costs for these three supply activities. In a similar way, the parameters e_1^Y through e_5^Y associated with the arcs $Y_1^{(t)}$, $Y_2^{(t)}$, $Y_3^{(t)}$, $Y_4^{(t)}$, $Y_5^{(t)}$, respectively, specify conversion efficiencies, while C_1^Y through C_5^Y represent the unit costs of the conversions. The shares of supplies $X_3^{(t)}$ and $X_2^{(t)}$ are bounded by a share-limit constraint, s_2 which is related to the electrical and thermal efficiencies. We assume, in this illustrative model, that the demand rates for electricity, heating and cooling are fixed over the ten periods considered and that the demand function is linear.

A complete list of the parameter values used in the illustrative model is given in Table 11.1, which is based on data and information in References 275,203.

This illustrative model is developed with the WATEMS. This WATEMS model is solved by a sequence of linear programs, which give an approximate solution to the model by maximizing the consumers' plus producers' surplus. The form of each linear program is as follows:

$$\text{Max.} \sum_{t=1}^{10}\sum_{i=1}^{3}(1+r)^{-t}P_1^{(t)}Q_2^{(t)} - \sum_{t=1}^{10}\sum_{i=1}^{3}(1+r)^{-t}C_i^X X_i^{(t)} - \sum_{t=1}^{10}\sum_{i=1}^{3}(1+r)^{-t}C_i^Z Z_i^{(t)}$$

s.t.

$$\tag{11.7}$$

Energy balances:

$$X_1^{(t)} + X_2^{(t)} = Y_1^{(t)} + Y_2^{(t)} + Y_3^{(t)}, \qquad t = 1,2,\ldots 10 \tag{11.8a}$$

$$X_3^{(t)} = Y_4^{(t)} + Y_5^{(t)}, \qquad t = 1,2,\ldots 10 \tag{11.8b}$$

$$Y_1^{(t)} = Q_1^{(t)}, \qquad t = 1,2,\ldots 10 \tag{11.8c}$$

$$e_2^Y Y_2^{(t)} + Y_4^{(t)} = Q_2^{(t)}, \qquad t = 1,2,\ldots 10 \tag{11.8d}$$

$$e_3^Y Y_3^{(t)} + e_5^Y Y_5^{(t)} = Q_3^{(t)}, \qquad t = 1,2,\ldots 10 \tag{11.8e}$$

Table 11.1 Parameter values for the illustrative model

Unit costs		
For electricity flows	$C_1^X = 7.0$ ¢/kWh(e)	$C_2^X = 5.0$ ¢/kWh(e)
For heat flows	$C_3^X = 2.0$ ¢/kWh(th)	
For energy conversions		
Electric heating		$C_2^Y = 0.5$ ¢/kWh(e)
Electric cooling		$C_3^Y = 3.0$ ¢/kWh(c)
Heat-driven absorption cooling		$C_5^Y = 4.0$ ¢/kWh(th)
Electricity to electricity		$C_1^Y = 0$ ¢/kWh(e)
Heat to heat		$C_4^Y = 0$ ¢/kWh(th)
Conversion efficiencies/COPs		
Electric heating		$e_2^Y = 0.98$
Electric cooling		$e_3^Y = 4.00$[1]
Heat-driven absorption cooling		$e_5^Y = 1.20$[1]
Electricity to electricity		$e_1^Y = 1$
Heat to heat		$e_4^Y = 1$
Heat/electricity product-ratio for cogeneration		
$s_2 = 0.90$ kWh(th)/kWh(e)		
Energy demands and price-based demand functions, per period[2]		
For electricity	$Q_1^{(t)} = 100$ kWh(e)	$Q_1^{(t)} = \{8.33[\text{kWh(e)}]^2/¢\}P_1^{(t)}$
For heating	$Q_2^{(t)} = 50$ kWh(th)	$Q_2^{(t)} = \{5.00[\text{kWh(e)}]^2/¢\}P_2^{(t)}$
For cooling	$Q_3^{(t)} = 50$ kWh(th)	$Q_3^{(t)} = \{6.25[\text{kWh(e)}]^2/¢\}P_3^{(t)}$

[1]These are COP values, based on a modern electric chiller and a double-effect absorption chiller.
[2]Prices for electricity, heating and cooling can be evaluated with the demand functions, as 12 ¢/kWh(e), 10 ¢/kWh(th) and 8 ¢/kWh(th), respectively.

Share constraint:

$$X_3{}^{(t)} = s_2 X_2{}^{(t)}, \qquad t = 1, 2, \ldots 10 \qquad (11.9)$$

Non-negativity:

$$X_i{}^{(t)} \geq 0, \qquad i = 1, 2, 3, \qquad t = 1, 2, \ldots 10 \qquad (11.10a)$$

$$Y_i{}^{(t)} \geq 0, \qquad i = 1, 2, 3, 4, 5, \qquad t = 1, 2, \ldots 10 \qquad (11.10b)$$

11.5.3 Resulting partial social welfare changes and improvements

The calculated results are summarized in Table 11.2, which shows the partial social welfare changes for Scenarios 1–3 and the percentage improvements for each

Table 11.2 Results for the illustrative model

Parameter	**Scenario**		
	1	**2**	**3**
Partial social welfare change, ΔW ($)	2,798	2,373	1,443
Improvement (%)	40.6	34.5	20.8

scenario relative to the BAU case. From the table, we see that the partial social welfare changes exhibit significant improvements for each scenario. The partial social welfare change increases by approximately 40 per cent for Scenario 1; the partial social welfare change exhibits reasonably large improvements compared to the BAU case even for the more realistic case, Scenario 2, where the time-lagged effect in new technology diffusion is considered. However, the partial social welfare change in Scenario 3, i.e., implementing the new technology with a three-period delay, only exhibits a 20 per cent increase. The results indicate that the partial social welfare change is very sensitive to new technology implementation delays.

11.6 Closure

Energy equilibrium models can aid in energy planning and decision-making. In this chapter, we use the energy equilibrium model to study conventional and cogeneration-based district energy systems for providing heating, cooling and electrical services in order not only to assess the potential economic and environmental benefits of utilizing cogeneration-based DE systems, but also to develop optimal configurations while accounting for factors such as economic and environmental impacts. The energy equilibrium model is formulated and solved here with software called WATEMS, which uses sequential nonlinear programming to calculate the intertemporal equilibrium of energy supplies and demands. An illustrative energy equilibrium model of conventional systems and cogeneration-based DE systems is developed within WATEMS to quantitatively and comparatively examine the economic impacts for a range of scenarios. According to the illustrative model, implementing cogeneration-based DE systems may provide potential economic benefits; a sensitivity analysis indicates that delays in implementing such systems can be costly.

Chapter 12

Developments and advances in technologies and systems for cogeneration and district energy

Overview

Developments and advances in technologies and systems for cogeneration and district energy, and related technologies such as trigeneration, multigeneration and distributed energy systems, are described, drawing extensively on recent and ongoing research and development activities throughout the world. Such advances help predict how the technologies will perform and be utilized in the future. Many of the advances involve renewable energy and advanced technologies such as energy storage, fuel cells and new HVAC devices, and utilize advanced methods such as exergy analysis. The material is loosely divided and organized for convenience, with the first part focusing on technical and related factors for cogeneration and extended cogeneration, heating and cooling, district energy, and integrated systems for cogeneration and district energy. The second part is centered on advances related to the economics of systems and technologies related to cogeneration and district energy, while the third part addresses environmental impact and climate change aspects of cogeneration and district energy and how the technologies can mitigate them. The final part describes advances regarding the optimization of systems related to cogeneration and district energy.

12.1 Introduction

Much research and development has been carried out and is ongoing in fields related to simple and extended cogeneration, heating and district heating, and cooling and district cooling. This work is leading to many advances in these field technologies that will likely affect how they perform and are utilized these in future. It is important for informed practitioners, designers and researchers to have an appreciation of these developments, especially since the technologies they deal with often have long life times, stretching into decades. Thus, it is prudent for decisions to be made based not only on what is available today, but also on what is likely to be available in the future.

Many of the advances involve renewable energy, which has been extensively covered [276], as well as energy storage [277], particularly thermal storage [278]. Also, the utilization of geothermal energy to provide sustainable heating and cooling using the ground has been described in detail [279]. Much of the research and development is aimed at methods and applications for HVAC [280].

It is noted that in much of this work, exergy analysis is employed [281], as are exergy-based economic methods [282] and exergy-based environmental and ecological methods [283]. That is, much of the past and present research related to technologies and systems for cogeneration and district energy, heating and district heating, and cooling and district cooling has utilized exergy-based methods. The extended use of exergy methods in the form of exergy-based economic and exergy-based environmental assessments has been particularly noteworthy.

Many of the advances described in this chapter are being spurred on by governments through incentives, taxes and policies. For instance, European Union governmental policies presently encourage the application of efficient and clean technologies in the energy sector by supporting distributed generation, cogeneration and trigeneration, as well as the exploitation of renewable sources [284].

Consequently, a selection of the important research and development and related work in the fields of cogeneration, heating and district heating, and cooling and district cooling are reviewed in this chapter, in order to provide the reader with an understanding of the present state of the art as well as expectations for future, in both the short and long terms. The material covered is outlined broadly in Figure 12.1.

For convenience and to help guide readers, the developments and advances described in this chapter are broken down into and organized in topical sections. But there is certainly an overlap in the areas, with many of the cited works easily fitting in numerous categories. So, the breakdown should only be viewed as approximate and done to facilitate the organization of the material.

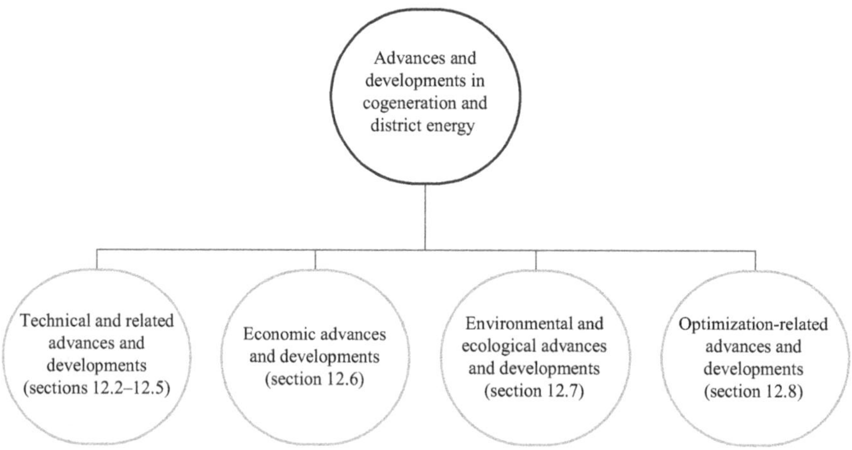

Figure 12.1 Outline of material covered in this chapter on advances and developments in cogeneration and district energy technologies

12.2 Cogeneration and extended cogeneration

Much research and development has been performed and is ongoing on cogeneration, as well as extensions of it such as trigeneration and multigeneration. We describe in this section a selection of the important work and advances in these fields.

12.2.1 Cogeneration

Proposals for improving conventional and novel cogeneration systems have been widely reported in recent years. For instance, exergetic performance analyses have been reported of various cogeneration systems for buildings by Kanoglu *et al.* [285], while the effects on cogeneration efficiency of hydrogen production through partial oxidation of natural gas over metal catalysts have been investigated by Le Corre *et al.* [286]. Many other informative recent examples are considered in this section.

Assessments of planning and operating issues for cogeneration systems have been investigated in recent years. Some informative recent examples are considered here. For example, a new cogeneration targeting procedure was recently proposed for integrating a gas-fired steam power plant with a total site utility by Abadi *et al.* [287]. Also, uncertainty-based operating strategies for cogeneration systems for heat and electricity have been described by Long *et al.* [288], in order to assist efforts at optimizing the operation of CHP systems based on criteria such as costs and/or energy use and/or emissions. The types of operating strategies typically considered include base loading, electric load following and thermal load following. The authors show that the uncertainties associated with operating strategy assessments usually are not considered in their selection, so they employ modular uncertainty analysis techniques to compare the uncertainties associated with the cost savings of different operating modes. They find that the choice among cogeneration operating strategies cannot always be clearly determined for power generation units with nominal sizes less than 60 per cent of the maximum electric load of a building due to overlapping uncertainty bands, and that ease of implementation and operation may be the main decision criteria in such cases. Furthermore, the impact of cogeneration planning restrictions on the efficiency of urban energy systems have been examined by Keirstead *et al.* [289] by considering the trade-offs between cogeneration plant size limitations and the performance of the overall energy system. They use a mixed integer linear programming model to evaluate urban energy system designs for a range of city sizes and technology scenarios, and show that the most cost-effective and energy-efficient scenarios require a mix of technology scales including cogeneration plants of appropriate size for the total urban demand. For cities with populations under 200,000 people, for instance, planning restrictions that prevent the use of cogeneration could lead to total system cost penalties of 2 per cent and energy-efficiency penalties of up to 24 per cent, relative to a boiler-only base case.

Many novel cogeneration systems, or systems using advanced energy technologies, have been proposed recently. Cogeneration systems have in some cases been combined with other advanced energy technologies, such as energy storage, fuel cells and renewable energy sources. Some informative recent examples are considered here. For example, the energy performance of a complex district heating system employing gas-driven cogeneration of heat and electricity, heat pumps and high-temperature aquifer TES has been assessed by Zeghici *et al.* [290]. This integrated system is proposed as a substitute for an old and inefficient district heating system for a condominium in Bucharest, Romania, and demonstrated to improve energy performance and CO_2 emissions significantly. Also, the use of solid oxide fuel cell technology for power generation has been reviewed, pointing out that the high-temperature exhaust gas from fuel cell can be utilized in cogeneration and trigeneration systems for heating and cooling [291], while energy and exergy analyses of the potential cogeneration efficiencies for fuel cells and other electricity generation devices have been compared [292]. Furthermore, electric boilers and heat pumps have been compared by Blarke [293] for distributed cogeneration, with the aim of avoiding intermittency issues with some energy systems, like those using renewable energy. He notes that some distributed operators in west Denmark are retiring cogeneration units or adding heat-only boilers as wind power penetration levels surpass 25 per cent, jeopardizing the system-wide energy, economic and environmental benefits cogeneration offers, and proposes distributed operators adapt their technology and operational strategies to achieve a better co-existence between cogeneration and wind power. He considers several options, including a new concept integrating a high pressure compression heat pump using low-temperature heat recovered from stack gases with an intermediate cold storage, to enable independent operation of heat pump and cogeneration system. Blarke finds that, although electric boilers avoid intermittency problems of distributed cogeneration, heat pumps can be more cost-effective, and that for higher gas/electricity price ratios compression heat pumps with intermediate thermal storages are superior for addressing intermittency issues while attaining efficient use of electricity in heating and cooling.

Many cogeneration systems have been proposed to mitigate or avoid the use of fossil fuels, by utilizing renewable energy sources. Some examples:

- The use of distributed cogeneration within an energy management strategy for mitigating fossil fuel consumption has been recently examined by Kikuchi *et al.* [294], in conjunction with reports by the United Nations Environment Program's International Resource Panel that quantitatively compare environmental and natural resource impacts and benefits of using demand-side efficient technologies for GHG mitigation to 2050. They conduct scenario analyses and show that a mismatch between the quality of energy produced, especially heat, or any benefits arising from scale from other energy technologies, can decrease the incentive to implement distributed energy technologies, and that the possibility exists to replace conventional single-generation technologies, such as boilers or power generators, with multigeneration technologies. They also demonstrate that cogeneration performance is very sensitive to changes in the grid power mix.

- A hybrid solar-fuel cell cogeneration system for residential applications has been assessed with energy and exergy analyses [295].
- For an integrated biomass-based cogeneration plant the effect of heat integration approach and temperature of torrefaction (a thermal pre-treatment technology that enhances biomass fuel characteristics) on performance has been evaluated by Sermyagina *et al.* [296]. They show that significant benefits can be achieved if torrefaction is combined with a large heat producer, by adding a torrefaction model into steam power plant simulations. For large plants, higher boiler loads are observed to capture a portion of heat so as to reduce the electricity output for all integration cases, and higher trigeneration efficiencies suggest that integration of torrefaction and cogeneration at full load could be beneficial. For small plants, free boiler capacity at reduced-load conditions satisfies the torrefaction heat demand and increases electricity generation for most integration cases. Integration method, process temperature and plant operation mode are concluded to be important factors for integrated cogeneration and torrefaction.
- A thermodynamic analysis of a novel multigeneration energy system based on heat recovery from a biomass cogeneration cycle has been carried out [297].

12.2.2 Trigeneration and multigeneration

Reviews of trigeneration and multigeneration have been reported in recent years. For instance, research and development recently carried out in the areas of trigeneration and polygeneration have been reviewed by Murugan and Horák [298], including studies on downsizing and improving the efficiency of cogeneration and trigeneration systems. Also, energy policies, prime movers, cooling technologies, configurations and operation strategies for trigeneration systems have been reviewed by Jradi and Riffat [299], focusing on recent developments in technology (prime movers, cooling devices, system configurations, fuels and renewable energy resources employed), support mechanisms and strategies to optimize system performance and improve overall efficiency. The study covers both conventional and commercial trigeneration technologies such as internal combustion engines, gas turbines and absorption cooling, as well as promising newer technologies still in the research and development phase such as ORCs, fuel cells and liquid desiccant cooling.

Analyses of the performance of trigeneration systems have also been reported in recent years. For example, a finite time analysis of a trigeneration cycle has been performed by Agnew *et al.* [300] to help avoid the challenge of many current trigeneration cycles exhibiting low thermal performance even when optimized for maximum useful output. The finite time analysis is based upon coupled power and refrigeration Carnot cycles, and is applicable to cycles that undergo isothermal heat transfer with the environment and are internally reversible (e.g., Stirling cycles). It is shown that a significantly higher energy utilization factor can be obtained with this type of cycle by treating the energy transferred during isothermal compression and expansion as useful products, making the energy utilization larger than the enthalpy drop of the cycle working fluid, and that the cycle has the highest energy utilization factor when energy is supplied from a low-temperature heat source, in which case the output is biased toward heating and cooling. In another investigation, energy and exergy efficiencies are determined of a building trigeneration system based on an internal combustion engine under two operational strategies by Espirito Santo [301]. Noting that trigeneration systems can be a good option for buildings with coincident energy demands and that building demand can vary widely with time of day, activity, occupancy and climate, the energy of the prime mover is compared with site energy demands to improve the design of cogeneration systems and raise their energy utilization factors. A computational hourly profile simulation methodology is developed that combines data from the literature and manufacturers, mathematical representations of physical phenomena and thermodynamic properties. The simulation methodology was used to predict the performance of a cogeneration concept under two operational strategies, and determined that the energy utilization factor varies between 65 and 81 per cent and the exergy efficiency between 35 and 38 per cent, and that – based on primary energy savings – the proposed cogeneration system can compete with high-efficiency thermal plants.

Microscale trigeneration, that is small-scale trigeneration typically with electrical capacities under 15 kW, has also been examined in recent years. For example, technologies, tools and trends for microscale trigeneration have been examined to help it decrease the environmental impact of energy sector while providing adequate standards of energy service, and thereby to contribute to energy sustainability [302]. Also, an analysis considering multiple perspectives of microscale trigeneration systems, as well as their role in the crowd energy concept, has been carried out [303]. In addition, a performance assessment was recently reported of cogeneration and trigeneration for small-scale applications and an index to assess economic feasibility introduced by Angrisani *et al.* [304]. This work commenced in International Energy Agency (IEA) Annex 54 (Integration of Micro-Generation and Related Energy Technologies in Buildings). Parts of the work were applied to small-scale commercial cogenerators, and sensitivity analyses were performed considering values for several European countries of electric efficiency, unitary natural gas and electricity prices, and emission factors.

Various novel trigeneration systems utilizing fossil fuels have been considered recently. The design and modeling of liquefied natural gas-fueled trigeneration plants, with electrical capacities of 1–10 MW, have been proposed and assessed for building applications in terms of efficiency and cost by Arsalis and Alexandrou [305]. They considered decentralized, liquefied natural gas-fueled trigeneration as an alternative to centralized electricity generation to improve efficiency and reduce operating costs. In the system, electricity is generated with a gas turbine and waste heat is recovered to satisfy the heating and cooling needs of nearby buildings, the high quality cooling energy in the LNG is used to cool the inlet air to the air compressor, and waste heat is recovered to provide cooling in the summer via double-effect absorption chillers and heating in winter. For a system with an electrical capacity of 10 MW, the net electrical efficiency was found to be 36.5 per cent, the primary energy ratio 90 per cent and the payback period four years, considering a lifecycle cost of 222 million euros and an investment cost of

13 million euros. The needs of 21,000 households can be met with this system compared to 12,000 households with an equivalent conventional system. Also, energy and exergy analyses were performed of a new compact trigeneration system incorporating an absorption chiller using an ammonia-water working fluid mixture and an internal combustion engine operating on liquefied petroleum gas by Pérez *et al.* [306]. The exergy efficiency of the trigeneration system was determined to be 51 per cent, while irreversibilities associated with combustion were seen to be the most significant contributor to the system inefficiency, representing 36 per cent of the total exergy input and 73 per cent of the total exergy destruction.

Trigeneration systems utilizing novel prime movers have been reported in recent years, with a particularly interesting example utilizing fuel cells. Energy and exergy assessments of a novel trigeneration system, based on a solid oxide fuel cell, a generator-absorber heat exchanger for cooling and a heat exchanger for heating, were carried for steady-state operation out by Ranjbar *et al.* [307]. The maximum energy efficiencies were found to be 79 per cent for the trigeneration system, 69 per cent for heating cogeneration, 58 per cent for cooling cogeneration and 46 per cent for electricity production, and the energy efficiency was found to be at least 33 per cent higher when using trigeneration compared to only the fuel cell system. But the highest corresponding trigeneration exergy efficiency was 47 per cent. The main exergy destructions occur in the air heat exchanger, the fuel cell and the after-burner. Particular attention was paid to the influences of two significant fuel cell parameters: current density and inlet flow temperature. As the fuel cell current density increases, the exergy efficiencies were found to decrease for the power cycle, cooling cogeneration, heating cogeneration and trigeneration and an optimal current density was observed to exist at which the net electrical power is a maximum. As fuel cell inlet flow temperature increases, the trigeneration energy and exergy efficiencies and net electrical power were seen to increase to a peak and then decrease.

Trigeneration systems utilizing renewable energy sources have also been reported in recent years. For example, an energy assessment of thermally driven solar cooling system configurations for an office building in Finland, in conjunction with district heating based on cogeneration, was reported by Reda *et al.* [308]. Solar thermal energy can of course also satisfy demands for space heating, domestic hot water and process heat. A dynamic simulation was employed through TRNSYS software, and two configurations of the solar cooling system were considered. In the first configuration only a storage tank can supply the chiller, while in the second the chiller can be supplied either by the tank or the solar collectors directly. System performance indexes, considering district heating as main building heating supply system in winter and as an auxiliary heating system for the chiller in summer, were evaluated as a function of the tank and solar thermal field sizes. The second configuration was found to exhibit better performance than the first, because of the versatility shown in summer and especially, when the solar irradiance is low. This study highlighted the potential of this technology in areas with cold climates supplied in conjunction with cogeneration and district heating. The solar cooling technology reduced energy use for heating and cooling and shaved summer electricity peaks. Also, biomass gasification cogeneration systems were reviewed and future perspectives provided by Ahrenfeldt *et al.* [309]. Cogeneration processes that produce heat and electricity, heat, electricity and fuel, and heat, electricity and fertilizer were compared, and thermal gasification of biomass was demonstrated to be flexible and efficient if used optimally. The gasification platforms considered included an up-draft gasifier with a gas engine and a fluidized-bed gasifier with a gas engine or a steam cycle, and two-stage down draft gasifiers with a gas engine, a microgas turbine, a solid oxide fuel cell and catalytic fuel synthesis.

Interesting trigeneration applications have been reported in recent years, for systems as small as buildings to as large as cities and regions. Some noteworthy examples are examined here. For instance, energy and exergy analyses of a high-efficiency engine trigeneration system for a hospital based on annual energy demand profiles was reported by Espirito Santo [310]. In this study an annual analysis of the engine trigeneration device as an integrated thermal system was performed using simulation to evaluate its performance. The simulation incorporates system and component characteristics, design assumptions and parameters, off-design operating conditions, energy demands profiles of the site and climatic data. The system was found to have an energy utilization factor between 58 and 77 per cent and an exergy efficiency between 35 and 41 per cent, and the primary energy savings analysis suggested that the proposed trigeneration system is superior to the best available technology used in central thermal plants. Also, the potential of trigeneration for Benevento, a small city in southern Italy, has been examined by Ascione *et al.* [311]. A geo-referenced energy model of the city is proposed, allowing evaluation and visualization of energy demands for space heating and cooling for each building. In addition, the electric demands of buildings are also available. The approach permitted the urban energy maps to be developed, which are useful for evaluating the present challenges and designing possible urban energy retrofits, e.g., the introduction of trigeneration.

Several multigeneration systems have been investigated in recent years. Examples of some of the systems proposed and assessed follow:

- a new renewable energy-based multigeneration system [312],
- a new solar based multigeneration system with hot and cold thermal storages and hydrogen production [313],
- a solar-biomass integrated cycle for multigeneration [314],
- an integrated biomass-based multigeneration energy system [315] and
- a novel solar and ocean thermal energy conversion based multigeneration system for coastal areas [316].

12.3 Heating and cooling

Much research and development has been performed and is ongoing on heating cooling technology. We now describe a selection of the important work and advances in these fields.

12.3.1 Heating

Although heating is often a mature and straightforward technology, innovations are nonetheless continually sought and studies have been performed on heating technology. From a general perspective, sustainable energy systems for building HVAC applications have been developed and analyzed [317]. More specifically, studies have been reported on double-pass photovoltaic-thermal solar air heaters, considering the effect of fins on the performance [318,319], and photovoltaic-thermal solar collectors for air heating have been critically reviewed [320]. Furthermore, solar water heaters with integrated collector-storage units have been proposed [321] and their performance compared considering various heat loss reduction strategies [322]. Also, the status of geothermal heat pump systems has been reviewed and compared with other heating options [323], and ground-source heat pumps for heating have been assessed parametrically considering a vapor compression cycle utilizing an economizer arrangement [324].

Much work has been reported on TES aimed at improving heating (and sometimes cooling) systems, often leading to advances. Systems and applications for TES have been covered extensively [325], but development is ongoing:

- Many advances have been made in sensible thermal storage. For example, natural and artificial methods for regeneration of heat resources for borehole heat exchangers to enhance the sustainability of underground thermal storages, and the modeling thermally interacting multiple boreholes, have both been recently reviewed [326,327]. The thermal interactions among multiple vertical ground heat exchangers for underground borehole TES have been compared analytically and numerically for systems of variable heating strength [328], and assessed with a three-dimensional analysis [329].
- Advances have also been reported for latent thermal storage. For example, phase change material solidification in a finned cylindrical shell TES has been investigated with an approximate analytical approach by Mosaffa *et al.* [330].
- Finally, advances have been made in thermochemical storage, which is at a much earlier stage of development. For instance, energy- and exergy-based assessments and comparisons have been reported for closed and open thermochemical energy storage systems [331], while some studies have focused on closed thermochemical energy storages [332,333].

12.3.2 Cooling

Much cooling technology is commercial and mature, yet innovations are nonetheless being made. For example, double-effect absorption refrigeration systems have been investigated in terms of performance characteristics by Garousi Farshi *et al.* [334], while the use of low-grade heat sources in combined ejector-double-effect absorption refrigeration systems has been studied [335] and combined ejector-absorption and single effect absorption refrigeration systems have been thermodynamically analyzed and compared [336]. Energy and exergy analyses of ammonia/salt absorption refrigeration systems have also been reported by Garousi Farshi *et al.* [337]. Also, the use of multiple phase-change material (PCM) thermal storage units for free cooling have been examined by to determine general thermal performance [338] and their thermodynamic behavior based on energy and exergy [339].

Many advances have been reported on cooling directly in conjunction with HVAC. For example, exergy analyses of psychometric processes for HVAC and refrigeration applications have been investigated generally [340], and advanced exergy analyses of air conditioning incorporating TES have been reported [341]. PCM solidification in a shell and tube finned thermal storage for air conditioning systems has been analytically modeled by Mosaffa *et al.* [342]. More broadly, these advances are often nowadays being examined in conjunction with the idea of achieving net-zero energy buildings and communities [343].

Advances in cooling (as well as heating in many instances) have been made by considering geothermal energy and underground thermal storage, often in conjunction with heat pumps. For example, comparative performance evaluations have been carried out of cascaded air-source hydronic heat pumps [344]. Analytical approaches to evaluating the effect of thermal interaction of geothermal heat exchangers on ground heat pump efficiency have been developed [345] and used to examine the thermal interactions among multiple vertical ground heat exchangers [346]. The role of energy storage in achieving net-zero energy buildings and communities has also been studied [343].

12.4 District energy

Increasing research and development has been devoted to district heating and district cooling in recent years, as well as to their combination in the form of district energy. For instance, numerous advanced district heating and cooling systems have recently been described [347]. Some of the important work and advances in these fields is described in this section.

Increasing research and development has been devoted to district heating recently. For example, strategies for implementing waste-to-energy via district heating have been reviewed and assessed by Pan *et al.* [348], in part to address simultaneously challenges related to energy demand, waste management and GHG emissions. Strategies were proposed to overcome technological, financial, institutional and regulatory barriers and technological supply chains such as green fuel pellets for heating supply and co-digestion of organic wastes for biogas production were reviewed and illustrated. Also, thermodynamic analyses and performance assessments of an integrated heat pump system for district heating applications have been reported [349], as have assessments of various configurations for multiple TESs in thermal networks [350].

Corresponding to the increased activity in district heating, increasing research and development has been observed into district cooling in recent years, sometimes taking quite novel approaches. For example, energy and exergy analyses of a novel power cycle using the cold of liquefied natural gas and low-temperature solar energy have been reported by Mehrpooya *et al.* [351], while deep borehole heat exchangers have been conceptually reviewed by Sapińska-Śliwa *et al.* [352].

Research and development on district energy, i.e., district heating and cooling, has also been examined with more urgency in recent years. For instance, to accelerate the advancement and implementation of advanced clean energy technologies, the use of a solar heating and cooling system with borehole TES for a community has been reviewed by Rad and Fung [353]. Solar district energy systems with integrated thermal storage, used either for heating or cooling, are considered. The potential of district heating and cooling integrated with borehole TES to reduce transmission and distribution heat losses within a community, to lower the required energy production and storage and to make the entire system techno-economically more attractive while maintaining its energy efficiency, is examined in relation to alternative systems. Also, district energy technology and potential enhancements have been reviewed [354], and the use of TES in a district energy application has been analyzed based on exergy [355], energy [356] and other factors [357].

12.5 Integrated systems for cogeneration and district energy

Much research and development has been performed and is ongoing on cogeneration-based district energy systems, due in large part to the benefits they provide. We now describe a selection of the important work and advances in these fields.

Some of the advances are restricted to cogeneration-based district heating. For instance, analyses of the location for peak heating in cogeneration-based district heating systems have been reported by Wang *et al.* [358]. Noting that cogeneration heat and electricity is the main technology for providing the base load of district heating in China, and that the technology is not efficient for providing the peak load (relative to a peaking boiler with high efficiency), the effect is examined of peak boiler location on the energy efficiency and economic performance of cogeneration-based district heating. The analysis considers the connection mode and control strategy for various peak heating locations, and the effect of peak boiler location on initial network investment and the cost of distributing thermal energy. When the objective is to place the peak boiler where the overall costs are smallest, it is found that the peaking boiler should be located at the cogeneration plant if that allows using cheaper electricity for distributing the heat, but that the peak boiler should be located closer to users with high-density heat loads if the market electricity price is used everywhere or if energy efficiency is a higher priority. Also, the development of a cogeneration-based district heating system for the new town of Parand, a residential area near Tehran, Iran to mitigate global warming has been examined by Mostafavi Tehrani *et al.* [359]. Thermal energy is recovered from the exhaust hot gases of an existing gas turbine power plant to meet the dynamic thermal energy requirements of the town, and to drive a steam turbine cycle. Two gas turbine power plants are located near the town, having electrical capacities of 954 MW and 790 MW. Two control methods are considered along with a thermal load following operation strategy: maximum power generation and minimum fuel consumption. For the best design from an environmental perspective, the primary energy supply reduction, carbon dioxide abatement and net present value were found to be 27 per cent, 2.6 million tons and 1,500 million dollars, respectively.

Some of the advances are considered here are more general, covering cogeneration-based district energy as well as trigeneration-based systems. For example, performance variations when renovating thermal power plants to trigeneration systems for district energy have been evaluated by Huseyin Erdem *et al.* [360]. Thermodynamic models are developed for eight thermal power plants for simulation purposes, and various performance criteria are considered: electrical power output, thermal efficiency, COP and comprehensive thermal efficiency (an efficiency that takes into account electricity, heating and cooling products of the trigeneration system). The comprehensive thermal efficiencies of the eight considered systems are determined to range from 49 per cent to 61 per cent in heating mode, with a slight decrease in generated electricity, suggesting that such modifications of an existing power plant can be beneficial in terms of efficiency and sustainability. Also, efficiency analyses of a cogeneration-based district energy have been reported based on energy [361] and exergy [362], while broader assessments of utility-level cogeneration-based district energy have been reported on an annual basis [363] and in cumulative form over a long time period [364].

12.6 Economics of systems and technologies related to cogeneration and district energy

Much research and development has been performed and is ongoing on the economics of technologies and systems related to cogeneration and district energy, including simple and extended cogeneration, heating and district heating, and cooling and district cooling. We now describe a selection of the important work and advances in these fields.

Some of this work involves exergy-based economic methods, which many consider to be an enhanced approach to energy economics. This field is discussed extensively elsewhere [282], with numerous applications in a wide variety of disciplines.

12.6.1 Economics of heating and cooling

Many advances and developments related to the economics of heating and cooling have occurred in recent years. For example, exergoeconomic analyses of double-effect absorption refrigeration systems have been performed by Garousi Farshi *et al.* [365], and exergoeconomic comparisons of double-effect and combined ejector-double-effect absorption refrigeration systems have been reported [366].

12.6.2 Economics of district energy

Numerous advances and developments related to the economics of heating and/or cooling have recently been reported. For example, an exergoeconomic analysis of a district heating system for geothermal energy was carried out using specific exergy cost method [367], in order to provide cost based information and suggests possible components and locations for them to improve cost effectiveness. The analysis was applied to the Afyon geothermal district heating system in Turkey and utilized operation data. The unit exergy cost of heat produced by the geothermal district heating system was found to average 5,620 $/h. System components were identified that should be modified due to their high total operating cost rate, relative cost difference and exergoeconomic factors. It was also observed that, when the well head temperature and ambient temperature increase, the heat production costs per unit exergy for all heat exchangers decrease due to the high exergy destruction cost rate of the system. Another exergoeconomic analysis of a geothermal district heating system along with a case study has been reported earlier [368].

12.6.3 Economics of cogeneration

Much research and development has been performed and is ongoing on the economics of cogeneration, and selected important examples of these activities, and the advances they have yielded, are described in this section.

Many economic studies have focused on methodology improvements and tool development for cogeneration systems analysis. Some examples follow:

- A decision support model for the economic evaluation of cogeneration systems that can assist energy planning and decision-making for cogeneration investments in an unstable energy market was proposed by Konstantakos *et al.* [369]. The model uses as variables state subsidies, natural gas and electricity selling price. A motivation for the work was the fact that many countries in Europe support cogeneration investments and/or provide enhanced selling tariffs for the electricity they produce. Five scenarios for Greece were considered. Using a sensitivity analysis with natural gas price and financial subsidy as variables, it was found that profits decrease as natural gas price increases, but that the investment remains viable for almost twice the natural gas price at the time of the study, suggesting that minor natural gas price fluctuations do not significantly affect the investment.
- The operational cost savings potential for two cogeneration units with differing operational strategies were evaluated based on a monthly power-to-heat ratio analysis, defined as the ratio of facility electric load to thermal load [370]. Noting that previous work has indicated that cogeneration system performance may be improved by limiting operation to months when the power-to-heat ratio is relatively low, the authors illustrated how the facility power-to-heat ratio can be used to establish the potential of a cogeneration system with varying operational strategies to reduce operational cost with respect to traditional cogeneration systems and conventional systems with separate heating and electricity generation. It was shown for eight separate buildings that the power-to-heat ratio parameter can predict the potential for a cogeneration system with varying operational strategies to reduce operating costs.
- Energy, environmental and economic dynamic performance assessments have been reported using simulation software for several residential micro-cogeneration systems fueled by natural gas (internal combustion engines and a reciprocating external combustion Stirling engine) [371]. Multi-family residences in Italy were considered. Auxiliary heat was supplied by a natural gas boiler and a heat storage tank was incorporated into the system. A sensitivity

analysis was performed considering hot water storage volumes ranging from 0.50 to 0.86 m^3 and set-point temperatures ranging from 55 to 60°C. The study utilized detailed dynamic models, developed within Annex 42 of the IEA and calibrated experimentally, for predicting the performance of the cogeneration devices. The proposed system was compared with a reference system comprised a natural gas-fired boiler for heating and the central electrical grid to permit a comparison of the cogeneration system and a system based on separate energy production. The proposed system was found to reduce primary energy consumption by up to 13 per cent, carbon dioxide equivalent emissions by up to 19 per cent, and operating costs by up to 21 per cent with respect to a conventional system. The greatest reductions were obtained with the storage tank having the lowest volume coupled with internal combustion engines when the set-point temperature was 55°C.

Other studies have focused on innovative cogeneration systems using integrated and advanced technologies and sometimes renewable energy sources. For example, energy and exergoeconomic assessments of a proposed cogeneration system for electricity and cooling based on a solid oxide fuel cell were carried out by Khani *et al.* [372]. Also, the techno-economic feasibility has been assessed for a biomass cogeneration plant based on an ORC [373]. Noting that biomass-fueled ORC power plants in a cogeneration topping layout have operated in central Europe since 2000 and that such plants usually operate in conjunction with district heating system proximate to the communities whose thermal and electric energy demands are to be satisfied, the energy performance was analyzed for various scenarios in this study. The economic analyses consider four typical organic fluids (two silicone oils, toluene and isopentane), subcritical and supercritical cycles, and the option for a recuperator, and treat plant size, biomass cost and annual operation schedule as important parameters. Heat and electricity are sold to a nearby village at competitive market rates, based on Spanish market prices, and subsidies were not considered for electricity. The technical analysis suggests that hexamethyldisiloxane use in recuperative cycles is superior in both the subcritical and the supercritical layouts, due to its advantageous characteristics in terms of health and environmental damage, reliability and efficiency. Supercritical cycles are found to be less profitable because their increased electric efficiency reduces the amount of useful heat produced, which is the main revenue source. It is demonstrated that the size of the plant can be set based on the fuel cost to achieve the same profitability, e.g., similar internal rates of return are obtained for a cogeneration plant of 1 MW electrical capacity using biomass costing 5.5 €/MWh and a cogeneration plant of 2 MW electrical capacity using biomass costing 15.5 €/MWh. It is determined that a 5 per cent IRR for subcritical recuperative plants requires an annual operation time of 2,750 h for a 2 MW (electric) plant using biomass costing 5.5 €/MWh and 5,500 h for a 1 MW (electric) plant using biomass costing 15.5 €/MWh. The above biomass costs are based on the thermal content of the fuel.

Some economic studies have focused on the use of cogeneration in a large region. for example, the potential for cogeneration from fossil fuels and nuclear energy using the electrical utility facilities of a province in Canada has been assessed based on energy, environmental, health and cost benefits [374].

12.6.4 Economics of trigeneration and multigeneration

Much research and development has been reported on the economics of trigeneration and multigeneration. Some interesting examples follow:

- Energy, environmental and economic dynamic performance assessments have been carried out for an innovative multi family residential building-integrated trigeneration system by Sibilio *et al.* [375]. The proposed system is based on a natural gas-fueled internal combustion engine micro-cogeneration unit for space heating and domestic hot water production. The cogeneration unit powers an electric air-cooled vapor compression water chiller for cooling. The trigeneration system feasibility was investigated with the dynamic simulation software TRNSYS and compared to conventional separate energy production processes, i.e., a natural gas-fired boiler coupled with the same air-cooled water electric chiller fed by the electric grid. The primary energy consumption, equivalent carbon dioxide emissions and operating costs of the proposed system were evaluated. A typical residential building in Italy is considered, and the analyses were performed for varying the climatic conditions as represented by three Italian cities: Palermo, Naples and Milan.
- The economics and energy management factors of a micro-trigeneration system have been assessed considering the use of solar photovoltaics and natural gas fuel (for the trigeneration system) [376]. The cooling, heating and electricity demands of a five-story residential building in Iran having a total floor area of 1,000 m^2 are considered, where the peak electrical demand is 48 kW and the respective heating and cooling demands are 100 kW and 50 kW. Two economic measures are considered: loss of power supply probability and levelized cost of energy. It is demonstrated that the present fuel subsidy diminishes the attractiveness of PV and hybrid PV/generator systems, and that the optimal operating strategy is affected by boiler and fuel prices and the number of solar collectors.
- Energy and exergy analyses and economic assessments have been reported of micro-cogeneration and micro-trigeneration cycles of a hybrid system operating on biomass and solar energy, and comprising an integrated ORC and vapor compression refrigeration cycle [377]. The biomass boiler and parabolic-trough solar collectors provide heat to the ORC via two intermediate-pressure water circuits. The ORC expander, the vapor compression refrigeration cycle compressor and the electrical generator are connected to the same shaft, and condensation occurs at a fixed pressure in a single condenser for both cycles. Heat from the condenser is used is used for hot water heating. In trigeneration mode (summer operation), some of the electricity from the ORC drives the vapor compression cycle compressor, while in cogeneration mode (winter operation) cooling is unnecessary so the vapor compression cycle is inactive. Subcritical operation pressures are considered for the organic working medium R245fa. For

an overall heat input rate of 50 kW and a cooling load of 5 kW (during summer), it is determined for a system having an evaporation temperature of 90°C and no superheating that the net electric efficiency is 2.4 per cent, with an electrical power output of 1.4 kW and a heat output rate of 53.5 kW, and that the exergy efficiency of the ORC is 7 per cent. The economic analysis considered a typical apartment block on an island in Greece and a parabolic-trough solar collector area of 50 m^2, and found that the reductions in fuel oil and electricity use yield an IRR of 12 per cent and a payback period of 7 years.

Increasing numbers of economic studies have considered multigeneration systems. For instance, a techno-economic assessment of a renewable energy-based integrated multigeneration system for green buildings was assessed recently [378]. Also, a techno-economic analysis for a natural gas-fired combined cycle plant for trigeneration of electricity, cooling and clean water using waste heat recovery was reported by Mohan *et al.* [379]. In that study, electricity was generated by a gas turbine power plant and a steam Rankine cycle, clean water by air gap membrane distillation, and cooling by a single stage vapor absorption chiller. The flue gases from a gas turbine power cycle provide thermal energy to the trigeneration system. Heat from the steam cycle condenser and excess heat from the flue gases drive the cooling and desalination cycles, which are optimized based on the cooling demands of the buildings. It was found that the energy efficiencies are 82–85 per cent for the trigeneration system (compared to 50–52 per cent for combined cycles), the use of waste heat recovery reduced trigeneration carbon dioxide emissions per MWh by 51.5 per cent, and the trigeneration system has a payback period is 1.4 years and a cumulative net present value of $66 million over the project life.

12.6.5 Economics of cogeneration-based district energy

Various developments and advances have been made on cogeneration-based district energy. For example, a techno-economic assessment of a natural gas combined cycle for combined heat and electricity integrated with District heating have been reported for the Netherlands [380], where space heating presently represents the greatest household energy use. Since methods to calculate energy savings from cogeneration systems differ widely, various methods to estimate primary energy savings relative to typical domestic gas-fired condensing boilers, including the method from the EU CHP Directive, are assessed and compared in the research. Using actual hourly cogeneration plant performance data, it is shown that supplying dwellings with district heat from a natural gas combined cycle for cogeneration reduces energy use, regardless of calculation method and for a wide range of efficiencies, and that CO_2 mitigation costs are acceptable from a social perspective (at discount rates up to 4 per cent, excluding fuel taxes) but not from a private perspective (at discount rates up to 10 per cent, including fuel taxes). In a different project, the cost of using power plant reject heat in low-temperature district heating and cooling networks has been evaluated by Colmenar-Santos *et al.* [381], in order to assess the economic impact of converting conventional stations to cogeneration plants connected to heat distribution and refrigeration systems. Annual capital and operating costs were determined for supplying cold and heat to urban areas for weather conditions typical northern, central and southern Europe – represented by Oldenburg–Wilhelmshaven, Bristol and Cartagena, respectively.

Various developments and advances have also been made on trigeneration-based district energy. A range of examples are now described.

Operating strategies for cogeneration and trigeneration for integrated communities including a residential district and a general hospital have been evaluated using RETScreen software [382]. In that study, the economic feasibility and CO_2 emissions was determined relative to energy systems that separately supply consumers with electricity, heating and cooling. It was found that equity payback period is 5.7 years, the electrical power generation capacity of the trigeneration system for the general hospital is 3.2 MW, and an electrical load-following operating strategy yielded the highest energy efficiency at 86 per cent. It was demonstrated that the simple payback period declines from 5.2 years to 3.0 and 2.1 years as the electricity cost rises by 10 and 20 per cent of the present price, and that the efficiency declines when the electrical generation capacity of the cogeneration system rises from 26 to 40 MW.

In another investigation, thermoeconomic and exergy analyses were carried out on a small-scale, decentralized trigeneration plant including a gas turbine cycle integrated with a cooling plant and a district energy network [383]. Noting that decentralization can reduce transmission and distribution losses and permit waste heat to be recovered for heating and/or cooling via absorption chillers, the study analyzed the merits of the proposed system at distant locations from centralized power plants. The system is fueled by liquefied natural gas and its cold is recovered in a liquefied natural gas regasification device that is coupled to the ambient air feed to the air compressor. In winter operation, recovered heat enters the district energy network, while in summer operation, recovered heat drives an absorption chiller to generate a cold medium that is distributed via the district energy network. It was determined that the average primary energy ratio of the proposed system is 0.91, the net electrical efficiency is 36 per cent in summer and 37 per cent in winter, and the average exergetic efficiency is 42 per cent, representing a 25 per cent improvement over the conventional system. The payback period was found to be reasonable at approximately 4 years, for a total initial cost of 17.1 million €.

In a different study, the feasibility and energy and economic benefits of converting an existing cogeneration system to trigeneration for sewage treatment plants in a subtropical area were assessed by Chen *et al.* [384]. A biogas-driven cogeneration system normally is used for both electricity generation and digester heating, but it yields a large amount of heat in summer in subtropical areas when the heating demand is low, creating the opportunity for using trigeneration for space cooling by adding an absorption chiller to the cogeneration scheme. Four trigeneration retrofit schemes were proposed with different types of absorption chillers. The hourly dynamic energy performances of the systems were simulated using Hong Kong weather data. A double-effect absorption chiller driven by 450°C flue gas was found to be the preferred retrofitting scheme in terms of highest energy efficiency (the annual average energy efficiency rose for it to 38 per cent from 21 per cent), shortest payback period, improved stability and easier control.

Some studies have considered multigeneration-based district energy. Two notable examples are as follows:

- Thermoeconomic analyses of the role of energy storage in a combined multigeneration and district system were carried out using the software W-ECoMP by Barberis *et al.* [385]. Data are taken from a smart multigeneration microgrid designed to satisfy the energy demands of the university campus of Savona, Italy. The system is composed of various cogeneration units (microgas turbines and an internal combustion engine) and renewable generators, as well as two auxiliary boilers (one of which is usually inactive). The generators are distributed around the campus and coupled to electrical and thermal storages (specifically hot and cold water tanks and electrical batteries). Prudent integration of the different storage systems with the several cogeneration units was required to determine the best management strategy and to follow both thermal and electrical loads while accounting for the significant differences between the two energy demand profiles. The analyses account for the temporal dependency of the energy demands over the year and off-design device behavior. For various scenarios, monetary savings and reductions in electrical energy purchases from the national grid are considered.
- Exergetic and exergoeconomic analyses of a novel hybrid solar-geothermal multigeneration system producing energy and water were performed using dynamic simulation model by Calise *et al.* [386]. The hybrid system supplies electricity, heating, cooling and fresh water for a small community, via district heating and cooling, and is composed of an ORC driven by medium-enthalpy geothermal energy and a parabolic trough collector solar field. Geothermal brine is also used for space heating and cooling, and to provide heat to a multi-effect distillation unit for desalinizing seawater. The overall simulation utilized TRNSYS software, and included control strategies for managing the system. The accounting of exergoeconomic costs provides a monetary value for all material and energy flows as well as a reasonable basis for price allocation. Both summer and winter operation are considered. It was demonstrated that global exergy efficiency varies between 40 and 50 per cent in heat recovery mode and between 16 and 20 per cent in cooling mode, and that the exergoeconomic costs are about 0.148–0.172 €/kWh for electricity, 0.188 €/kWh of exergy for chilled water, 0.017 €/kWh of exergy for cooling water and 0.58 €/kWh of exergy for desalinated water.

12.7 Reducing environmental impact and climate change with cogeneration and district energy

Much research and development has been performed and is ongoing on the mitigation of environmental impact and climate change that is possible with systems for cogeneration (simple and extended), district energy and related technologies, including heating and cooling systems. We describe a selection of the important work and advances in these fields in this section. It is noted that much of that work involves exergy methods, which have recently been increasingly linked with environment and ecology approaches in order to provide enhanced approaches to environmental and ecological management [283].

12.7.1 Heating, cooling and district energy

Advances have been made on reducing environmental impact and climate change through the use of systems for heating and cooling as well as district energy. For example, exergoeconomic and environmental analyses of a CO_2/NH_3 cascade refrigeration system equipped with different types of flash tank intercoolers were performed by Mosaffa *et al.* [387], while an enviro-economic function for assessing energy resources for district energy systems has been recently proposed [388]. In an interesting study for a specialized application, the potential for improving the energy efficiency of HVAC systems at hospitals has been assessed considering economic and environmental aspects [389]. The study was motivated by the fact that for the country considered hospitals account for 6 per cent of the total energy consumption in the utility buildings sector and HVAC systems account for most of the electricity use at hospitals. Payback periods were evaluated for HVAC-related measures involving the installation of cogeneration units, trigeneration units, advanced chillers, new burners, improved heat exchangers and steam traps, and variant refrigerant flow technology was shown to enable notable energy efficiency and cost savings relative to conventional HVAC systems.

12.7.2 Cogeneration and related systems

Noteworthy developments have been made on reducing environmental impact and climate change through the use of cogeneration and related systems. For example, exergy-based methods for allocating carbon dioxide emissions from cogeneration systems have been developed [390], as pointed out in Chapter 11. Also, at the utility scale, the energy and environmental benefits of regional cogeneration has been evaluated for electrical utilities operating on nuclear energy and coal [391], and reductions in energy use and environmental emissions achievable with utility-based cogeneration have been determined for Ontario, the largest province in Canada [392]. In a study focused on methodology, environmental performance assessments based on local and global approaches and focusing on air pollutants have been developed for cogeneration systems by Bianchi *et al.* [393]. The "Avoided Heat Generator" method was argued to be an appropriate approach for local-scale environmental impact evaluation of cogeneration, as it calculates the pollutant emissions reductions due to cogeneration operation, taking into account the emissions of a heater that provides the equivalent thermal power of the cogeneration prime mover. Methods using the "Pollutant Saving Index" were argued as suitable for global-scale environmental impact estimates. In the study, environmental benefits in terms of reduced NO_x, CO and CO_2 emissions were evaluated numerically for several cogeneration systems using different technologies and electric power capacities, in order to provide an example of the two methods and their potential for use in cogeneration authorization procedures.

12.7.3 Trigeneration and related systems

Many interesting advances and developments have been made on reducing environmental impact and climate change through the use of trigeneration and related systems. For instance, GHG emission and exergoenvironmental analyses have been recently reported for a trigeneration system [394]. Some of the other significant studies in these fields are described here.

Some of the environmental studies focus on the use of trigeneration in conjunction with renewable energy, such as biomass and solar energy. Some examples:

- Environmental assessments of trigeneration systems based on biomass combustion relative to conventional generation systems have been reported by Maraver *et al.* [395]. Noting that trigeneration based on biomass combustion have demonstrated their benefits in some operating conditions but that their environmental and energy performance does not always exceed that of conventional standalone generation systems, biomass-based trigeneration plants are evaluated with the lifecycle assessment methodology to provide environmental guidelines. The primary energy savings ratio was evaluated as a contrasting methodology. Various sizes of cogeneration and cooling units were considered. Based on the lifecycle assessments, small cooling-to-heating ratios make trigeneration plants environmentally superior to conventional standalone generation, while high trigeneration cooling-to-heating ratios have the opposite effect. The use of the primary energy savings ratio alone was shown to be inadequate for assessing the steady-state performance of a trigeneration plant because in some circumstances it may misleadingly indicate there are no environmental benefits of using trigeneration.
- Energy and environmental/ecological assessments, integrated with exergy methods, have been made for a system integrating biogas trigeneration and solar photovoltaics [396]. The trigeneration system is based on an internal combustion engine and an adsorption chiller. The total and relative primary energy savings of the biogas trigeneration system aided by a PV system are assessed, as are the reductions in emissions of total GHGs. The algorithm used considers actual hourly data over the year of operation and accounts for the variability of solar energy availability and the daily variability in biogas availability. It also allocates fuel between useful products based on exergy.

Some of the environmental studies focus on the use of trigeneration in conjunction with advanced energy technologies, such as fuel cells and ORCs. For example, an exergoenvironmental analysis of trigeneration using an ORC has been reported [397]. In another study, GHG emission and exergy analyses were carried out of a novel integrated trigeneration system driven by a solid oxide fuel cell and including a generator-absorber heat exchanger absorption refrigeration system [398]. Various performance parameters were evaluated and four operating cases were considered: electrical power generation, electrical power and cooling cogeneration, electrical power and heating cogeneration and trigeneration. It was found that a maximum enhancement of 46 per cent could be achieved in exergy efficiency when the fuel cell was used as the primary mover for trigeneration compared to using the fuel cell on a standalone basis, that the main irreversibilities were associated with the air heat exchanger, the fuel cell and the afterburner, and that the unit CO_2 emission were double when only electricity is generated compared to when the system operates in a trigeneration mode.

12.7.4 Cogeneration-based district energy systems

Developments have been made on the environmental aspects of cogeneration-based district energy. For example, the potential Environmental and Health Benefits of exploiting Utility-Based Cogeneration and District Cooling in the province of Ontario, Canada have been analyzed [399]. Also, indexes for evaluating the potential advantages of cogeneration from the perspectives of energy, exergy, economics and environmental impact – in terms of pollutant emissions – have been investigated by Torchio [400]. Noting that legislative constraints usually are based on energy efficiency, primary energy savings and GHG reductions, the author suggests that other criteria should also be accounted for to better evaluate cogeneration, in terms of both individual cogeneration units and cogeneration systems, the latter including all the cogeneration units and auxiliary plants necessary to satisfy demand. Several indexes were proposed to evaluate the cogeneration potential, based on energy, exergy, environmental aspects (global and local) and economics, and applied to a cogeneration-based district heating system.

12.8 Optimization of systems related to cogeneration and district energy

Much research and development has been performed and is ongoing on the optimization of technologies and systems related to cogeneration and district energy, including simple and extended cogeneration, heating and district heating, and cooling and district cooling. We now describe a selection of the important work and advances in these fields.

Much of this work utilizes general methods for optimizing energy systems, which have been described in detail recently [401], although technology-specific methods are also employed in many studies.

12.8.1 Optimization of distributed energy systems

Distributed energy systems often incorporate or are integrated with technologies and systems related to cogeneration and district energy, including simple and extended cogeneration, heating and district heating, and cooling and district cooling, and have been the subject of much optimization activity in recent years. An important selection of optimization studies pertaining to distributed energy systems and the advances they yielded are described in this section.

Many of these studies relate to the optimal design and operation of distributed energy systems. Some interesting examples of this work and the enhancements to which it has led follow:

- Optimal options for natural gas-based distributed energy systems for four building categories (residential buildings, hotels, offices and mixed buildings) in Dalian, China have been examined by Li *et al.* [402], motivated largely by the fact that current energy policy in China encourages liquefied natural gas applications. Basic and improved distributed energy systems cases were developed and evaluated in terms of such measures as energy savings, exergy efficiency, annual total cost savings and environmental impacts. It was shown that in summer no energy savings are achieved for all types of buildings using basic distributed energy systems but savings occur for some improved distributed energy systems (e.g., one involving hybrid cooling); energy savings are obtained for all types of buildings in winter; the exergy efficiencies of distributed energy systems are lower in summer than winter due to the large cooling demands and low COP of absorption chillers; and distributed energy systems decrease annual total costs and CO_2 emissions for all examined cases compared with a reference system.
- A mixed integer linear programming model for the optimal design of district-scale distributed energy resource systems has been developed by Yang *et al.* [403], focusing on systems in which energy is produced outside energy-consuming buildings and conveyed to the buildings through energy distribution networks. The model simultaneously optimizes operation strategies, location/site for energy generation, and synthesis factors such as structure of energy distribution networks and type, capacity and number of devices, and accounts for the discrete nature of equipment capacities, equipment part-load behavior and the effect of ambient temperature on gas turbine performance. The objective function is the system total annual cost (investment, maintenance and operating). An application to an urban area in Guangzhou, China demonstrated significant economic benefits for the proposed distributed energy resource system over a conventional energy system. In a related work, the optimal design of distributed energy resource systems coupled with energy distribution networks for heat, cool and electricity has been investigated using a superstructure-based mixed integer linear programming model [404].
- A mixed integer linear programming tool called Distributed Energy System Optimal Design was recently developed to optimally design a distributed energy system for heating, cooling and electricity in an urban neighborhood [284].

Numerous studies on the optimization of distributed energy systems are applied to the use of renewable energy or other free energy sources. For instance, a thermoeconomic multi-objective optimization of a novel biomass-based integrated energy system was reported recently [405], while the optimization of the thermal performance of free cooling systems using enhanced latent heat thermal storage has been investigated [406]. Furthermore, the effect of various economic support policies on the optimal synthesis and operation of a distributed energy supply system with renewable energy sources was examined for an industrial area in northeast Italy by

Casisi *et al.* [407]. In that study, the social cost in terms of economic investment needs were investigated in parallel with the corresponding reductions in CO_2 emissions and overall energy use by utilizing a mixed integer linear programming optimization procedure developed to evaluate the effect of economic support policies on the optimal configuration and operation of a distributed energy supply system. The objective function to be minimized is the total annual cost for owning, operating and maintaining the overall energy system. It is demonstrated that a proper mix of renewable energy technologies and cogeneration systems is often included in the optimal solution, depending on the quantity and nature of the support policies.

12.8.2 Optimization of cogeneration systems

Many efforts have been made to optimize cogeneration systems for electricity and heat. Some of these have focused on methodologies and tools. For instance, multi-step and mixed integer linear programming approaches for cogeneration systems optimization have been compared [408]. Other studies have concentrated on the use of renewable energy in cogeneration, such as a multi-objective optimization of a solar-hybrid cogeneration cycle was carried out by Soltani *et al.* [409]. Two other important studies relating to optimizing cogeneration systems that incorporate renewable energy follow:

- The modeling and optimization of the smart hybrid renewable energy for communities have been investigated by Wang *et al.* [410], where "smart" is taken to imply high security and reliability, enhanced market service, low environmental impacts, reduced costs and greater use of renewable energy. The smart hybrid renewable energy system proposed considers heating, cooling and electricity markets in large communities and utilizes renewable energy, cogeneration and energy storage. A planning model for cogeneration is developed along with a linear programming algorithm to optimize the system over a weekly period. Three weeks typical of cold, warm and mid-season weather are assessed, demonstrating the superior efficiency and flexibility of the developed modeling and optimization method for smart hybrid renewable energy systems.
- A lifecycle multi-objective economic and environmental assessment has been carried out for distributed generation in buildings based on cogeneration and solar photovoltaic and thermal technologies, and compared with conventional boilers [411]. The life-cycle model combines distributed generation and conventional sources to determine the cost and environmental impacts of meeting the building energy demands over a defined planning period. The electricity generation mix in Portugal were considered and the Pareto optimal frontiers derived, showing the trade-offs between various environmental impacts (non-renewable cumulative energy demand, GHG emissions, acidification, eutrophication) and cost to meet the energy demand of a commercial building. It was shown that a specific design and operational strategy is appropriate for the energy system depending on the distributed generation objective (reduce cost or environmental impact), and that strategies to minimize each type of environmental impact and the associated costs trade-offs can be determined using the solutions on the Pareto optimal frontiers.

12.8.3 Optimization of extended cogeneration systems

Efforts have been made to optimize extended cogeneration systems, such as trigeneration for electricity, heat and cool, and other multigeneration systems. Some of these are described here.

Numerous reviews have been reported on the optimization of trigeneration systems, speaking holistically to the complexity and importance of the topic. One survey of the state of the art for trigeneration systems focused on the distinct advantages of high efficiency and economic savings plus reduced GHG emissions [412]. That review contained four parts: system and component development and working schemes; prime mover and thermally activated facilities; management, control, system optimization and sizing; and applications in countries and development barriers. Another review of trigeneration and multigeneration optimization was premised in part on the challenges of modeling and sizing these systems, due to the high number of inputs and computational complexity [413]. A taxonomy was constructed and novel self-organizing map used to cluster the material based on optimization technique. The crowd of clusters demonstrates widely used optimization techniques and gaps in research. Another trigeneration review centered on performance improvement and optimization [414], and covered relevant energy and exergy methods, system optimization, performance improvement, systems development and emerging trends in trigeneration technologies. Finally, challenges and opportunities regarding the use of multigeneration in moving to energy sustainability were reviewed from optimization viewpoints by Rong and Lahdelma [415], stressing that efficient decision support tools for sustainable multigeneration systems are needed.

Some studies have dwelled on details involved in the optimization of trigeneration systems. Selected examples of these are covered here:

- The optimization was reported of mixed working fluids (20 zeotropic mixtures in total) for a novel trigeneration system based on ORC and heat pumps [416], focusing on optimal component concentration. In the research, a system model was considered with idealized operating conditions and a geothermal water heat source at a temperature of 95°C and a mass flow is 40 t/h. The heating fluid is heated to 45°C when the ambient temperature is −5°C, and the refrigerating fluid is cooled to 0°C when the ambient temperature is 35°C. It was shown that R141b/R134a, R141b/R152a and R123/152a exhibit higher COP's and exergy efficiencies than the other working fluids, and that a mixture of dry and wet working fluids is more suitable for the system. The system attains higher energy output if zeotropic mixtures have a higher proportion of wet rather than dry working fluids.
- The influence of plant scheme and boundary conditions on the optimization of trigeneration systems by mathematical programming was examined by Piacentino *et al.* [417]. Noting that it is often difficult to identify optimal design and operation strategies since economic viability is influenced by endogenous factors (e.g., energy load profiles) and exogenous factors (i.e., external factors such as energy prices and support mechanisms), an enhanced decision tool was

proposed in that study that simultaneously optimizes plant layout, sizes of the main components and operation strategy. An analysis was performed for a specific building in the hotel sector to identify the most promising plant configuration, in terms of type of cogeneration unit (microturbine or diesel oil/natural gas-fueled reciprocate engine) and absorption chiller. The sensitivity was assessed of the tax exemption for the fuel consumed in "efficient cogeneration" mode, and the dynamic behavior of the prime mover and its ability to rapidly adjust to follow energy loads.

- Noting that current trigeneration cycles exhibit low thermal performance even when optimized for maximum useful output, the maximum output from a trigeneration cycle was examined by Agnew *et al.* [418]. Trigeneration based on coupled Stirling cycles was considered. It was shown that a significantly higher energy utilization factor is possible by considering the energy transferred during isothermal compression and expansion as useful products, allowing the energy utilization to exceed the enthalpy drop of the working fluid. The cycle was shown to work best with a low-temperature heat source and to have the highest energy utilization factor when the three outputs were approximately equal.

Some studies have been carried out on various details involved in optimizing trigeneration systems using renewable energy forms such as biomass and advanced energy technologies such as energy storage. Selected examples are covered here. For instance, a thermoeconomic optimization using an evolutionary algorithm of a trigeneration system driven by a solid oxide fuel cell was reported by Sadeghi *et al.* [419]. In another investigation, a simple, rigorous trigeneration plant model for designing, comparing, optimizing and simulating small to medium scale trigeneration configurations based on biomass gasification was proposed by Puig-Arnavat *et al.* [420]. The model includes five configurations and various components (gasifier, internal combustion engine, and single- and double-effect absorption chillers), and enables the outputs to be evaluated for various biomass types, operating conditions and configurations. The model was applied to a plant to be built in Cerdanyola del Vallès near Barcelona, Spain. In a different study, a hybrid optimization model for biomass trigeneration system combined with pit TES was proposed by Dominković *et al.* [421], aimed at managing properly excess heat production in trigeneration and thereby increasing the power plant annual efficiency. A double piping district heating and cooling network in the residential area without industry consumers was considered, allowing simultaneous heating and cooling transport and making it suitable for assessing different regions. The degree-hour method was used to find hourly heating and cooling energy demands. The system covered annual heating and cooling energy needs, electricity was taken to be transferable to the grid. The system was modeled in MATLAB® and a hybrid optimization model was used to maximize the net present value. The system economics were found to be favorable when the economy-of-scale of both the power plant and pit TES were utilized, and the pit TES was seen to be useful for storing energy and peak shaving.

Some studies on the optimization of other more complex multigeneration systems. For example, multi-objective optimizations of various multigeneration energy systems have been reported [422,423,424]. Other selected examples follow:

- A multi-objective evolutionary-based optimization was performed on a new multigeneration energy system based on a microgas turbine, a dual pressure heat recovery steam generator, an absorption chiller, an ejector refrigeration cycle, a domestic water heater and a proton exchange membrane electrolyzer [425]. The system produces electricity, heating, cooling, hot water and hydrogen. Energy, exergy and environmental factors are considered. The multi-objective optimization is based on a fast and elitist non-dominated sorting genetic algorithm (NSGA-II). The two objective functions considered were system total cost rate (i.e., the costs associated with fuel, components and environmental impact), and the system exergy efficiency. The total cost rate of the system was minimized and the system exergy efficiency maximized.
- The performance and economic viability of small-scale multigeneration technologies for cost-optimal and zero-energy office buildings was assessed by Mohamed *et al.* [426], following the Energy Performance of Buildings Directive comparative framework methodology. The local and global cost-optimal solutions for an office building in Helsinki, Finland were determined. The results based on considering over 2,300 cases showed that (1) the global cost-optimum occurs for the ground-source heat pump with free ground cooling; (2) biomass-based cogeneration is economically viable only with high overall efficiency and low power-to-heat ratio; (3) biomass-based trigeneration does not exhibit economic or environmental advantages over biomass-based cogeneration due to the significantly higher investment and operational costs; (4) fossil fuel-based cogeneration with high operational costs is the least beneficial option economically and environmentally and (5) extending the cost optimal solutions with photovoltaic panels yields a net-zero energy office building with minimum life-cycle costs.
- The promotion of multigeneration for buildings through sector- and user-oriented eligibility criteria based on high-efficiency cogeneration has been investigated based on the current European framework [427], noting that such criteria can favor or discourage market growth. Economic and energetic factors were considered, and the optimal layout and operation strategy were determined for three distinct buildings: a large hotel, a hospital and an office building. It was shown that the availability of free space to install a large heat storage moderately influences the amount of electricity eligible to obtain dispatch priority.

12.8.4 Optimization of district energy systems

Much research and development has been performed and is ongoing on the optimization of technologies and systems related to district heating and cooling. We describe below some of the noteworthy work in these fields.

Some optimization studies in this area have focused on design, operating strategies and planning. For example, an optimization approach for district heating strategic network design has been proposed [428]. Noting that the use of optimization for the design of such networks is motivated by the high infrastructure cost and made challenging by the technical characteristics and size of actual applications, a mathematical approach was developed to support district heating system planning. An optimal set of new users to be connected to an existing thermal network is selected so as to maximize revenues and minimize infrastructure and operational costs. In another interesting example, a multi-objective, multi-period optimization of the design and operating strategy of district energy systems was investigated by Fazlollahi *et al.* [429]. It was shown via a case study that selecting adequate resources, centralized and decentralized conversion technologies and distribution networks can reduce environmental impacts by 50–65 per cent and total annual costs by 22–27 per cent, and that the efficiency of a system integrating cogeneration, endogenous resources and waste heat recovery is 75 per cent.

Other optimization studies relating to district energy have focused on integrating it with renewable and alternative resources. The latter includes MSW, the disposal of which has proven controversial in many regions and countries because of disagreement regarding what constitutes appropriate waste management policies and technologies. Energy recovery is one approach for disposing of MSW, especially since it contains a considerable amount of biowaste and thus may be considered a type of renewable energy. The overall efficiency of using MSW cogeneration or trigeneration is typically high. Some interesting optimization studies on district energy using renewable and alternative resources follow:

- An optimization model has been reported for the combination of a base-load waste-to-energy facility using MSW and peak-load biomass boiler for district energy applications [430], in part to support decision-making on issues relating to energy recovery from MSW. The model optimizes system specifications such as waste-to-energy facility capacity, biomass boiler capacity and facility location, and determines the quantity of each fuel source that should be used annually to maximize the financial yield of the investment.
- The optimal integration of solar energy in a district heating network has been investigated [431], accounting for the variable and non-controllable nature of solar energy. An optimization procedure was been developed to determine the dispatching strategy for the different power sources present in the network. The optimization procedure was shown to be useful at the planning level, as a means to ascertain the best sizing proportions of solar and conventional sources and for defining the optimal capacity of storage.

12.8.5 Optimization of systems integrating cogeneration or trigeneration and district energy

12.8.5.1 Optimization of systems integrating cogeneration and district energy

Many studies have focused on the optimization of cogeneration-based district energy systems and the various details involved in that activity.

Some of these studies have been aimed at design and operation in the context of application type. For example, an economic and environmental optimization model has developed for the design and operation of a cogeneration systems distributed in an urban area [432]. The mixed integer linear programming model permits the optimal design and operation for an urban area where buildings are equipped with small-size cogeneration plants and connected by a district heating network. A multi-objective function was used to account for capital and operating costs as well as CO_2 emissions. A case study for a residential and tertiary district in Arenzano, a city in the north of Italy was examined. In another study, a multi-objective optimization of a distributed energy supply system for an industrial area, based on a mixed integer linear programming model, was reported by Buoro *et al.* [433]. The supply system includes small-scale cogeneration systems, a large centralized solar plant, a thermal storage and a district heating network connecting them and users. The optimization determines the optimal system structure and component sizing as well as the optimal operation strategy. The objective function is a variable linear combination of the annual cost for owning, maintaining and operating the overall system and its carbon dioxide emissions. The ensuing Pareto Front highlights possible improvements in both economic and environmental terms. For a specific case study it was found that the solar plant coupled with an optimal thermal storage permits environmental and economic goals to be satisfied.

Other studies on optimizing cogeneration-based district energy systems have focused on the use of renewable energy resources. Some examples:

- The modeling and optimization of cogeneration-based district heating system with renewable energy production and energy storage was investigated by Wang *et al.* [434]. An optimization method was developed that seeks to minimize the overall costs of the net acquisition for heat and electricity in a deregulated power market and thereby assist planning and operation. The method was demonstrated for a solar thermal plant and a TES. It was shown that in the future a higher share of renewable energy will likely be used and that a larger TES will likely be necessary, because of both the higher renewable energy share and increasing fluctuations in cogeneration loads.
- A distributed cogeneration system with solar district heating was optimized by Buoro, Pinamonti and Reini [435], with the aim of identifying the optimal energy production system and the optimal operation strategy to satisfy the energy demands in an industrial area. The energy supply system includes a district heating network, a solar thermal plant with long-term TES, a set of natural gas cogeneration units, boilers and compression chillers. The required heat can be produced by the solar thermal plant or cogeneration or conventional boilers. The decision variable set of the optimization procedure includes the component sizes, the solar field size and the thermal energy recovered in the storage. The optimization algorithm is based on a mixed integer linear programming model that minimizes the total annual cost (capital, maintenance and operating) of the overall energy supply system, and can determine both the economic and the environmental benefits and the share of the thermal demand covered by renewable energy for optimal solutions. It was shown by considering various system configurations that the minimum value of the average useful heat costs is achieved when cogenerators, district heating network, solar field and thermal storage are present in the energy supply system and optimized consistently.

12.8.5.2 Optimization of systems integrating extended cogeneration and district energy

Numerous studies focus on the optimization of district energy integrated with extended cogeneration systems, such as trigeneration-based district energy systems, and many of the relevant factors involved in such optimization efforts.

Some of these studies focus on the optimization of the use of renewable energy sources and advanced energy technologies in extended cogeneration systems integrated with district energy. For example, the optimal integration of a natural gas cogeneration plant as a utility producer for a residential district has been described, as has the potential for combining cogeneration with photovoltaic electricity generation [436]. Utilizing residential energy demand data from a smart-grid demonstration project in Austin, Texas, an optimal operating strategy for an integrated cogeneration/solar utility and the impact of photovoltaic generation on plant operation and operating profit were determined. The system was found to be advantageous for a residential district in a hot climate. In a different study, a methodology has been developed to optimize the size in terms of electric power capacity of a cogeneration plant based on a biomass-fired ORC and connected to an existing district heating network, so as to maximize profitability, and applied to two locations in Spain with different climates [437]. In that assessment, subsidies were not included, hourly thermal energy demand profiles in Spain were utilized and partial load operation of the organic Rankine plant was allowed to increase operation time. For a population of 10,000–20,000, the plant electrical capacity was found to between 2 and 9 MW and the IRR between 6 and 18 per cent. It was also demonstrated that cooling is only worthwhile within trigeneration in locations with very warm summers and at full-load operation mode, and that the optimal size of the plant is smaller for trigeneration than cogeneration.

Other studies address specific details involved in the optimization of extended cogeneration systems integrated with district energy. Selected examples of these are covered here:

- The optimization model described previously for using MSW and biomass for district energy [430] was utilized to examine the use of energy sources from rural areas, including MSW and agricultural residue biomass via trigeneration. A case study application for a rural area in Greece was considered and demonstrated that using mixed MSW and biomass can provide a positive financial yield of investment. In addition, a sensitivity analysis identified the most significant parameters affecting the optimum solution to be interest rate, investment cost and heating oil price. The sensitivity analysis was enhanced with stochastic analysis to determine the effect of parameter volatility on the robustness of solution obtained.
- A method for energy optimization with reliability analysis of a trigeneration and district heating system on urban scale has been applied to the area of Arci, near Rome, Italy by Salata *et al.* [438]. The work was motivated by the need of trigeneration and district heating for reliability to avoid service interruptions, often by building in redundancies. The method was developed for a residential context, and compares optimized energy solutions with conventional ones. The results suggested changes to improve system availability.

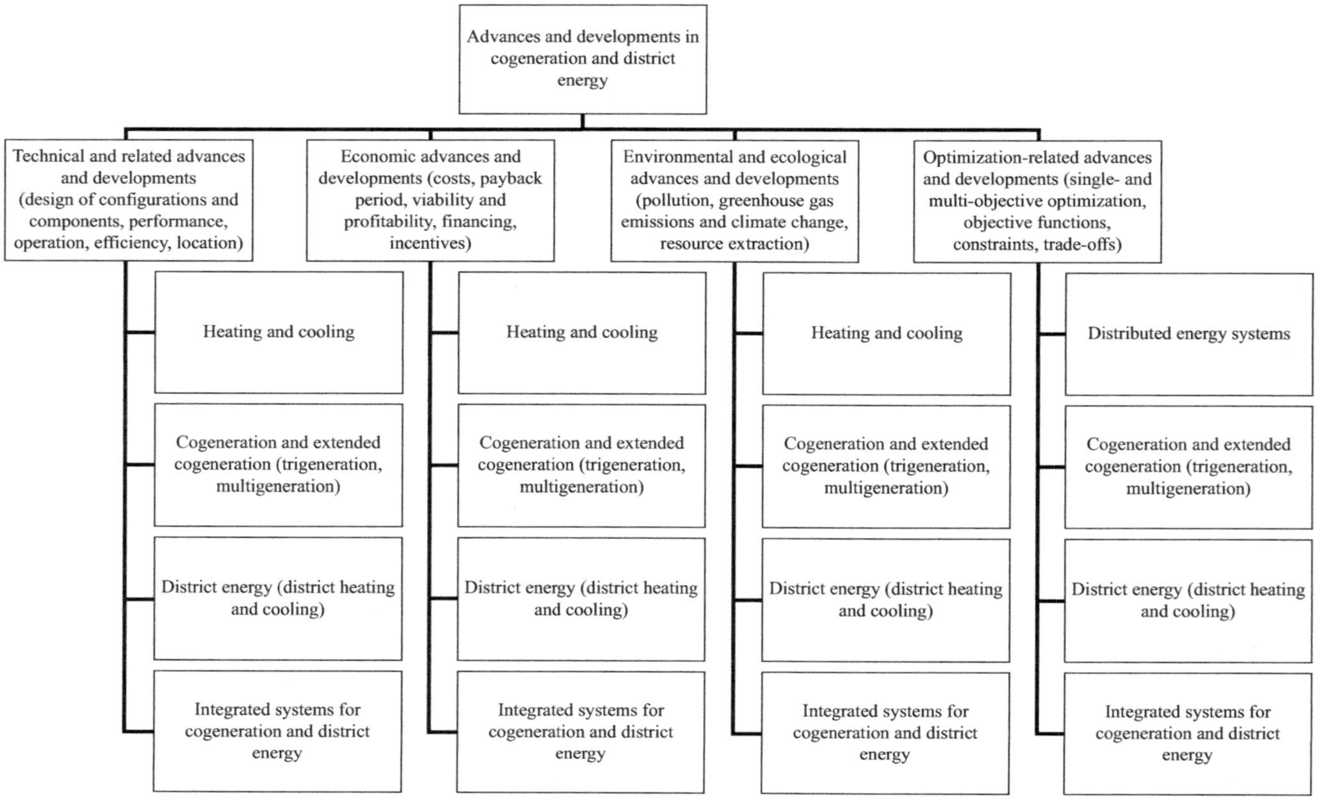

Figure 12.2 Summary of advances and developments in cogeneration and district energy technologies covered in this chapter, showing breakdowns into the main systems and technologies that are integrated to form cogeneration-based district energy systems as well as relevant related systems

- The operational optimization of a complex trigeneration system connected to a district heating and cooling network has been investigated [439]. This work was carried out to address the fact that the high number of technologies available to produce electricity, heating and cooling and the many ways of integrating them make it difficult to select the optimal configuration, and that the high variability in energy demand makes selection of the optimal operational strategy difficult. The optimization models were used to perform economic, energetic and environmental assessments of a new multigeneration system, composed of high-efficiency natural gas cogeneration engines, thermal cooling facilities and a district heating and cooling network of four tubes, in Cerdanyola del Vallès, Spain. The system was shown to be able to reduce primary energy consumption and CO_2 emissions by up to 24 per cent.
- The optimal design and operation of district heating and cooling networks with trigeneration systems was studied for a residential complex by Ameri and Besharati [440]. The method utilized is based on a mixed integer linear programming model, which minimizes the initial and operating costs of the energy supply system to determine component capacities and the optimal operation strategy. It was applied to determine the optimal capacity and operation of seven trigeneration systems in the heating and cooling network of Shahid Beheshti Town, a residential district east of Tehran, Iran. The energy supply system includes district heating and cooling networks, gas turbines as the prime mover for cogeneration, photovoltaic systems, boilers and absorption and compression chillers, so the required heat is obtained from natural gas cogeneration and auxiliary boilers. Four scenarios were considered, and the reductions in costs and CO_2 emissions were determined for each relative using boilers to produce heat and the grid for electricity.

12.9 Closure

A summary of the advances and developments in cogeneration and district energy technologies covered in this chapter is provided in Figure 12.2, loosely organized by technical and related factors, economics, environmental impact and climate change, and optimization activities. Breakdowns into the main systems and technologies that are integrated to form cogeneration-based district energy systems are shown in that figure, as are relevant related systems. Many of the advances involve renewable energy and advanced technologies, and advanced methods and techniques such as exergy analysis. The developments and advances described draw on recent and ongoing global activities, and may assist efforts to predict how the technologies will perform and be utilized in the future.

Chapter 13
Closing and future considerations

Overview

The material covered in this book on cogeneration and district energy systems, as well as their modeling, analysis and optimization, is summarized. Several closing remarks are provided, highlighting key points raised in the book that readers should be aware of and take into account in designs and applications of cogeneration and/or district energy systems. The closing remarks also draw on the insights of the authors and include the advantages of cogeneration and district energy systems over conventional heating and cooling systems in terms of factors such as efficiency, reliability and safety, reduced environmental impact and economics. The importance is emphasized of modeling activities, including characterizing configurations and components of cogeneration and district energy systems, analysis based on thermodynamic, economic and environmental factors, and optimization using appropriate objective functions and constraints. Finally, the authors speculate on future issues and considerations relating to the topics covered in this book, based on material included in this book and in the literature as well as their own insights and views. This includes views on the significant likelihood of increasing use in the future of systems that integrate district energy systems with cogeneration plants for electricity and heat, due to the significant technical, environmental and economic benefits.

13.1 Summary

Cogeneration and district energy systems, as well as cogeneration-based district energy systems are described in this book, along with methods to model, analyze and optimize them. The motivation, rationale and objectives for the book are outlines, and the approaches and methodologies employed are detailed throughout.

The manner in which cogeneration and district energy systems can be integrated into more complex cogeneration-based district energy systems helps to summarize the coverage of the book. Such a summary is presented in diagrammatic form in Figure 13.1, which is a modified form of the illustration of cogeneration and district energy technologies in Figure 1.1. It is seen that cogeneration can involve either the simple production of electricity and thermal energy in the form of heat or electricity and thermal energy in the form of cool. When combined, cogeneration can involve the production of electricity and heat and cool, i.e., trigeneration. Similarly, district

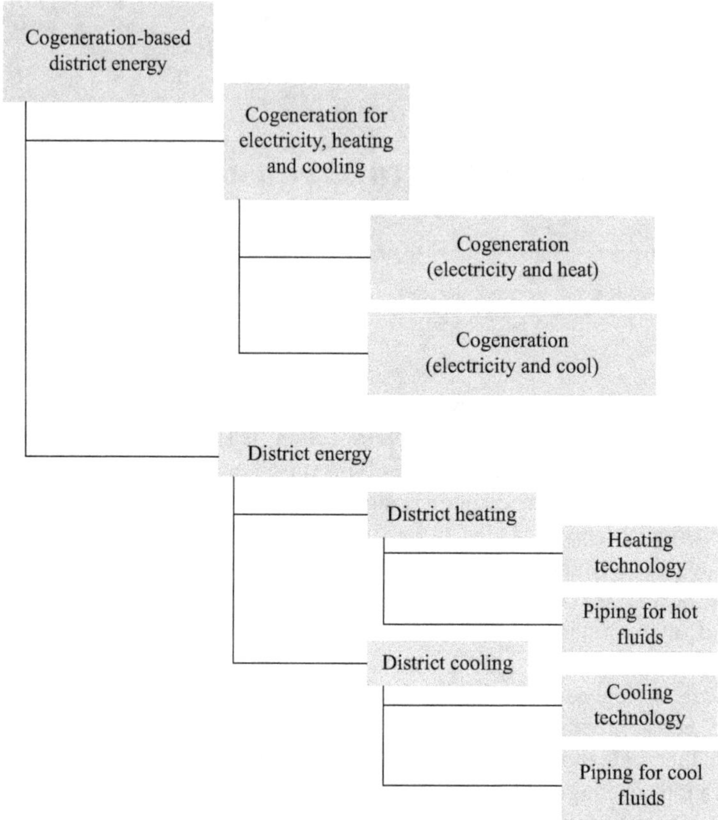

Figure 13.1 Breakdown of cogeneration-based district energy into its main component parts, cogeneration and district energy, and their components

energy represents the combination of district heating and district cooling, each of which involves technology for heating and cooling, respectively, and piping technology for hot and cool fluids, respectively. The integration of these technologies leads to the formation of cogeneration-based district energy, which is presented at the top of the diagram.

Relevant background material on thermodynamic analysis is presented, with a strong focus on exergy analysis. This material forms the basis of much of the material in the book on modeling, as well as a key input to the material on analysis and optimization.

Modeling, analysis and optimization activities are described, with a focus on cogeneration-based district energy systems. Models and descriptions are presented for the principal technologies relevant to the book: cogeneration or CHP production, heating and district heating, chilling and district cooling, and configurations for integrating cogeneration and district energy technologies.

For illustration and to help demonstrate the merits of different integrated systems for cogeneration and district energy, several such systems are compared and contrasted, in a variety of ways.

Given the growing concerns with environmental issues in general and climate change in particular, the environmental impact of cogeneration systems are described with a focus on the allocation wastes for cogeneration systems, and the contributions to climate change of cogeneration are discussed with an emphasis on carbon dioxide emissions.

The economics of cogeneration and district energy technologies are essential to any engineering activities involving systems that use them separately or in an integrated manner. Hence, relevant economic factors and methods are discussed in the book. Correspondingly, appropriate optimization procedures are described.

Drawing on global research and development activities, advances in technologies and systems for cogeneration and district energy and related technologies are described, focusing on technical and related factors, economics, environmental impact and climate change, and optimization activities.

Several closing remarks and future considerations can be made, based on material covered in the book, other information sources and the insights of the authors. These are provided in the remainder of this chapter.

13.2 Closing remarks

The authors provide the following closing remarks based on their own insights and by drawing from the materials that are included in this book as well as information available in the literature and elsewhere.

Cogeneration facilities and technologies can provide heating, cooling and electricity generation with many advantages over conventional and separate systems for these services, including much higher efficiency and much lower environmental impacts.

District energy (i.e., district heating and/or district cooling) technologies can provide heating and cooling services for communities with many advantages over separate heating and cooling systems in each building, including higher efficiencies, lower environmental impacts, and better economics and reliability.

Cogeneration facilities can be beneficially integrated with district energy technologies, for the provision of heating, cooling and electricity generation. The benefits include higher efficiencies and lower environmental impacts than non-integrated systems.

Appropriate modeling, analysis and optimization work is extremely useful for cogeneration systems, district energy systems and integrated systems for cogeneration-based district energy. These activities help provide both theoretical and practical understanding of the characteristics and behaviors of the technologies, and are particularly important for the more complex cogeneration-based district energy systems. Appropriate modeling, analysis and optimization work also is essential for the design and improvement of such systems, and for their expanded usage.

Numerous configurations for integrated cogeneration-based district energy systems are possible, many of which are considered and examined in this book. These configurations can be optimized, while accounting for factors such as efficiency, economics, environmental impact, reliability and safety. However, the "optimum" system is dependent on the objectives and constraints considered by a designer.

Exergy analysis (or second-law analysis), an advanced thermodynamic method for assessing and comparing the efficiency and performance of energy systems, permits more uniform and consistent evaluations and comparisons to be made of cogeneration systems, district energy systems and integrated systems for cogeneration-based district energy, compared to other analysis techniques. The incorporation of exergy analysis into the tools developed and activities carried out in this study can help ensure that the better systems for using energy resources are identified without the lengthy and inefficient process of waiting for them to be sorted out by competitive economic success in the marketplace.

The information provided and design tools described in this book can serve to help industry develop better (in terms of efficiency, effectiveness, economics, environmental impact, etc.) overall designs of integrated cogeneration-based district energy systems for the provision of heating, cooling and electrical services. This remark applies to both new and retrofit situations, and to both urban and, to a lesser degree, rural environments.

The development of "better" designs (as described in the previous point) may permit an increased number of cogeneration systems, district energy systems and integrated systems for cogeneration-based district energy to be developed and implemented in many communities, regions and countries, and, consequently, more of the potential benefits of such systems to be realized in those jurisdictions.

The main factors that need to be considered for the successful development and implementation of cogeneration-based district energy systems are summarized pictorially in Figure 13.2.

13.3 Future considerations

Based on their own thoughts and views, the authors speculate on future issues and considerations relating to the topics covered in this book. These future considerations are also informed by material included in this book as well as information available in the literature and other sources.

Assessment, comparison and optimization activities for cogeneration systems and district energy systems as well as for integrated systems for cogeneration and district energy can benefit from using the advanced approaches and methods presented in this book, particularly those related to exergy analysis, in order to realize as many benefits as possible from such systems. Thus, use of these assessment, comparison and optimization activities is likely to expand and become increasingly necessary in the future.

Integrated cogeneration-based district energy systems should be considered where they are utilizable in new and retrofit situations, and implemented in those

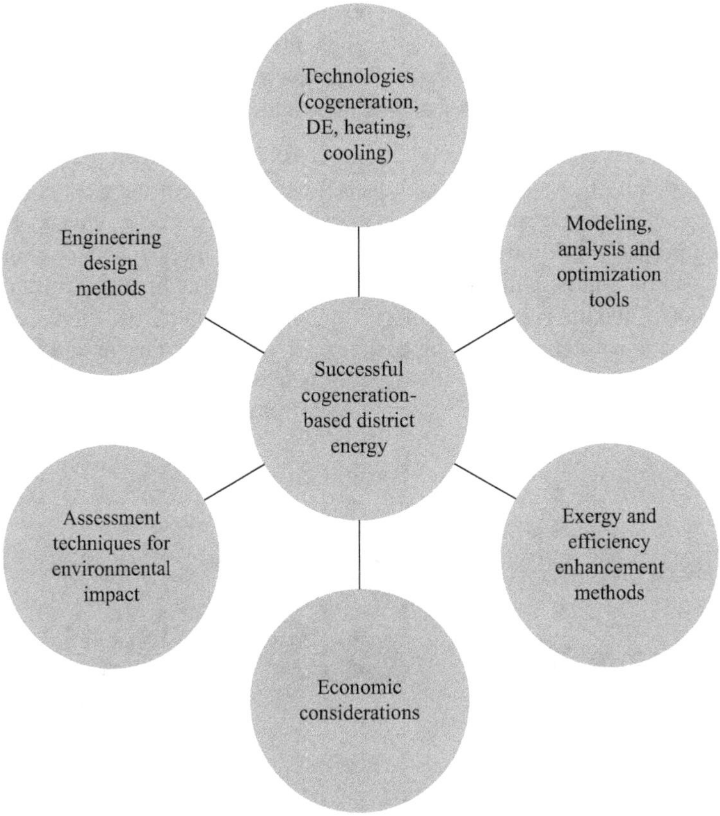

Figure 13.2 Illustration of some of the main factors to be considered for successful cogeneration-based district energy systems

situations where these systems are appropriate and advantageous. Advantageousness can be defined from many perspectives, such as efficiency, effectiveness, economics and environmental impact, but regardless of which perspective is or perspectives are adopted, this point is likely to remain valid and become increasingly important in the future.

Governments should consider offering incentives (or enhancing existing incentives) to promote the development and use of cogeneration systems and district energy systems and especially integrated systems for cogeneration and district energy, so as to assist in permitting all of the potential benefits of such systems to be realized.

Meaningful information on cogeneration systems, district energy systems and integrated systems for cogeneration-based district energy, such as is provided in this book, should be communicated appropriately to all possible user groups and interested parties. In particular, this information should be conveyed to all relevant industries and government organizations, and incorporated into the academic curricula at appropriate educational institutions (e.g., engineering departments).

Further work is merited and should be undertaken to advance the technologies of cogeneration and district energy as well as integrated systems for cogeneration-based district energy. In particular, the following activities seem required and thus are likely to form the basis of research in the future:

- The accuracy and generality of the methods for assessment, comparison and optimization of cogeneration and district energy systems as well as integrated systems for cogeneration-based district energy should be enhanced. Many assumptions and simplifications are utilized in present day work, and these can likely to be improved upon, to enrich benefits such as financial returns.
- The tools and techniques developed should be applied to more real-world systems, to show how they can assist in all aspects of design and to increase confidence in them and thereby expand their utilization.

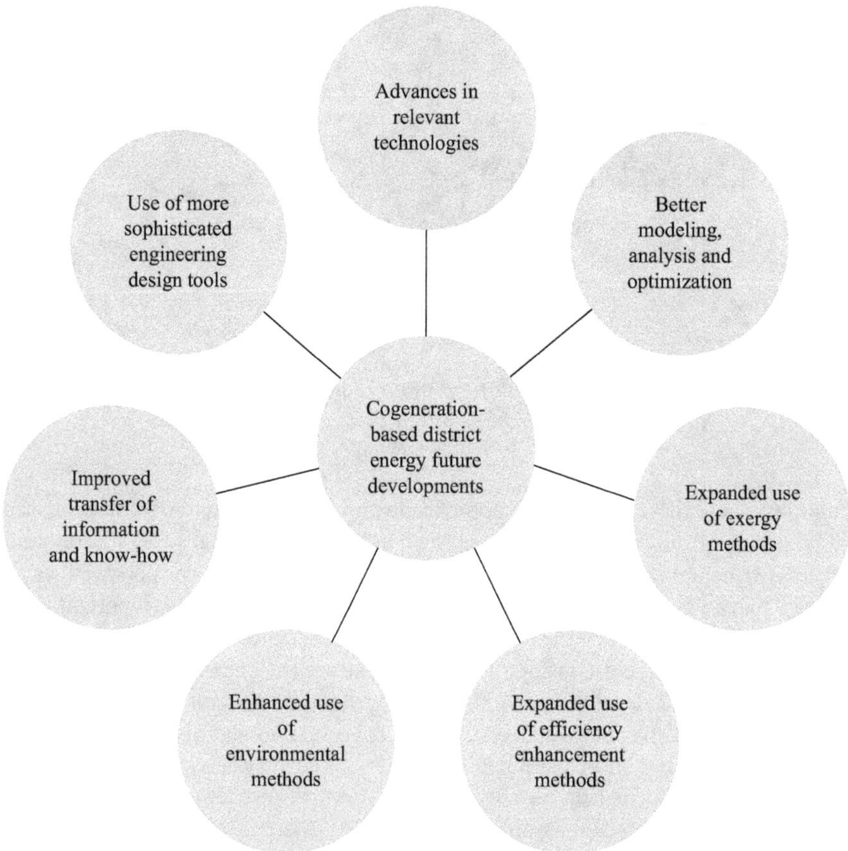

Figure 13.3 Expected future developments in cogeneration-based district energy systems, focusing on advances in the technology and developments that will support achieving such advances

- Several aspects of the material covered in this book should be extended in the future. For instance, efforts are likely to be expended in the future on enhancing the portions of the work related to exergy analysis to make it more user-friendly and attractive, especially to industry. Efforts are also likely to be directed toward developing more focused and specific design tools and methodologies that are applicable for a range of locations.

The main expected future developments in cogeneration-based district energy systems are summarized in Figure 13.3. In this figure, the focus is on advances in the technologies involved (cogeneration, district energy, heating, cooling) and developments in methods and tools that will support achieving such advances.

The benefits of the expected future developments in Figure 13.3 in cogeneration-based district energy systems are great in number and significant. These anticipated benefits are summarized in Figure 13.4, and involve items of value to consumers, industry providers, governments and society. The benefits of future developments in cogeneration-based district energy systems are expected to contribute in key value areas like economics, environmental stewardship, efficiency, safety and reliability.

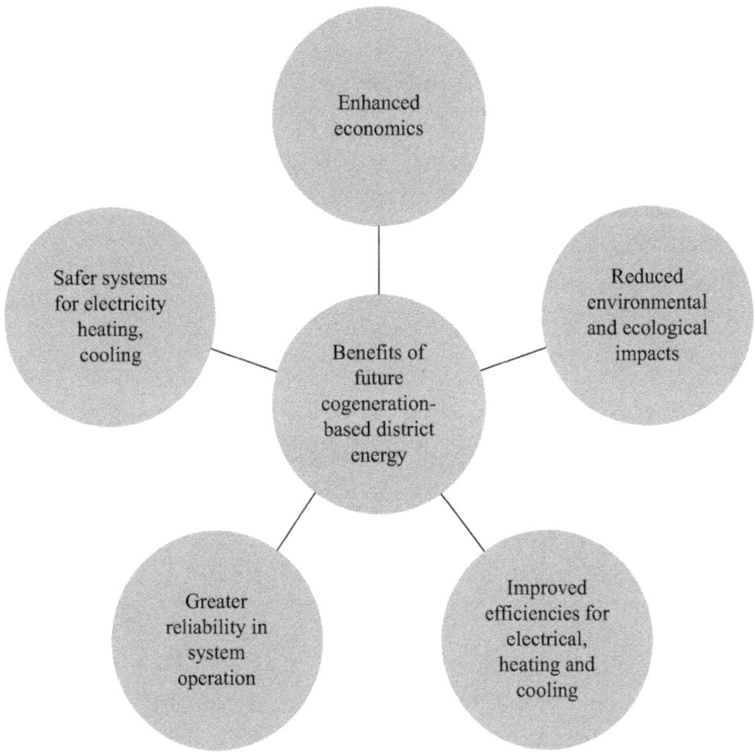

Figure 13.4 Benefits of expected future developments in cogeneration-based district energy systems

References

1 Dincer I, Rosen MA. Exergy: energy, environment and sustainable development. 2nd ed. London: Elsevier; 2013.

2 Wark Jr K, Richards DE. Thermodynamics. McGraw-Hill. 6th ed. Toronto, ON: McGraw-Hill; 1999.

3 Reynolds WC, Perkins HC. Engineering thermodynamics. 2d ed. Toronto, ON: McGraw-Hill; 1977.

4 Kotas TJ. The exergy method of thermal plant analysis. London: Paragon Publishing; 2012.

5 Moran, MJ. Availability analysis: a guide to efficient energy use. New York: American Society of Mechanical Engineers; 1989.

6 Szargut J, Morris DR, Steward FR. Exergy analysis of thermal, chemical and metallurgical processes. New York: Hemisphere; 1988.

7 Galletti A, Gusso R, Stecco SS. Energy–exergy analysis of an actual regenerative cogenerating gas turbine plant. In: Gaggioli RA, editor. Computer-aided engineering of energy systems. AES-vols. 2–3. New York: American Society of Mechanical Engineers; 1986. p. 69–77.

8 Habib MA. First and second law analysis of cogeneration plants. In: Reistad GM, Gaggioli RA, Bejan A, Tsatsaronis G, editors. Thermodynamics and energy systems – fundamentals. AES-vol. 24. New York: American Society of Mechanical Engineers; 1991. p. 31–42.

9 Rosen MA. Comparison based on energy and exergy analyses of the potential cogeneration efficiencies for fuel cells and other electricity generation devices. Int J Hydrogen Energy 1990;15: 267–74.

10 Rosen MA, Scott DS. Energy and exergy analyses of selected production processes for hydrogen and electricity from coal. In: Wepfer WJ, Tsatsaronis G, Bajura RA, editors. Thermodynamic analysis of chemically reactive systems. AES-Vol. 4. New York: American Society of Mechanical Engineers; 1988. p. 13–21.

11 Rosen MA, Dimitriu, J, Horozak, DA. Thermodynamic assessment of a pressurized fluid-bed combustion process for electricity generation. Proceedings of the 10th international Pittsburgh coal conference, Pittsburgh, PA, 1993.

12 Rosen MA, Dimitriu J, Brygadyr, R. Assessment of clean coal technologies. Final Report for Ontario Ministry of Environment and Energy. Toronto: Ryerson Polytechnic University; 1993.

13 Kumar A, Kachhwaha SS, Mishra RS. Thermodynamic analysis of a regenerative gas turbine cogeneration plant. J Sci Ind Res 2010;69(03):225–31.
14 Ghaebi H, Amidpour M, Karimkashi S, Rezayan O. Energy, exergy and thermoeconomic analysis of a combined cooling, heating and power (CCHP) system with gas turbine prime mover. Int J Energy Res 2011;35:697–709.
15 Khaliq A, Dincer I. Energetic and exergetic performance analyses of a combined heat and power plant with absorption inlet cooling and evaporative aftercooling. Energy 2011;36:2662–70.
16 Khaliq A, Agrawal BK, Kumar R. First and second law investigation of waste heat based combined power and ejector-absorption refrigeration cycle. Int J Refrig 2012;35:88–97.
17 Khaljani M, Khoshbakhti Saray R, Bahlouli K. Comprehensive analysis of energy, exergy and exergo-economic of cogeneration of heat and power in a combined gas turbine and organic Rankine cycle. Energy Conv Manag 2015;97:154–65.
18 Ozgener L, Hepbasli A, Dincer I. Effect of reference state on the performance of energy and exergy evaluation of geothermal district heating systems: Balcova example. Build Environ 2006;41:699–709.
19 Fonseca JGS, Schneider PS. Simulation of a thermal power plant with district heating: Comparative results of 5 different codes. Energy 2006; 31:1955–68.
20 Gaggioli RA, Petit PJ. Use the second law first. Chemtech 1977;7:496–506.
21 Sato N. Chemical energy and exergy: an introduction to chemical thermodynamics for engineers. Oxford: Elsevier; 2005.
22 Szargut J. Exergy method: technical and ecological applications. Southampton: WIT Press; 2005.
23 Hevert HW, Hevert SC. Second law analysis: an alternative indicator of system efficiency. Energy Int J 1980;5:865–73.
24 Moran MJ, Sciubba E. Exergy analysis: principles and practice. J Eng Gas Turb Power 1994;116:285–90.
25 Sciubba E. From engineering economics to extended exergy accounting: a possible path from monetary to resource-based costing. J Ind Ecol 2004;8 (4):19–40.
26 Moran MJ, Shapiro HN, Boettner DD, Bailey MB. Fundamentals of engineering thermodynamics. 8th ed. New York: Wiley; 2014.
27 Szargut J. International progress in second law analysis. Energy 1980;5:709–18.
28 Edgerton RH. Available energy and environmental economics. DC Heath, Toronto; 1982.
29 Rosen MA. Second law analysis: approaches and implications. Int J Energy Res 1999;23:415–29.
30 Sciubba E. Exergy as a direct measure of environmental impact. In: Aceves SM, *et al.* editors. Proceedings of ASME advanced energy systems division. AES-Vol. 39. New York: ASME; 1999. p. 573–81.
31 Ayres RU, Ayres LW, Martinas K. Exergy, waste accounting, and life-cycle analysis. Energy 1998;23:355–63.

32 Connelly L, Koshland CP. Two aspects of consumption: using an exergy-based measure of degradation to advance the theory and implementation of industrial ecology. Resour Conserv Recycling 1997;19:199–217.
33 Creyts JC, Carey VP. Use of extended exergy analysis as a tool for assessment of the environmental impact of industrial processes. In: Ramalingam ML, editor. Proceedings of ASME advanced energy systems division, AES-Vol. 37. New York: ASME; 1997. p. 129–37.
34 Berthiaume R, Bouchard C, Rosen MA. Exergetic evaluation of the renewability of a biofuel. Exergy Int J 2001;1:256–68.
35 Crane P, Scott DS, Rosen MA. Comparison of exergy of emissions from two energy conversion technologies, considering potential for environmental impact. Int J Hydrogen Energy 1992;17:345–50.
36 Rosen MA, Dincer I. On exergy and environmental impact. Int J Energy Res 1997;21:643–54.
37 Rosen MA, Dincer I. Exergy analysis of waste emissions. Int J Energy Res 1999;23:1153–63.
38 Rosen MA, Dincer I. Exergy as the confluence of energy, environment and sustainable development. Exergy Int J 2001;1:3–13.
39 Gunnewiek LH, Rosen MA. Relation between the exergy of waste emissions and measures of environmental impact. Int J Environ Pollution 1998;10:261–72.
40 Daniel JJ, Rosen MA. Exergy-based environmental assessment of life cycle emissions for various automobiles and fuels. Proceedings of CSME forum, Kingston, Ontario; 21–24 May 2002. Sec. 20, Paper 4. p. 1–14.
41 Cownden R, Nahon M, Rosen MA. Exergy analysis of a fuel cell power system for transportation applications. Exergy Int J 2001;1:112–21.
42 Rosen MA. Evaluation of energy utilization efficiency in Canada using energy and exergy analyses. Energy Int J 1992;17:339–50.
43 Rosen MA. Thermodynamic investigation and comparison of selected production processes for hydrogen and hydrogen-derived fuels. Energy Int J 1996;21:1079–94.
44 Rosen MA. Regional cogeneration via electrical utilities operating on nuclear energy and coal: energy and environmental benefits. Trans CSME 2009;33(1):131–43.
45 Rosen MA. Energy, environmental, health and cost benefits of cogeneration from fossil fuels and nuclear energy using the electrical utility facilities of a province. Energy Sustain Dev 2009;13(1):42–50.
46 Rosen MA, Dincer I. Energy and exergy analyses of thermal energy storage systems. In: Thermal energy storage systems and applications. London: Wiley; 2002. p. 411–510 [chapter 10].
47 Rosen MA, Horazak DA. Energy and exergy analyses of PFBC power plants. In: Alvarez Cuenca M, Anthony EJ, editors. Pressurized fluidized bed combustion. London: Chapman and Hall; 1995. p. 419–48 [chapter 11].
48 Rosen MA. Comparison based on energy and exergy analyses of the potential cogeneration efficiencies for fuel cells and other electricity generation devices. Int J Hydrogen Energy 1990;15:267–74.

49 Rosen MA, Le M. Efficiency analysis of a process design integrating cogeneration and district energy. In: Duncan AB, Fiszdon J, O'Neal D, Den Braven K, editors. Proceedings of ASME advanced energy systems division, AES-Vol. 36. New York: ASME; 1996. p. 473–80.

50 Tsatsaronis G. Definitions and nomenclature in exergy analysis and exergoeconomics. Energy 2007;32:249–53.

51 Kotas TJ, Raichura RC, Mayhew YR. Nomenclature for exergy analysis. In: Moran MJ, Sciubba E, editors. Second law analysis of thermal systems. New York: American Society of Mechanical Engineers; 1987. p. 171–6.

52 Wepfer WJ, Gaggioli RA. Reference datums for available energy. In: Gaggioli RA, editor. Thermodynamics: second law analysis. ACS Symposium Series 122. Washington, DC: American Chemical Society; 1980. p. 77–92.

53 Sussman MV. Second law efficiencies and reference states for exergy analysis. Proceedings of the 2d world congress of chemical engineering. Montreal, QC: Canadian Society of Chemical Engineers; 1981. p. 420–21.

54 Ahrendts J. Reference states. Energy Int J 1980;5:667–78.

55 Baehr HD, Schmidt EE. Definition und berechnung von brennstoffexergien (definition and calculation of fuel exergy). Brennst-Waerme-Kraft 1963;15:375–81.

56 Rodriguez LSJ. Calculation of available-energy quantities. In: Gaggioli RA, editor. Thermodynamics: second law analysis. ACS Symposium Series 122. Washington, DC: American Chemical Society; 1980. p. 39–60.

57 Szargut J. Grenzen fuer die anwendungsmoeglichkeiten des exergiebegriffs (limits of the applicability of the exergy concept). Brennst-Waerme-Kraft 1967;19:309–13.

58 Bosnjakovic E. Bezugszustand der exergie eines reagiernden systems (reference states of the exergy in a reacting system). Forsch Ingenieurw 1963;20:151–2.

59 Szargut J, Valero A, Stanek W, Valero A. Towards an international legal reference environment. Proceedings of 18th international conference on efficiency, cost, optimization, simulation and environmental impact of energy systems (ECOS), NTNU, Trondheim, Norway, Vol. 1; June 2005. p. 409–17.

60 Rosen MA, Dincer I. Effect of varying dead-state properties on energy and exergy analyses of thermal systems. Int J Thermal Sci 2004;43:121–33.

61 Utlua Z, Hepbasli A. Energetic and exergetic assessment of the industrial sector at varying dead (reference) state temperatures: a review with an illustrative example. Renew Sustain Energy Rev 2008;12:1277–301.

62 Aljundi IH. Energy and exergy analysis of a steam power plant in Jordan. Appl Thermal Eng 2009;29:324–8.

63 Etele J, Rosen MA. Sensitivity of exergy efficiencies of aerospace engines to reference environment selection. Exergy Int J 2001;1:91–9.

64 Lior N, Zhang N. Energy, exergy, and Second Law performance criteria. Energy 2007;32:281–96.

65 Keenan JH, Chao J, Kayle J. Gas tables, international version: properties of air products of combustion and component gases compressible flow functions. Malabar, FL: Krieger Publishing; 1992.

66 1967 Steam tables. London: Edward Arnold Publishers; 1967.
67 Chase MW, Daives CA, Downey JR, Frurip DJ, McDonald RA, Syverud AN, editors. JANAF thermochemistry tables. 3rd ed. Washington, DC: American Chemical Society and American Institute of Physics; 1985.
68 CFCC Developmental Program, General Electric Co. Commercial Plant Concept. FE-2357–42; March 1981.
69 Bayrak M, Gungor A. Fossil fuel sustainability: exergy assessment of a cogeneration system. Int J Energy Res 2011;35:162–8.
70 Tsatsaronis G, Tawfik T, Lin L, Gallaspy DT. Exergy analysis of an IGCC design configuration for plant Wansley. In: Bajura RA, von Spakovsky MR, Geskin ES, editors. Analysis and design of energy systems: thermodynamic analysis of industrial processes, AES-Vol. 10-3. New York: American Society of Mechanical Engineers, 1989. p. 21–9.
71 Tsatsaronis G, Tawfik T, Lin L, Gallaspy DT. Exergetic comparison of two KRW-based IGCC power plants. In: Reistad GM, Moran MJ, Wepfer WJ, Lior N, editors. Second law analysis: industrial and environmental applications, AES-Vol. 25/HTD-Vol. 191. New York: American Society of Mechanical Engineers; 1991. p. 7–18.
72 Alefeld G. What are thermodynamic losses and how to measure them? A future for energy: proceedings of the Florence world energy research symposium, Firenze, Italy, 1990. p. 271–9.
73 Boehm R, Spask M. First and second law comparisons of high efficiency Brayton cycles. In: Analysis and design of advanced energy systems: applications. Vol. 3–2. Boston, MA: American Society of Mechanical Engineers; 1987. p. 83–8.
74 Dunbar WR, Gaggioli RA, Lior N. The component equations of energy and exergy. In: Thermodynamics and energy systems – fundamentals. Vol. 24. Atlanta, GA: American Society of Mechanical Engineers; 1991. p. 21–30.
75 Dunbar WR, Lior N. A breakdown of the exergy losses in combustion. A future for energy: proceedings of the Florence world energy research symposium, Firenze, Italy; 1990. p. 347–58.
76 Gaggioli RA, editor. Analysis of energy systems-design and operation. Vol. 1. Miami Beach, FL: American Society of Mechanical Engineers; 1985.
77 Gaggioli RA, editor. Efficiency and costing: second law analysis of processes. Washington, DC: American Chemical Society; 1983.
78 Habib MA. First and second law analysis of cogeneration plants. Thermodynamics and energy systems – fundamentals. Vol. 24. Atlanta, GA: American Society of Mechanical Engineers; 1991. p. 31–42.
79 Rosen MA. Evaluation of energy utilization efficiency in Canada using energy and exergy analyses. Energy Int J 1992;17:339–50.
80 Moran MJ, Sciubba E, editors. Second law analysis of thermal systems: 4th international symposium on second law analysis of thermal systems, Rome, Italy; 1987.
81 Moran MJ. Second law analysis. What is the state of the art? A future for energy: proceedings of the Florence world energy research symposium, Firenze, Italy; 1990. p. 249–60.

82 Morris DR, Steward PR, Szargut J. Exergy analysis of thermal, chemical, and metallurgical processes. New York: Springer-Verlag; 1988.
83 Tsatsaronis G, Pisa J, Lin L, Tawfik T. Optimization of an IGCC power plant. In: RF Boehm, editor. Thermodynamics and the design, analysis, and improvement of energy systems (Parts I and II). AES-Vol. 27/HTD-Vol. 228. New York: American Society of Mechanical Engineers; 1992. p. 37–67.
84 Rosen MA, Scott DS. On the sensitivities of energy and exergy analyses to variations in dead-state properties. In: Moran MJ *et al.*, editor. Analysis and design of advanced energy systems. New York: American Society of Mechanical Engineers; 1987. AES-Vol. 3–1, p. 23–32.
85 Rosen MA. The relation between thermodynamic losses and capital costs for a modern coal-fired electrical generating station. In: Tsatsaronis G *et al.*, editor. Computer-Aided Energy-Systems Analysis. New York: American Society of Mechanical Engineers; 1990. AES-Vol. 21, p. 69–78.
86 Rosen MA. Assessment of an integrated coal-gasification combined-cycle (IGCC) power plant using energy and exergy analyses. In: RF Boehm, editor. Thermodynamics and the Design, Analysis and Improvement of Energy Systems. New York: American Society of Mechanical Engineers; 1992. AES-Vol. 27/HTD-Vol. 228, p. 101–9.
87 Barclay FJ. Combined power and process – an exergy approach. London: Mechanical Engineering Publications Ltd.; 1995.
88 Rosen MA, Horazak DA. Energy and exergy analysis of PFBC power plants. In: Alvarez Cuenca M, Anthony EJ, editors. Pressurized fluidized bed combustion. London: Chapman and Hall; 1995. Chapter 11, p. 419–48.
89 Kouremenos DA, Rogdakis ED, Antonopoulos KA. Cogeneration with combined gas and aqua-ammonia absorption cycles. In: RJ Krane, editor. Thermodynamics and the design, analysis, and improvement of energy systems. New York: American Society of Mechanical Engineers; 1994. AES-Vol. 33, p. 231–8.
90 Karakas A, Egrican N, Uyger S. Second-law analysis of solar absorption-cooling cycles using lithium bromide/water and ammonia/water as working fluids. Appl Energy 1990;37:169–87.
91 Ayres RU, Turton H, Casten T. Energy efficiency, sustainability and economic growth. Energy 2007;32:634–48.
92 Çakir U, Çomakli K, Yüksel F. The role of cogeneration systems in sustainability of energy. Energy Convers Manag 2012;63:196–202.
93 MacLaren Engineers Inc. Cogeneration sourcebook. Report for Ontario Ministry of Energy, Toronto; 1988.
94 Calise F, d'Accadia MD, Vicidomini M, Scarpellino M. Design and simulation of a prototype of a small-scale solar CHP system based on evacuated flat-plate solar collectors and Organic Rankine Cycle. Energy Convers Manag 2015;90:347–63.
95 1991 Cogeneration Database. Computer database compiled by publishers of Cogeneration & Small Power Monthly. Atlanta: Association of Energy Engineers; 1991.
96 Office of Energy Efficiency & Renewable Energy. CHP Installation Database. <https://doe.icfwebservices.com/chpdb/> [accessed 05.04.16].

97 CIEEDAC Canadian Cogeneration Database. <https://www.cieedac.sfu.ca/db_cogen_new/> [accessed 05.04.16].

98 Nelson RM, Kainlauri EO. District heating with cogeneration. ASHRAE J 1989;31:20–2.

99 MacRae KM. Realizing the benefits of community integrated energy systems. Calgary, AB: Canadian Energy Research Institute; 1992.

100 Spiewak SA. Cogeneration & small power production manual. 3rd ed. Lilburn, GA: Fairmont Press; 1991.

101 Orlando JA. Cogeneration planner's handbook. Lilburn, GA: Fairmont Press; 1991.

102 Limaye DR. Planning cogeneration systems. Lilburn, GA: Fairmont Press; 1985.

103 Horlock JH. "Some practical CHP schemes" in Cogeneration: combined heat and power (CHP). Oxford: Pergamon; 1987. p. 158–74.

104 International Energy Agency (IEA). Cogeneration and renewables, solution for a low carbon energy future; 2011.

105 Guinn GR. Planning cogeneration systems. ASHRAE J 1991;33(2):18–22.

106 Hennefonh JC, Todd DM. Cogeneration on a large scale. Mech Eng 1988;110(8):51–3.

107 FVB/Eltec. Potential heat production from existing and future electric generating plants. Report No. 91 02-G-806 for Canadian Electrical Association; 1993.

108 Collins S. The perfect match: combined cycles and district heating/cooling. Power 1991;135(4):117–20.

109 Hufford PE. Absorption chillers improve cogeneration energy efficiency. ASHRAE J 1992;34(3):46–53.

110 Collet PJ. Cogeneration in combination with heat pumps. ASHRAE J 1989;31(2):30–4.

111 Kavvadias KC, Tosios AP, Maroulis ZB. Design of a combined heating, cooling and power system: Sizing, operation strategy selection and parametric analysis. Energy Convers Manag 2010;51:833–45.

112 O'Brian JM, Bansal PK. Modelling of cogeneration systems Part 1: Historical perspective. Proc Inst Mech Eng A: J Power Energy 2000;214(2):115–24.

113 Hawkes A, Staffell I, Brett D, Brandon N. Fuel cells for micro-combined heat and power generation. Energy Environ Sci 2009;2(7):729–74.

114 Adam A, Fraga ES, Brett DJL. Options for residential building services design using fuel cell based micro-CHP and the potential for heat integration. Appl Energy 2015;138:685–94.

115 Dong L, Liu H, Riffat S. Development of small-scale and micro-scale biomass-fuelled CHP systems – a literature review. Appl Thermal Eng 2009; 29, (11–12):2119–26.

116 De Paepe M, D'Herdt P, Mertens D. Micro-CHP systems for residential applications. Energy Convers Manag 2006;47:3435–46.

117 Qui G, Liu Q, Riffat S. Expanders for micro-CHP systems with organic Rankine cycle. Appl Thermal Eng 2011;31(16):3301–7.

118 Kong XQ, Wang RZ, Huang XH. Energy efficiency and economic feasibility of CCHP driven by stirling engine. Energy Convers Manag 2004;45:1433–42.
119 Klein M. The need for standards to promote low emission, high efficiency gas turbine facilities. Proceedings of the ASME international gas turbine and aeroengine congress and exhibition, Indianapolis, Indiana, Paper 99-GT-405, 7–10 June; 1999. p. 1–10.
120 Klein M. Full fuel cycle emissions estimations. Proceedings of the 13th symposium on industrial applications of gas turbines, Banff, Alberta, Paper 99-IAGT-402; 13–15 October 1999. p. 1–22.
121 Klein M. Brief descriptions of commercial and institutional CHP systems in Canada. Report by Oil, Gas and Energy Branch, Environment Canada; 1999 (estimated).
122 Klein M. Canadian gas turbine cogeneration plants. Report by Oil, Gas and Energy Branch, Environment Canada; 1999.
123 Klein M. High efficiency combined heat and power facilities: benefits and barriers. Proceedings of the Canadian technology development conference, Toronto, Canada; 2001. p. 1–20.
124 Nyboer J, Groves S, Baylin-Stern A. A review of existing cogeneration facilities in Canada: 2013. Report by Canadian Industrial Energy End Use Data and Analysis Centre, Simon Fraser University, Burnaby, BC, Canada; April 2013.
125 MacRae KM. Realizing the benefits of community integrated energy systems. Calgary, Alberta: Canadian Energy Research Institute; 1992.
126 Rogner HH. Clean energy services without pain: district energy systems. Energy Stud Rev 1993;5:114–20.
127 FVB/Eltec potential heat production from existing and future electric generating plants. Report No. 9102 G 806 for Canadian Electrical Association; 1993.
128 MacLaren Engineers Inc. Cogeneration sourcebook. Report for Ontario Ministry of Energy, Toronto; 1988.
129 Henneforth JC, Todd DM. Cogeneration on a large scale. Mech Eng 1988 (August);110:51–3.
130 Acres International Limited. Cogeneration potential in Ontario and barriers to its development. Report for Ontario Ministry of Energy, Toronto; 1987.
131 Horlock JH. Some practical CHP schemes. In: Cogeneration: combined heat and power (CHP). Oxford: Pergamon; 1987. p. 158–74 [chapter 5].
132 Rosen MA. Energy utilization efficiency in a macrosystem (Ontario): evaluation and improvement through cogeneration. Proceedings of international symposium on CO2 fixation and efficient utilization of energy, Tokyo; 29 November–1 December 1993. p. 17–26.
133 Rosen MA. Assessment of various scenarios for utility-based cogeneration in Ontario. Energy Int J 1994;19:1143–9.
134 Rosen MA. Reductions in energy use and environmental emissions achievable with utility-based cogeneration: simplified illustrations for Ontario. Appl Energy 1998;61:163–74.

135 Rosen MA, Le MN, Dimitriu J. Efficiency improvement for large energy systems through utility-based cogeneration and district energy: investigation for Ontario, Canada. World Energy Syst Int J 1997;1:56–63.
136 Hart DR, Rosen MA Environmental and health benefits of utility-based cogeneration in Ontario, Canada. Energy Environ 1994;5:363–78.
137 Rosen MA, Le M. Assessment of the potential cumulative benefits of applying utility-based cogeneration in Ontario. Energy Stud Rev 1994;6:154–63.
138 Sherwood M, Rosen MA. Efficiency comparison of several alternatives for cogeneration-based district heating and cooling. Proceedings of symposium on thermal and fluids engineering at the CSME forum; Hamilton, Ontario, 7–9 May 1996. p. 293–8.
139 Simpson J, Rosen MA. District energy and cogeneration systems: modelling, simulation and analysis. Proceedings of symposium on thermal and fluids engineering at the CSME Forum, Hamilton, Ontario; 7–9 May 1996. p. 275–82.
140 Gowrishankar V, Angelides C, Druckenmiller H. Combined heat and power systems: Improving the energy efficiency of our manufacturing plants, buildings, and other facilities. Natural Resources Defense Council (NRDC); April 2013.
141 Midwest CHP Application Centre. Rochester wastewater reclamation plant. <http://www.rochestermn.gov/home/showdocument?id=2549> [accessed 12.05.16].
142 Norton B. Solar process heat: distillation, agriculture, drying and industrial uses. In: Gordon JM, editor. Solar energy: the state of the art. London: Routledge; 2013. p. 477–96.
143 Garg HP. Advances in solar energy technology; Volume 2: industrial applications of solar energy. London: Springer; 2012.
144 Hess S. Solar thermal process heat (SPH) generation. In: Stryi-Hipp G, editor. Renewable heating and cooling. Cambridge: Woodhead Publishing; 2016.
145 Kalogirou S. The potential of solar industrial process heat applications. Appl Energy 2003;76:337–61.
146 Sharaf OZ, Orhan MF. Concentrated photovoltaic thermal (CPVT) solar collector systems: Part I – fundamentals, design considerations and current technologies. Renew Sustain Energy Rev 2015;50:1500–65.
147 United States Environmental Protection Agency (EPA). Renewable industrial process heat. <https://www.epa.gov/rhc/renewable-industrial-process-heat> [accessed 07.05.16].
148 Chapas RB, Colwell JA. Industrial technologies program research plan for energy-intensive process industries. Prepared by Pacific Northwest National Laboratory for the US Department of Energy. <http://www1.eere.energy.gov/manufacturing/pdfs/itp_research_plan.pdf>; 2007 [accessed 07.05.16].
149 International Renewable Energy Agency (IRENA). Solar heat for industrial processes. IEA-ETSAP and IRENA technology brief. <http://www.irena.org/DocumentDownloads/Publications/IRENA_ETSAP_Tech_Brief_E21_Solar_Heat_Industrial_2015.pdf>; January 2015 [accessed 07.05.16].

150 Brunner C, Slawitsch Kanellina Giannakopoulou B, Schnitzer H. Industrial process indicators and heat integration in industries. IEA SHC Task 33 and SolarPACES Task IV Joanneum Research, Graz, Austria; 2008. <http://archive.iea-shc.org/publications/downloads/Process_Integration_Booklet.pdf> [accessed 07.05.16].

151 Office of Energy Efficiency & Renewable Energy. Industrial process heating – technology assessment. <http://energy.gov/eere/amo/process-heating-systems> [accessed 05.04.15].

152 Persson U, Werner S. Heat distribution and the future competitiveness of district heating. Appl Energy 2011;88:568–76.

153 Lund H, Moller B, Mathiesen BV, Dyrelund A. The role of district heating in future renewable energy systems. Energy 2010;35:1381–90.

154 District Energy Association. History of the district energy industry. <http://www.districtenergy.org/history-of-the-district-energy-industry> [accessed 07.05.16].

155 Huculak M, Jarczewski W, Dej M. Economic aspects of the use of deep geothermal heat in district heating in Poland. Renew Sustain Energy Rev 2015;49:29–40.

156 Flynn C, Siren K. Influence of location and design on the performance of a solar district heating system equipped with borehole seasonal storage. Renew Energy 2015;81:377–388.

157 Drake Landing Solar Community. <http://www.dlsc.ca/> [accessed 05.09.15].

158 Sibbitt B, McClenahan D, Djebbar R, Thornton J, Wong B, Carriere J, Kokko J. The performance of a high solar fraction seasonal storage district heating system – five years of operation. Energy Proc 2012;30:856–65.

159 Natural Resources Canada. Drake Landing Solar Community, <http://www.nrcan.gc.ca/media-room/backgrounders/2012/3299#archived> [accessed 07.05.16].

160 Leidos Canada. Drake landing solar community. Annual Performance Monitoring Report for 2012–2013, prepared for Natural Resources Canada; February 2014. <http://www.dlsc.ca/reports/JUL2015/2012-2013%20Annual%20Report%20v07%20FINALpdf.> [accessed 07.05.16].

161 Solar District Heating. <http://solar-district-heating.eu/> [accessed 05.09.15].

162 Sunstore4. <http://sunstore4.eu/background/project-brief/> [accessed 08.09.15].

163 International Energy Agency. Key World Energy Statistics; 2014.

164 Metro Vancouver. <http://www.metrovancouver.org/> [accessed 18.05.16].

165 International Energy Agency (IEA). Linking heat and electricity systems, co-generation and district heating and cooling solutions for a clean energy future; 2014. <http://www.iea.org/publications/freepublications/publication/LinkingHeatandElectricitySystems.pdf> [accessed 10.05.16].

166 Marstal Fjernvarme. Summary technical description of the SUNSTORE 4 plant in Marstal; 2013. <http://www.solarmarstal.dk/media/2854117/summary-technical-description-marstal.pdf> [accessed 10.05.16].

167 Marstal Fjernvarme. <http://www.solarmarstal.dk/english/> [accessed 10.05.16].

168 Koreneff G, Lehtilä A, Hurskainen M. Nordic heating technology solution pathways. Report VTT-V-89025-14/ 101843-15; 2015.
169 Rolfsman B. Combined heat-and-power plants and district heating in a deregulated electricity market. Appl Energy 2004;78:37–52.
170 Colen HR. HVAC systems evaluation. New York: RS Means; 1990.
171 Harris NC. Modern air conditioning practice. 3rd ed. New York: Gress Division, McGraw-Hill; 1983.
172 Puetzer K. Handbook of HVAC design. New York: McGraw-Hill; 1990.
173 Dalkilic AS, Wongwises S. A performance comparison of vapour-compression refrigeration system using various alternative refrigerants. Int Commun Heat Mass Transfer 2010;37:1340–9.
174 Bilir N, Kursad Ersoy H. Performance improvement of the vapour compression refrigeration cycle by a two-phase constant area ejector. Int J Energy Res 2009;33:469–80.
175 Ahamed JU, Saidur R, Masjuki HH. A review on exergy analysis of vapor compression refrigeration system. Renew Sustain Energy Rev 2011;15: 1593–1600.
176 Yan J, Caia W, Lina C, Lib C, Li Y. Experimental study on performance of a hybrid ejector-vapor compression cycle. Energy Convers Manag 2016;113: 36–43.
177 Xing M, Yan G, Yu J. Performance evaluation of an ejector subcooled vapor-compression refrigeration cycle. Energy Convers Manag 2015;92:431–6.
178 Azzouz K, Leducq D, Gobin D. Performance enhancement of a household refrigerator by addition of latent heat storage. Int J Refrig 2008;31(5): 892–901.
179 Chua KJ, Chou SK, Yang WM. Advances in heat pump systems: a review. Appl Energy 2010;87(12):3611–24.
180 Skagestad B, Mildenstein P. District heating and cooling connection handbook. International Energy Agency. <http://www.districtenergy.org/> [accessed 05.04.16].
181 Climate Control. World's largest district cooling plant opens in The Pearl-Qatar. <http://climatecontrolme.com/2010/11/worlds-largest-district-cooling-plant-opens-pearl-qatar/> [accessed 08.05.16].
182 Eggen G, Vangsnes G. Heat pump for district cooling and heating at OSLO Airport Gardermoen. Proceedings of the IEA eighth heat pump conference, Las Vegas; 2006.
183 Enwave, Deep Lake Water Cooling. Keeping Toronto Cool. <http://www.districtenergy.org/assets/CDEA/Case-Studies/Enwave-case-history-Toronto7-19-07.pdf> [accessed 05.04.15].
184 GBE Factory. World's largest solar cooling plant with solar flat plate collectors in operation. <http://www.gbefactory.eu/worlds-largest-solar-cooling-plant-in-operation/> [accessed 05.04.15].
185 Climespace. District cooling. <http://www.climespace.fr/en/district-cooling/> [accessed 11.05.15].

186 Climespace. Cooling in the heart of the city. <file:///C:/Users/100400687/Downloads/CLIMESPACE-PLAQUETTE-CORPO-ENGLISH_VF.pdf> [accessed 11.05.16].

187 Di Cecca A, Benassis F, Poeuf P. Energy storage: The Parisian district cooling system. Available on Energy Learning. <http://www.energy-learning.com/index.php/archive/94-energy-storage-the-parisian-district-cooling-system> [accessed 11.05.16].

188 United Nations Environment Programme (UNEP). District energy in cities, Paris case study. <http://districtenergyinitiative.org/case-studies/PARIS-CASE-STUDY.pdf> [accessed 11.05.16].

189 Herold KE, Radermacher R. Absorption heat pumps. Mech Eng 1989;111(8):68–72.

190 Spurr M, Larsson I. Integrating district cooling with combined heat and power. International Energy Agency Report No. 1996: Nl, by Resource Efficiency Inc.; 1996.

191 Grossman G, Zaltash A, Adcock W, DeVault RC. Simulating a 4-effect absorption chiller. ASHRAE J 1995;37:45–53.

192 Ziegler F, Alefeld G. Comparison of multi-effect absorption cycles. Proceedings of the international absorption heat pump conference, New Orleans; January 1995. American Society of Mechanical Engineers, AES-Vol. 31. p. 19–21.

193 Chicco G, Mancarella P. Trigeneration primary energy saving evaluation for energy planning and policy development. Energy Policy 2007;35:6132–44.

194 Hernandez-Santoyoa J, Sanchez-Cifuentes A. Trigeneration: an alternative for energy savings. Appl Energy 2003;76:219–27.

195 Fumo N, Chamra LM. Analysis of combined cooling, heating, and power systems based on source primary energy consumption. Appl Energy 2010;87:2023–30.

196 Clark S, Mardiat E, Swinson SK. A Texas-sized district energy system: the energy behind what's next at the Texas Medical Center. District Energy, International District Energy Association (IDEA) 2011; Third quarter:12–17.

197 International District Energy Association (IDEA). Award winning CHP system serves Texas Medical Center; 2015. <http://www.districtenergy.org/blog/2015/09/03/award-winning-chp-system-serves-texas-medical-center-2/> [accessed 12.05.16].

198 US Department of Energy, Energy Efficiency & Renewable Energy. Recovery act case study. <https://www1.eere.energy.gov/manufacturing/rd/pdfs/teco_chp__recoveryact_casestudy.pdf> [accessed 12.05.16].

199 Garcia J, Mardiat E. Combined heat and power in the Texas Medical Center. Proceedings combined heat and power conference and trade show, Houston, TX; 2012.

200 Solar Energy Research Institute. Economic aspects of district heating and cooling. Heating/Piping/Air Cond 1987;59(9):47–56.

201 DeVault RC, Marsala J. Ammonia-water triple-effect absorption cycle. ASHRAE Trans 1990;96:676–82.

202 Bulgan AT. Optimization of the thermodynamic model of aqua-ammonia absorption refrigeration systems. Energy Convers Manag 1995;36:135–43.

203 Bejan A, Tsatsaronis G, Moran M. Thermal and design optimization. New York: John Wiley & Sons; 1996.

204 HVAC Systems and Equipment. ASHRAE Handbook. American Society of Heating, Refrigerating and Air-Conditioning Engineers, Inc.; 1996.

205 Uglietto SR. District heating and cooling conversion of buildings. ASHRAE Trans 1987;93:2096–106.

206 Knebel DA, Houston S. Thermal storage retrofit. ASHRAE J 1989;31(5):34–42.

207 Porter RW. Economic distribution distance for cogenerated district heating and cooling. Energy 1985;10(7):851–9.

208 Lee TR, Chatterjee AK. Integrated cogeneration and chiller system study. Energy Eng 1990;87(5):42–8.

209 Ozgener O, Hepbasli A, Dincer I, Rosen MA. Modelling and assessment of ground-source heat pump systems using exergoeconomic analysis for building applications. Proceedings building simulation: ninth international building performance simulation association conference, Montreal, QC; 2005. p. 915–20.

210 Ozgener O, Hepbasli A. Exergoeconomic analysis of a solar assisted ground-source heat pump greenhouse heating system. Appl Thermal Eng 2005;25: 1459–71.

211 Ucar A, Inalli M. Exergoeconomic analysis and optimization of a solar-assisted heating system for residential buildings. Build Environ 2008;41:1551–56.

212 Ozgener O, Hepbasli A, Ozgener L. A parametric study on the exergoeconomic assessment of a vertical ground-coupled (geothermal) heat pump system. Build Environ 2007;42:1503–09.

213 Sayyaadi H, Hadaddi Amlashi E, Amidpour M. Multi-objective optimization of a vertical ground source heat pump using evolutionary algorithm. Energy Convers Manag 2009;50:2035–46.

214 Ozturk IT, Karabay H, Bilgen E. Thermo-economic optimization of hot water piping systems: a comparison study. Energy 2006;31:2094–107.

215 Ozgener L, Hepbasli A, Dincer I, Rosen MA. Exergoeconomic analysis of geothermal district heating systems: a case study. Appl Thermal Eng 2007;2:1303–10.

216 Khir T, Jassim RK, Zaki GM. Application of exergoeconomic techniques to the optimization of a refrigeration evaporator coil with continuous fins. J Energy Resour Technol 2007;129:266–77.

217 Jassim RK, Khir T, Ghaffour N. Thermoeconomic optimization of the geometry of an air conditioning precooling air reheater dehumidifier. Int J Energy Res 2005;30:237–58.

218 Jassim RK, Khir T, Habeebullah BA, Zaki GM. Exergoeconomic optimisation of the geometry of continuous fins on an array of tubes of a refrigeration air cooled condenser. Int J Exergy 2005;2:146–71.

219 Jassim RK, Khir T. Exergoeconomic optimisation of an air-conditioning rotary regenerator: effect of matrix thermal conductivity on its performance. Int J Exergy 2004;1:215–36.

220 Kan X, Chen L, Sun F, Wu F. Exergoeconomic performance optimization for a steady-flow endoreversible refrigerator cycle model including five typical cycles. Int J Low-Carbon Technol 2010;5:74–80.
221 Sahoo PK, Misra RD, Gupta A. Exergoeconomic optimisation of an aqua-ammonia absorption refrigeration system. Int J Exergy 2004;1:82–93.
222 Palacios Bereche R, Gonzales Palomino R, Nebra SA. Thermoeconomic analysis of a single and double-effect LiBr/H_2O absorption refrigeration system. Int J Thermodyn 2009;12:89–96.
223 Piacentino A, Cardona F. Scope-oriented thermoeconomic analysis of energy systems. Part I: looking for a non-postulated cost accounting for the dissipative devices of a vapour compression chiller. Is it feasible? Appl Energy 2010;87:943–56.
224 Piacentino A, Cardona F. Scope-oriented thermoeconomic analysis of energy systems. Part II: formation structure of optimality for robust design. Appl Energy 2010;87:957–70.
225 Ma K, Chen L, Sun F. Profit rate performance optimization for a generalized irreversible combined refrigeration cycle. Sadhana 2009;34:851–64.
226 Erlach B, Tsatsaronis G, Cziesla F. A new approach for assigning costs and fuels to cogeneration products. Int J Appl Thermodyn 2001;4:145–56.
227 Tsatsaronis G, Pisa J. Exergoeconomic evaluation and optimization of energy systems: application to the CGAM problem. Energy 1994;19:287–321.
228 Silveira JL, Tuna CE. Thermoeconomic analysis method for optimization of combined heat and power systems: Part I. Prog Energy Combus Sci 2003;29:479–85.
229 Silveira JL, Tuna CE. Thermoeconomic analysis method for optimization of combined heat and power systems. Part II. Prog Energy Combus Sci 2004;30:673–78.
230 Balli O, Aras H, Hepbasli A. Exergoeconomic analysis of a combined heat and power (CHP) system. Int J Energy Res 2008;32:273–89.
231 Tsatsaronis G, Moran MJ. Exergy-aided cost minimization. Energy Convers Manag 1997;38:1535–42.
232 Abusoglu A, Kanoglu M. Exergetic and thermoeconomic analyses of diesel engine powered cogeneration. Part 1: formulations and Part 2: application. Appl Thermal Eng 2009;29:234–49.
233 Abusoglu A, Kanoglu M. Exergoeconomic optimization of a diesel-engine-powered cogeneration plant. Energy Fuels 2009;23:1977–89.
234 Sugimoto K, Fujii T, Ohta J. An appraisal method of exergy cost minimisation for co-generation systems. Int J Exergy 2006;3;255–71.
235 Kim S-M, Oh S-D, Kwon Y-H, Kwak H-Y. Exergoeconomic analysis of thermal systems. Energy 1998;23:393–406.
236 Aras H, Balli O. Exergoeconomic analysis of a combined heat and power system with the micro gas turbine. Energy Explor Exploit 2008;26:53–70.
237 Luz-Silveira J, Beyene A, Leal EM, Santana JA, Okada D. Thermoeconomic analysis of a cogeneration system of a university campus. Appl Thermal Eng 2002;22:1471–83.

238 Ozgur Colpan C, Yesin T. Energetic, exergetic and thermoeconomic analysis of Bilkent combined cycle cogeneration plant. Int J Energy Res 2006; 30:875–94.
239 Colpan CO, Yesin T. Thermodynamic and thermoeconomic comparison of combined cycle cogeneration systems. Int J Exergy 2006;3:272–90.
240 Lian ZT, Chua KJ, Chou SK. A thermoeconomic analysis of biomass energy for trigeneration. Appl Energy 2010;87:84–95.
241 Strickland C, Nyboer J. Cogeneration potential in Canada: Phase 2. Report for Natural Resources Canada, by MK Jaccard and Associates; 2002.
242 Strickland C, Nyboer J. A review of existing cogeneration facilities in Canada. Report by Canadian Industrial Energy End-Use Data and Analysis Center, Simon Fraser University; 2002.
243 Upton B. Letter of 16 November 2001 to P Bhatia, World Resources Institute, Washington, DC, from B Upton, Senior Research Engineer, National Council for Air and Stream Improvement, Inc., West Coast Regional Center, Corvallis, OR; 2001.
244 Phylipsen GJM, Blok K, Worrell E. Handbook on international comparisons of energy efficiency in the manufacturing industry. The Netherlands: Department of Science, Technology and Society, Utrecht University; 1998.
245 European Parliament. Directive 2004/8/EC of the European Parliament and of the Council of 11 February 2004 on the Promotion of Cogeneration Based on a Useful Heat Demand in the Internal Energy Market and Amending Directive 92/42/EEC. Off J Eur Union 2004;L52(47):50–60.
246 CEN/CENELEC Workshop Agreement Manual for Determination of Combined Heat and Power (CHP). CWA 45547, European Committee for Standardization (CEN) and European Committee for Electrotechnical Standardization (CENELEC), Brussels; 2004.
247 Verbruggen A. What's needed next to refine the EU directive on cogeneration regulation. Electricity J 2007;20(2):63–70.
248 Verbruggen A. Quantifying combined heat and power (CHP) activity. Int J Energy Technol Policy 2007;5(1):17–35.
249 Rosen MA. Comparison based on energy and exergy analyses of the potential cogeneration efficiencies for fuel cells and other electricity generation devices. Int J Hydrogen Energy 1990;15:267–74.
250 Rosen MA. Regional cogeneration via electrical utilities operating on nuclear energy and coal: energy and environmental benefits. Trans Canadian Soc Mech Eng 2009;33(1):131–43.
251 Rosen MA. Energy, environmental, health and cost benefits of cogeneration from fossil fuels and nuclear energy using the electrical utility facilities of a province. Energy Sustain Dev 2009;13(1):42–50.
252 Consumers Gas. Cogeneration at the University of Toronto. Information sheet; 1995 [estimated].
253 Wiggin M. Cornwall, Ontario District Heating System. Information sheet, CANMET Energy Technology Centre, Natural Resources Canada; 1997 [estimated].

254 Harrell G. Steam System Survey Guide. Report ORNL/TM-2001/263 for Oak Ridge National Laboratory; 2002.
255 Samuelson PA. Foundations of economic analysis. Cambridge, MA: Harvard University Press; 1983.
256 Ahn, BH. Computation of market equilibria for policy analysis. New York: Garland Publishing; 1979.
257 Daniel TE, Goldberg HM. Dynamic equilibrium energy modelling: the Canadian BALANCE model. Oper Res 1981;29(5):829–52.
258 Wu Y, Fuller JD. Convergence of multi-period equilibrium calculation, with geometric distributed lag demand. Oper Res Lett 1991;10:197–202.
259 Wu YJ, Fuller JD. Introduction of geometric, distributed lag demand into energy-process models. Energy Int J 1995;20(7):647–56.
260 Berger C, Dubois R, Haurie A, Lessard E, Loulou R, Waaub JP. Canadian MARKAL: an advanced linear programming system for energy and environmental modelling. Inf Syst Oper Res 1992;30:222–39.
261 Manne AS. On the formulation and solution of economic equilibrium models. In: Manne AS, editor. Economic equilibrium: model formulation and solution. Amsterdam: North-Holland; 1985. p. 1–22.
262 Nagumey A. Computational comparison of spatial price equilibrium methods. J Regional Sci 1987;27(1):55–76.
263 Stone JC. Sequential optimization and complementarity techniques for computing economic equilibria. In: Manne AS, editor. Economic equilibrium: model formulation and solution. Amsterdam: North-Holland; 1985. p. 173–91.
264 Wu YJ, Fuller JD. An algorithm for the multi-period market equilibrium model with geometric, distributed lag demand. Oper Res 1996;44(6):1002–12.
265 Ahn BH, Hogan WW. On convergence of the PIES algorithm for computing equilibria. Oper Res 1982;30(2):281–300.
266 Manne AS, Richels RG. Buying greenhouse insurance: the economic costs of carbon dioxide emission limits. Cambridge, MA: MIT Press; 1992.
267 Nordhaus WD. To slow or not to slow: the economics of the greenhouse effects. Economics J 1991;101:920–37.
268 Goto N, Sawa T. An analysis of the macroeconomic costs of various emissions control policies in Japan. Energy J 1993;14(1):83–110.
269 Chung W, Wu YJ, Fuller JD. Dynamic energy and environment equilibrium model, for the assessment of CO_2 emission control in Canada and the USA. Energy Economics 1997;19(1):103–24.
270 Murtagh BA, Saunders MA. MINOS 5.0 user's guide. Technical Report SOL 83-20, Systems Optimization Laboratory, Department of Operations Research, Stanford University; 1983.
271 Fuller JD. In: Lev B, Bloom JA, Murphy FH, Gleit AS, editors. Strategic planning in energy and natural resources. Amsterdam: Elsevier; 1987.
272 Fuller JD, Luthra D. Model formulation with the Waterloo Energy Modelling System (WATEMS). Energy 1990;15(5):413–25.
273 Harberger AC. Three basic postulates for applied welfare economics: an interpretive essay. J Economic Lit 1971;9(3):785–97.

274 Wu YJ, Chung W. Assessing the control of energy-related CO_2 emissions with a dynamic energy process model. Energy 1996;22(7):693–704.

275 Spurr M, Larsson I. Integrating district cooling with combined heat and power. International Energy Agency Report, No. 1996: N1; 1996.

276 Rosen MA, editor. Renewable energy. London: Routledge/Taylor and Francis (part of series: Critical Concepts in Natural Resources); 2015.

277 Rosen MA, editor. Energy storage. Hauppauge, NY: Nova Science Publishers; 2012.

278 Rosen MA, editor. Energy storage. London: Routledge/Taylor and Francis (part of series: Critical Concepts in Natural Resources); 2015.

279 Rosen MA, Koohi-Fayegh S. Geothermal energy: sustainable heating and cooling using the ground. London: Wiley; 2016.

280 Dincer I, Rosen MA. Exergy analysis of heating, refrigerating and air conditioning: methods and applications. Oxford: Elsevier; 2015.

281 Dincer I, Rosen MA. Exergy: energy, environment and sustainable development. 2nd ed. Oxford: Elsevier; 2013.

282 Rosen MA. Economics and exergy: an enhanced approach to energy economics. Hauppauge, NY: Nova Science Publishers; 2011.

283 Rosen MA. Environment, ecology and exergy: enhanced approaches to environmental and ecological management. Hauppauge, NY: Nova Science Publishers; 2012.

284 Bracco S, Dentici G, Siri S. DESOD: a mathematical programming tool to optimally design a distributed energy system. Energy 2016;100:298–309.

285 Kanoglu M, Dincer I, Rosen MA. Exergetic performance analysis of various cogeneration systems for buildings. ASHRAE Trans 2007;113(2): 105–12.

286 Le Corre O, Saikaly K, Rosen MA. Effects on CHP plant efficiency of H_2 production through partial oxydation of natural gas over two group VIII metal catalysts. Int J Hydrogen Energy 2012;37(13):10380–89.

287 Abadi SK, Manesh MHK, Rosen MA, Amidpour M, Hamedi MH. Integration of a gas fired steam power plant with a total site utility using a new cogeneration targeting procedure. Chin J Chem Eng 2014;22(4):455–68.

288 Long WC, Luck R, Mago PJ. Uncertainty based operating strategy selection in combined heat and power systems. Appl Thermal Eng 2016;98: 1013–24.

289 Keirstead J, Samsatli N, Shah N, Weber C. The impact of CHP (combined heat and power) planning restrictions on the efficiency of urban energy systems. Energy 2012;41(1):93–103.

290 Zeghici RM, Damian A, Frunzulică R, Iordache F. Energy performance assessment of a complex district heating system which uses gas-driven combined heat and power, heat pumps and high temperature aquifer thermal energy storage. Energy Buildings 2014;84:142–151.

291 Choudhury A, Chandra H, Arora A. Application of solid oxide fuel cell technology for power generation—a review. Renew Sustain Energy Rev 2013;20:430–42.

292 Rosen MA. Comparison based on energy and exergy analyses of the potential cogeneration efficiencies for fuel cells and other electricity generation devices. Int J Hydrogen Energy 1990;15(4):267–74.
293 Blarke MB. Towards an intermittency-friendly energy system: comparing electric boilers and heat pumps in distributed cogeneration. Appl Energy 2012;91(1):349–65.
294 Kikuchi Y, Kanematsu Y, Sato R, Nakagaki T. Distributed cogeneration of power and heat within an energy management strategy for mitigating fossil fuel consumption. J Ind Ecol 2016;20:289–303.
295 Hosseini M, Dincer I, Rosen MA. Hybrid solar-fuel cell combined heat and power systems for residential applications: energy and exergy analyses. J Power Sources 2013;221:372–80.
296 Sermyagina E, Saari J, Zakeri B, Kaikko J, Vakkilainen E. Effect of heat integration method and torrefaction temperature on the performance of an integrated CHP-torrefaction plant. Appl Energy 2015; 149:24–34.
297 Soltani R, Dincer I, Rosen MA. Thermodynamic analysis of a novel multigeneration energy system based on heat recovery from a biomass CHP cycle. Appl Thermal Eng 2015;89:90–100.
298 Murugan S, Horák B. Tri and polygeneration systems – a review, renewable and sustainable. Energy Rev 2016;60:1032–51.
299 Jradi M, Riffat S. Tri-generation systems: energy policies, prime movers, cooling technologies, configurations and operation strategies. Renew Sustain Energy Rev 2014;32:396–415.
300 Agnew B, Walker S, Ng B, Tam ICK. Finite time analysis of a tri-generation cycle. Energies 2015;8(6):6215–29.
301 Espirito Santo DB. Energy and exergy efficiency of a building internal combustion engine trigeneration system under two different operational strategies. Energy Buildings 2012;53:28–38.
302 Sonar D, Soni SL, Sharma D. Micro-trigeneration for energy sustainability: technologies, tools and trends. Appl Thermal Eng 2014;71(2):790–6.
303 Parantapa S, Meftah N, Pfafferott J. Multiperspective analysis of microscale trigeneration systems and their role in the crowd energy concept. J Electron Sci Technol 2015;3:003.
304 Angrisani G, Akisawa A, Marrasso E, Roselli C, Sasso M. Performance assessment of cogeneration and trigeneration systems for small scale applications. Energy Convers Manag. Available online 11 April 2016, ISSN 0196-8904. <http://dx.doi.org/10.1016/j.enconman.2016.03.092>.
305 Arsalis A, Alexandrou A. Design and modeling of 1–10 MWe liquefied natural gas-fueled combined cooling, heating and power plants for building applications. Energy Buildings 2015;86:257–67.
306 Pérez NP, Sadamitsu MT, Silveira JL, *et al.* Energetic and exergetic analysis of a new compact trigeneration system run with liquefied petroleum gas. Energy 2015;90(2):1411–19.

307 Ranjbar F, Chitsaz A, Mahmoudi SMS, Khalilarya S, Rosen MA. Energy and exergy assessments of a novel trigeneration system based on a solid oxide fuel cell. Energy Convers Manag 2014;87:318–27.
308 Reda F, Viot M, Sipilä K, Helm M. Energy assessment of solar cooling thermally driven system configurations for an office building in a Nordic country. Appl Energy 2016;166:27–43.
309 Ahrenfeldt J, Thomsen TP, Henriksen U, Clausen LR. Biomass gasification cogeneration – a review of state of the art technology and near future perspectives. Appl Thermal Eng 2013;50(2):1407–17.
310 Espirito Santo DB. An energy and exergy analysis of a high-efficiency engine trigeneration system for a hospital: a case study methodology based on annual energy demand profiles. Energy Buildings 2014;76:185–98.
311 Ascione F, Canelli M, De Masi RF, Sasso M, Vanoli GP. Combined cooling, heating and power for small urban districts: an Italian case-study. Appl Thermal Eng 2014;71(2):705–13.
312 Malik M, Dincer I, Rosen MA. Development and analysis of a new renewable energy-based multi-generation system. Energy 2015;79:90–9.
313 Almahdi M, Dincer I, Rosen MA. A new solar based multigeneration system with hot and cold thermal storages and hydrogen production. Renewable Energy 2016;91:302–14.
314 Khalid F, Dincer I, Rosen MA. Energy and exergy analyses of a solar-biomass integrated cycle for multi-generation. Solar Energy 2015;112:290–9.
315 Ahmadi P, Dincer I, Rosen MA. Development and assessment of an integrated biomass-based multi-generation energy system. Energy Int J 2013;56:155–66.
316 Ahmadi P, Dincer I, Rosen MA. Performance assessment of a novel solar and ocean thermal energy conversion based multigeneration system for coastal areas. ASME J Solar Energy Eng 2014;137(1):011013-011013-12.
317 Khalid F, Dincer I, Rosen MA. Development and analysis of sustainable energy systems for building HVAC applications. Appl Thermal Eng 2015;87:389–401.
318 Rosen MA, Kumar R. Performance of a photovoltaic/thermal solar air heater: effect of vertical fins on a double pass system. Int J Energy Environ Eng 2011;2(4):1–12.
319 Kumar R, Rosen MA. Performance evaluation of a double pass PV/T solar air heater with and without fins. Appl Thermal Eng 2011;31(8–9):1402–10.
320 Kumar R, Rosen MA. A critical review of photovoltaic-thermal solar collectors for air heating. Appl Energy 2011;88(11):3603–14.
321 Kumar R, Rosen MA. Integrated collector-storage solar water heater with extended storage unit. Appl Thermal Eng 2011;31(2–3):348–54.
322 Kumar R, Rosen MA. Comparative performance investigation of integrated collector-storage solar water heaters with various heat loss reduction strategies. Int J Energy Res 2011;35:1179–87.
323 Self SJ, Reddy BV, Rosen MA. Geothermal heat pump systems: status review and comparison with other heating options. Appl Energy 2013;101:341–48.

324 Self SJ, Reddy BV, Rosen MA. Ground source heat pumps for heating: parametric energy analysis of a vapor compression cycle utilizing an economizer arrangement. Appl Thermal Eng 2013;52(2):245–54.

325 Dincer I, Rosen MA. Thermal energy storage: systems and applications. 2nd ed. London: Wiley.

326 Sliwa T, Rosen MA. Natural and artificial methods for regeneration of heat resources for borehole heat exchangers to enhance the sustainability of underground thermal storages: a review. Sustainability 2015;7(10):13104–25.

327 Koohi-Fayegh S, Rosen MA. A review of the modelling thermally interacting multiple boreholes. Sustainability 2013;5(6):2519–36.

328 Koohi-Fayegh S, Rosen MA. An analytical approach to evaluating the effect of thermal interaction of geothermal heat exchangers on ground heat pump efficiency. Energy Convers Manag 2014;78:184–192.

329 Koohi-Fayegh S, Rosen MA. Three dimensional analysis of the thermal interaction of multiple vertical ground heat exchangers. Int J Green Energy 2015;12(11):1144–50.

330 Mosaffa AH, Talati F, Rosen MA, Basirat Tabrizi, H. Phase change material solidification in a finned cylindrical shell thermal energy storage: an approximate analytical approach. Thermal Sci 2013;17(2):407–18.

331 Haji Abedin A, Rosen MA. Closed and open thermochemical energy storage: energy- and exergy-based comparisons. Energy 2012;41(1):83–92.

332 Haji Abedin A, Rosen MA. Thermodynamic analyses of closed thermochemical energy storages. Int J Exergy 2012;10(2):125–41.

333 Haji Abedin A, Rosen MA. Assessment of a closed thermochemical energy storage using energy and exergy methods. Appl Energy 2012;93:18–23.

334 Garousi Farshi L, Seyed Mahmoudi SM, Rosen MA, Yari, M. A comparative study on the performance characteristics of double effect absorption refrigeration systems. Int J Energy Res 2012;36(2):182–92.

335 Garousi Farshi L, Seyed Mahmoudi SM, Rosen MA, Yari M. Use of low grade heat sources in combined ejector-double effect absorption refrigeration systems. Proc Inst Mech Engrs Part A: J Power Energy 2012;226(5): 607–22.

336 Garousi Farshi L, Mosaffa AH, Infante Ferreira CA, Rosen MA. Thermodynamic analysis and comparison of combined ejector-absorption and single effect absorption refrigeration systems. Appl Energy 2014;133: 335–46.

337 Garousi Farshi L, Infante Ferreira CA, Seyed Mahmoudi SM, Rosen MA. First and second law analysis of ammonia/salt absorption refrigeration systems. Int J Refrig 2014;40:111–21.

338 Mosaffa AH, Infante Ferreira CA, Talati F, Rosen MA. Thermal performance of a multiple PCM thermal storage unit for free cooling. Energy Convers Manag 2013;67:1–7.

339 Mosaffa AH, Garousi Farshi L, Infante Ferreira CA, Rosen MA. Energy and exergy evaluation of a multiple-PCM thermal storage unit for free cooling applications. Renewable Energy 2014;68:452–58.

340 Kanoglu M, Dincer I, Rosen MA. Exergy analysis of psychrometric processes for HVAC&R applications. ASHRAE Trans 2007;113(2):172–80.
341 Mosaffa AH, Garousi Farshi L, Infante Ferreira CA, Rosen MA. Advanced exergy analysis of an air conditioning system incorporating thermal energy storage. Energy Int J 2014;77:945–52.
342 Mosaffa AH, Talati F, Basirat Tabrizi H, Rosen MA. Analytical modeling of PCM solidification in a shell and tube finned thermal storage for air conditioning systems. Energy Buildings 2012;49:356–61.
343 Rosen MA. Net-zero energy buildings and communities: potential and the role of energy storage. J Power Energy Eng 2015;3(4):470–74.
344 Soltani R, Dincer I, Rosen MA. Comparative performance evaluation of cascaded air-source hydronic heat pumps. Energy Convers Manag 2015;89:577–87.
345 Koohi-Fayegh S, Rosen MA. An analytical approach to evaluating the effect of thermal interaction of geothermal heat exchangers on ground heat pump efficiency. Energy Convers Manag 2014;78:184–92.
346 Koohi-Fayegh S, Rosen MA. Examination of thermal interaction of multiple vertical ground heat exchangers. Appl Energy 2012;97:962–69.
347 Wiltshire R, editor. Advanced district heating and cooling (DHC) systems. Cambridge: Woodhead Publishing; 2015.
348 Pan SY, Du MA, Huang IT, Liu IH, Chang E, Chiang PC. Strategies on implementation of waste-to-energy (WTE) supply chain for circular economy system: a review. J Cleaner Prod 2015;108(A):409–21.
349 Soltani R, Dincer I, Rosen MA. Thermodynamic analysis and performance assessment of an integrated heat pump system for district heating applications. Appl Thermal Eng 2015;89:833–42.
350 Rezaie B, Rosen MA, Reddy BV. Configurations for multiple thermal energy storages in thermal networks. Int J Process Syst Eng 2015;3(1/2/3): 191–206.
351 Mehrpooya M, Moftakhari Sharifzadeh MM, Rosen MA. Energy and exergy analyses of a novel power cycle using the cold of LNG (liquefied natural gas) and low-temperature solar energy. Energy 2016;95:324–45.
352 Sapińska-Śliwa A, Rosen MA Gonet A, Śliwa T. Deep borehole heat exchangers: a conceptual review. Int J Air-Cond Refrig 2016;24(1):1630001.
353 Rad FM, Fung AS. Solar community heating and cooling system with borehole thermal energy storage – review of systems. Renew Sustain Energy Rev 2016;60:1550–61.
354 Rezaie B, Rosen MA. District heating and cooling: review of technology and potential enhancements. Appl Energy 2012;93:2–10.
355 Rezaie B, Reddy BV, Rosen MA. Exergy analysis of thermal energy storage in a district energy application. Renew Energy 2015;74:848–54.
356 Rezaie B, Reddy BV, Rosen MA. Energy analysis of thermal energy storages with grid configurations. Appl Energy 2014;117:54–61.
357 Rezaie B, Reddy BV, Rosen MA. Assessment of the utilization of thermal energy storage in district energy systems. Int J Energy Environ Economics 2012;20(5):395–414.

358 Wang H, Lahdelma R, Wang X, Jiao W, Zhu C, Zou P. Analysis of the location for peak heating in CHP based combined district heating systems. Appl Thermal Eng 2015;87:402–11.
359 Mostafavi Tehrani SS, Saffar-Avval M, Mansoori Z, *et al.* Development of a CHP/DH system for the new town of Parand: an opportunity to mitigate global warming in Middle East. Appl Thermal Eng 2013;59(1–2): 298–308.
360 Erdemb HH, Akkaya AV, Dagdas A, *et al.* Renovating thermal power plant to trigeneration system for district heating/cooling: evaluation of performance variation. Appl Thermal Eng 2015;86:35–42.
361 Rosen MA, Le MN, Dincer I. Efficiency analysis of a cogeneration and district energy system. Appl Thermal Eng 2005;25(1):147–59.
362 Rosen MA, Le MN, Dincer I. Exergetic analysis of cogeneration-based district energy systems. Proc Inst Mech Eng Part A: J Power Energy 2004;218 (6):369–75.
363 Rosen MA. Assessment of various scenarios for utility-based cogeneration in Ontario. Energy Int J 1994;19(11):1143–49.
364 Rosen MA, Le M. Assessment of the potential cumulative benefits of applying utility-based cogeneration in Ontario. Energy Stud Rev 1994;6 (2):154–63.
365 Garousi Farshi L, Mahmoudi SMS, Rosen MA, Yari M, Amidpour M. Exergoeconomic analysis of double effect absorption refrigeration systems. Energy Convers Manag 2013;65:13–25.
366 Garousi Farshi L, Mahmoudi SMS, Rosen MA. Exergoeconomic comparison of double effect and combined ejector-double effect absorption refrigeration systems. Appl Energy 2013;103:700–11.
367 Alkan MA, Keçebaş A, Yamankaradeniz N. Exergoeconomic analysis of a district heating system for geothermal energy using specific exergy cost method. Energy 2013;60:426–34.
368 Ozgener L, Hepbasli A, Dincer I, Rosen MA. Exergoeconomic analysis of geothermal district heating systems: a case study. Appl Thermal Eng 2007;27 (8–9):1303–10.
369 Konstantakos V, Pilavachi PA, Polyzakis A, Theofylaktos C. A decision support model for combined heat and power economic evaluation. Appl Thermal Eng 2012;42:129–35.
370 Knizley A, Mago PJ, Tobermann J. Evaluation of the operational cost savings potential from a D-CHP system based on a monthly power-to-heat ratio analysis. Cogent Eng 2015;2(1):1004994.
371 Rosato A, Sibilio S, Ciampi G. Energy, environmental and economic dynamic performance assessment of different micro-cogeneration systems in a residential application. Appl Thermal Eng 2013;59(1–2):599–617.
372 Khani L, Mahmoudi SMS, Chitsaz A, Rosen MA. Energy and exergoeconomic evaluation of a new power/cooling cogeneration system based on a solid oxide fuel cell. Energy 2016;94:64–77.

373 Uris M, Linares JI, Arenas E. Techno-economic feasibility assessment of a biomass cogeneration plant based on an Organic Rankine Cycle. Renew Energy 2014;66:707–13.

374 Rosen MA. Energy, environmental, health and cost benefits of cogeneration from fossil fuels and nuclear energy using the electrical utility facilities of a province. Energy Sustain Dev 2009;13(1):42–50.

350 Sibilio S, Rosato A, Ciampi G, Scorpio M, Akisawa A. Building-integrated trigeneration system: energy, environmental and economic dynamic performance assessment for Italian residential applications. Renew Sustain Energy Rev. Available online 18 March 2016, ISSN 1364-0321. <http://dx.doi.org/10.1016/j.rser.2016.02.011>

376 Baniasad Askari I, Oukati Sadegh M, Ameri M. Energy management and economics of a trigeneration system considering the effect of solar PV, solar collector and fuel price. Energy Sustain Dev 2015;26:43–55.

377 Karellas S, Braimakis K. Energy–exergy analysis and economic investigation of a cogeneration and trigeneration ORC–VCC hybrid system utilizing biomass fuel and solar power. Energy Convers Manag 2016;107:103–13.

378 Khalid F, Dincer I, Rosen MA. Techno-economic assessment of a renewable energy based integrated multigeneration system for green buildings. Appl Thermal Eng 2016;99:1286–94.

379 Mohan G, Dahal S, Kumar U, Martin A, Kayal H. Development of natural gas fired combined cycle plant for tri-generation of power, cooling and clean water using waste heat recovery: techno-economic analysis. Energies 2014, 7 (10):6358–81.

380 Klaassen RE, Patel MK. District heating in the Netherlands today: a techno-economic assessment for NGCC-CHP (Natural Gas Combined Cycle combined heat and power). Energy 2013;54:63–73.

381 Colmenar-Santos A, Rosales-Asensio E, Borge-Diez D, Collado-Fernández E. Evaluation of the cost of using power plant reject heat in low-temperature district heating and cooling networks. Appl Energy 2016;162:892–907.

382 Choi J, Yun R. Operation strategy and parametric analysis of a CHP and a tri--generation system for integrated community. Int J Air-Cond Refrig 2015;23(01).

383 Arsalis A, Alexandrou A. Thermoeconomic modeling and exergy analysis of a decentralized liquefied natural gas-fueled combined-cooling–heating-and-power plant. J Nat Gas Sci Eng 2014;21:209–20.

384 Chen Y, Zhang T, Yang H, Peng J. Study on energy and economic benefits of converting a combined heating and power system to a tri-generation system for sewage treatment plants in subtropical area. Appl Thermal Eng 2016;94:24–39.

385 Barberis S, Rivarolo M, Traverso A, Massardo AF, Thermo-economic analysis of the energy storage role in a real polygenerative district. J Energy Storage 2016;5:187–202.

386 Calise F, Dentice d'Accadia M, Macaluso A, Piacentino A, Vanoli L. Exergetic and exergoeconomic analysis of a novel hybrid solar–geothermal

polygeneration system producing energy and water. Energy Convers Manag 2016;115:200–20.
387 Mosaffa AH, Garousi Farshi L, Infante Ferreira CA, Rosen MA. Exergoeconomic and environmental analyses of a CO_2/NH_3 cascade refrigeration system equipped with different types of flash tank intercoolers. Energy Convers Manag 2016;117:442–53.
388 Rezaie B, Reddy BV, Rosen MA. An enviro-economic function for assessing energy resources for district energy systems. Energy 2014;70:159–64.
389 Teke A, Timur O. Assessing the energy efficiency improvement potentials of HVAC systems considering economic and environmental aspects at the hospitals. Renew Sustain Energy Rev 2014;33:224–35.
390 Rosen MA. An exergy-based method for allocating carbon dioxide emissions from cogeneration systems. Int J Exergy 2009;6(1):1–14.
391 Rosen MA. Regional cogeneration via electrical utilities operating on nuclear energy and coal: energy and environmental benefits. Trans Can Soc Mech Eng 2009;33(1):131–43.
392 Rosen MA. Reductions in energy use and environmental emissions achievable with utility-based cogeneration: simplified illustrations for Ontario. Appl Energy 1998;61(3):163–74.
393 Bianchi M, Branchini L, De Pascale A, Peretto A. Application of environmental performance assessment of CHP systems with local and global approaches. Appl Energy 2014;130:774–82.
394 Ahmadi P, Rosen MA, Dincer I. Greenhouse gas emission and exergo-environmental analyses of a trigeneration energy system. Int J Greenhouse Gas Control 2011;5(6):1540–49.
395 Maraver D, Sin A, Sebastián F, Royo J. Environmental assessment of CCHP (combined cooling heating and power) systems based on biomass combustion in comparison to conventional generation. Energy 2013;57:17–23.
396 Gazda W, Stanek W. Energy and environmental assessment of integrated biogas trigeneration and photovoltaic plant as more sustainable industrial system. Appl Energy 2016;169:138–49.
397 Ahmadi P, Dincer I, Rosen MA. Exergo-environmental analysis of an integrated organic Rankine cycle for trigeneration. Energy Convers Manag 2012;64:447–53.
398 Chitsaz A, Mahmoudi SMS, Rosen MA. Greenhouse gas emission and exergy analyses of an integrated trigeneration system driven by a solid oxide fuel cell. Appl Thermal Eng 2015;86:81–90.
399 Hart DR, Rosen MA. Environmental and health benefits of utility-based cogeneration and district cooling in Ontario, Canada. Energy Int J 1996;21 (12):1135–46.
400 Torchio MF. Energy–exergy, environmental and economic criteria in combined heat and power (CHP) plants: indexes for the evaluation of the cogeneration potential. Energies 2013;6:2686–708.
401 Dincer I, Rosen MA, Ahmadi P. Optimization of energy systems. London: Wiley; 2016.

402 Li M, Mu H, Li N, Li H, Gui S, Chen X. Optimal option of natural-gas district distributed energy systems for various buildings. Energy Buildings 2014;75:70–83.

403 Yang Y, Zhang S, Xiao Y. An MILP (mixed integer linear programming) model for optimal design of district-scale distributed energy resource systems. Energy 2015;90(2):1901–15.

404 Yang Y, Zhang S, Xiao Y. Optimal design of distributed energy resource systems coupled with energy distribution networks. Energy 2015;85:433–48.

405 Ahmadi P, Dincer I, Rosen MA. Thermoeconomic multi-objective optimization of a novel biomass-based integrated energy system. Energy Int J 2014;68:958–70.

406 Mosaffa AH, Infante Ferreira CA, Rosen MA, Talati F. Thermal performance optimization of free cooling systems using enhanced latent heat thermal storage unit. Appl Thermal Eng 2013;59(1–2):473–79.

407 Casisi M, De Nardi A, Pinamonti P, Reini M. Effect of different economic support policies on the optimal synthesis and operation of a distributed energy supply system with renewable energy sources for an industrial area. Energy Convers Manag 2015;95:131–9.

408 Bischi A, Pérez-Iribarren E, Campanaria S, *et al.* Cogeneration systems optimization: comparison of multi-step and mixed integer linear programming approaches. Int J Green Energy. doi:10.1080/15435075.2016.1161635

409 Soltani R, Mohammadzadeh Keleshtery P, Vahdati M, Khoshgoftar Manesh MH, Rosen MA, Amidpour M. Multi-objective optimization of a solar-hybrid cogeneration cycle: application to CGAM problem. Energy Convers Manag 2014;81:60–71.

410 Wang H, Abdollahi E, Lahdelma R, Jiao W, Zhou Z. Modelling and optimization of the smart hybrid renewable energy for communities (SHREC). Renewable Energy 2015;84:114–123.

411 Safaei A, Freire F, Henggeler Antunes C. A life cycle multi-objective economic and environmental assessment of distributed generation in buildings. Energy Convers Manag 2015;97:420–27.

412 Liu M, Shi Y, Fang F. Combined cooling, heating and power systems: a survey. Renew Sustain Energy Rev 2014;35:1–22.

413 Ünal A N, Ercan S, Kayakutlu G. Optimisation studies on tri-generation: a review. Int J Energy Res 2015;39:1311–34.

414 Cho H, Smith AD, Mago P. Combined cooling, heating and power: a review of performance improvement and optimization. Appl Energy 2014;136:168–85.

415 Rong A, Lahdelma R. Role of polygeneration in sustainable energy system development challenges and opportunities from optimization viewpoints. Renew Sustain Energy Rev 2016;53:363–72.

416 Li Z, Li W, Xu B. Optimization of mixed working fluids for a novel trigeneration system based on organic Rankine cycle installed with heat pumps. Appl Thermal Eng 2016;94:754–62.

417 Piacentino A, Gallea R, Cardona F, Lo Brano V, Ciulla G, Catrini P. Optimization of trigeneration systems by mathematical programming: influence of plant scheme and boundary conditions. Energy Convers Manag 2015;104:100–14.
418 Agnew B, Walker S, Ng B, Tam ICK. Maximum output from a tri-generation cycle. Appl Thermal Eng 2015;90:1015–20.
419 Sadeghi M, Chitsaz A, Mahmoudi SMS, Rosen MA. Thermoeconomic optimization using an evolutionary algorithm of a trigeneration system driven by a solid oxide fuel cell. Energy 2015;89:191–204.
420 Puig-Arnavat M, Bruno JC, Coronas A. Modeling of trigeneration configurations based on biomass gasification and comparison of performance. Appl Energy 2014;114:845–56.
421 Dominković DF, Ćosić B, Bačelić Medić Z, Duić N. A hybrid optimization model of biomass trigeneration system combined with pit thermal energy storage. Energy Convers Manag 2015;104:90–99.
422 Ahmadi P, Rosen MA, Dincer I. Multi-objective exergy-based optimization of a polygeneration energy system using an evolutionary algorithm. Energy Int J 2012;46(1):21–31.
423 Ahmadi P, Dincer I, Rosen MA. Multi-objective optimization of a novel solar-based multigeneration energy system. Solar Energy 2014;108:576–91.
424 Ahmadi P, Dincer I, Rosen MA. Performance assessment and optimization of a novel integrated multigeneration system for residential buildings. Energy Buildings 2013;67:568–78.
425 Ahmadi P, Dincer I, Rosen MA. Thermodynamic modeling and multi-objective evolutionary-based optimization of a new multigeneration energy system. Energy Convers Manag 2013;76:282–309.
426 Mohamed A, Hamdy M, Hasan A, Sirén K. The performance of small scale multi-generation technologies in achieving cost-optimal and zero-energy office building solutions. Appl Energy 2015;152:94–108.
427 Piacentino A, Barbaro C, Cardona F. Promotion of polygeneration for buildings applications through sector- and user-oriented "high efficiency CHP" eligibility criteria. Appl Thermal Eng 2014;71(2):882–94.
428 Bordin C, Angelo Gordini, Daniele Vigo. An optimization approach for district heating strategic network design. Eur J Oper Res 2016;252(1): 296–307.
429 Fazlollahi S, Becker G, Ashouri A, Maréchal F. Multi-objective, multi-period optimization of district energy systems: IV – a case study. Energy 2015;84:365–81.
430 Rentizelas AA, Tolis AI, Tatsiopoulos IP. Combined municipal solid waste and biomass system optimization for district energy applications. Waste Manag 2014;34(1):36–48.
431 Carpaneto E, Lazzeroni P, Repetto M. Optimal integration of solar energy in a district heating network. Renew Energy 2015;75:714–21.
432 Bracco S, Dentici G, Siri S. Economic and environmental optimization model for the design and the operation of a combined heat and power distributed generation system in an urban area. Energy 2013;55:1014–24.

433 Buoro D, Casisi M, De Nardi A, Pinamonti P, Reini M. Multicriteria optimization of a distributed energy supply system for an industrial area. Energy 2013;58:128–37.

434 Wang H, Yin W, Abdollahi E, Lahdelma R, Jiao W. Modelling and optimization of CHP based district heating system with renewable energy production and energy storage. Appl Energy 2015;159:401–21.

435 Buoro D, Pinamonti P, Reini M. Optimization of a distributed cogeneration system with solar district heating. Appl Energy 2014;124:298–308.

436 Ondeck AD, Edgar TF, Baldea M. Optimal operation of a residential district-level combined photovoltaic/natural gas power and cooling system. Appl Energy 2015;156:593–606.

437 Uris M, Linares JI, Arenas E. Size optimization of a biomass-fired cogeneration plant CHP/CCHP (combined heat and power/combined heat, cooling and power) based on Organic Rankine Cycle for a district network in Spain. Energy 2015;88:935–45.

438 Salata F, Vollaro AL, Vollaro RL, Mancieri L. Method for energy optimization with reliability analysis of a trigeneration and teleheating system on urban scale: a case study. Energy Buildings 2015;86:118–36.

439 Ortiga J, Bruno JC, Coronas A. Operational optimisation of a complex trigeneration system connected to a district heating and cooling network. Appl Thermal Eng 2013;50(2):1536–42.

440 Ameri M, Besharati Z. Optimal design and operation of district heating and cooling networks with CCHP systems in a residential complex. Energy Buildings 2016;110:135–48.

Index

Page numbers in **bold** refer to pages where the definition of the word or more detailed information about it is provided. Page numbers in *italic* refer to pages where the word appears in tables or figures.

www.ingramcontent.com/pod-product-compliance
Lightning Source LLC
LaVergne TN
LVHW011912200926
848484LV00002B/601

* 9 7 8 1 7 8 5 6 1 1 2 6 1 *